GW00693888

AVIAN SURVIVORS

AVIAN SURVIVORS

THE HISTORY AND BIOGEOGRAPHY OF PALEARCTIC BIRDS

CLIVE FINLAYSON

T & AD POYSER
London

To my son Stewart,
gifted with the ability to feel the spirit of the wild

Published 2011 by T & AD Poyser, an imprint of Bloomsbury Publishing Plc, 36 Soho Square, London W1D 3QY

ISBN (print) 978-0-7136-8865-8
ISBN (e-pub) 978-1-4081-3732-1
ISBN (e-pdf) 978-1-4081-3731-4

A CIP catalogue record for this book is available from the British Library

This book is produced using paper that is made from wood grown in managed sustainable forests. It is natural, renewable and recyclable. The logging and manufacturing processes conform to the environmental regulations of the country of origin.

Commissioning Editor: Nigel Redman
Project Editor: Jim Martin

Design by Julie Dando at Fluke Art

Printed in China by South China Printing Company

10 9 8 7 6 5 4 3 2 1

Cover art by Tim Worfolk

Front: Gorham's Cave, Gibraltar, 40,000 years ago. Several miles inland, the cave lay above a plain of scrub not dissimilar to the Coto Doñana today, with Crag Martins roosting in the cave, Azure-winged Magpies foraging from bush to bush and Lammergeiers soaring above.

Back: Gorham's Cave today. Crag Martins still roost in the cave in good numbers, but the plain has been submerged beneath the sea for millennia.

Contents

Preface

This book owes its existence to a man and to a bird. Both influenced me at key moments of my life and in fundamental ways that shaped my thoughts about the distribution of birds in the Palearctic and beyond: why is one species here and not somewhere else? Why are there more species in one family than in another? How do similar species manage to live side by side without one ousting the other? Why do some birds migrate long distances while others hardly move from the territory in which they were born? These questions are not new, and there is a huge body of ecological and ornithological literature that has tried to answer them, with varying degrees of success. Ian Newton's recent volumes (Newton 2003, 2008) provide excellent summaries of the literature as it applies to birds. So this book is not another synthesis of what has been written. It is, instead, a new approach that owes its origins to the inspirational works of the late Reg Moreau, the man I alluded to in the first sentence of this book.

I became familiar with Moreau's work very early in my career, as an undergraduate in 1973. His seminal book *The Palaearctic-African Bird Migration Systems*, published posthumously in 1972, was the culmination of years of field work, research and papers in journals such as the *Ibis*. It was, in the words of David Lack, 'his best work', and it is still the best ornithological book that I have ever read, and the one that has the central place in my library. I recommend it to everyone interested in bird migration for its clarity, insight, and as a limitless source for reflection. And it is also a fine example of how to write science in a clear, concise and engaging way.

Moreau's work was a catalyst to me. He raised my childhood passion for birds and migration to a new level, and I have not recovered from it since. My research path has taken me in many different directions in the intervening thirty-seven years, but I always end up leafing excitedly through the pages of his final work as though it was the first time. I ask for forgiveness from readers for submitting them now to a brief history of my own career but I feel it is essential, not out of vanity or self-importance, but because it provides an explanation for the reasons why I have written this book.

I grew up close to birds; the phenomenon of bird migration was ever-present, as I watched the annual movements of all kinds of birds crossing the Strait of Gibraltar between Europe and Africa, just as the Reverend John White, Gilbert White of Selborne's brother, had done in the same spot two hundred years before (White, 1789). Bird migration stimulated my interest in nature and pointed me in the direction of zoology as a subject of study. On completion of my zoology degree at the University of Liverpool I was lucky to be awarded the first David Lack studentship by the British Ornithologists' Union and, in 1976, I went to read for a DPhil at the Edward Grey Institute of Field Ornithology of the University of Oxford. Here I was to study the ecology of closely related species, attempting to answer the question of how two or more similar species could co-exist in the same area without one outcompeting the other. It was a natural consequence of David Lack's work on competition and his 1971 book *Ecological Isolation in Birds*. And it was also natural that the subjects of study should be swifts; after all, David Lack had dedicated much time to the study of the Common Swift *Apus apus* in the tower of the University's Natural History Museum (Lack, 1956).

As I was familiar with the swifts in Gibraltar, where I was born and grew up, it followed that this should be the place of study and I would compare two similar species – the Common Swift with the Pallid Swift *A. pallidus.* Unhappy with this species pair alone, I decided to add a comparison of two *Sylvia* warblers that also nested in Gibraltar – Sardinian Warbler *Sylvia melanocephala* and Blackcap *S. atricapilla.* This was an interesting combination as these Blackcaps were quite unlike their northern counterparts – dull with short, rounded wings, they lived in the Olive maquis scrub and not in forest, and they shared space with Sardinian Warblers, the archetypal Mediterranean scrub warbler. My results revealed differences in food, behaviour and life cycle in each of the two pairs but I was never happy that I had resolved the questions that I had set out to answer. I had found differences but the real question was whether or not these differences mattered – were they of a scale that allowed the species pairs to coexist?

These were the days before the personal computer, so complicated multivariate statistical analyses required the negotiation of computer time, and the results came out in lengthy printouts that needed acres of space to read and hours of peace to assimilate. Oh, what I would have given for a Windows-based package and a laptop then! But even the results, using the latest available analyses, simply explained proportions of the observed variation and left a lot to future interpretation or to dismissal as noise. The acid test of interspecific competition, to put two species in a controlled environment and alter variables one by one, might have worked for the Russian scientist Georgyi Frantsevitch Gause and his tiny protozoans in 1934, but could you even begin to think of replicating his experiments with swifts?

But my biggest misconception, and one that I should have expected had I read Moreau more carefully, was to think that I could explain everything that I saw without reference to history. And I was not alone in this. Ecologists looked at the present and thought that they could explain it in splendid isolation or, at least, even if history was acknowledged it was not that important or easy to submit to the scrutiny of statistics. So, for better or worse, I came away thinking that I had some kind of handle on what was going on between Common and Pallid Swifts and why they could live side-by-side. It would be another two decades before the penny would start to drop.

History is full of contingent events that matter and they happen at many scales, affecting individuals at one end and, less often, species at the other. The extinction of the dinosaurs at the Cretaceous-Tertiary (K/T) boundary is an extreme example of historical contingency affecting an entire lineage. Contingent events also shape and steer people's lives, and one such occasion manoeuvred mine in a new direction. It was in 1989 that I first met Chris Stringer and Andy Currant, palaeontologists at the Natural History Museum in London, who were visiting Gibraltar prospecting for cave sites that had shown evidence of occupation by Neanderthals when excavated in the 1950s. I got involved with them, fortuitously, because they wanted to see these caves, which were on military land, and I knew them well because I used to ring hundreds of Crag Martins *Ptyonoprogne rupestris* there every winter, when more than 3,000 birds used the caves to roost. Little did I know then that beneath the sandy slopes by the cave entrances on which I placed my mist nets were rich archaeological and palaeontological deposits going back to the time of the Neanderthals.

This was the start of my involvement with the world of the remote past. I need to impose on the reader's patience a little longer by describing this world of caves and bones, because it gives some kind of context to the words that will compose this book. Often at the end of a day of excavation inside Gorham's Cave, the main site that we have been working in Gibraltar for over two decades now, I would stand at the cave entrance and let my colleagues go ahead. Once on my own I would stare at the dark Mediterranean Sea and feel the past. I would be standing on eighteen metres of sediment accumulated over the past 60,000 years. In the sediment were the bones of countless birds, mammals, amphibians, reptiles and fish that had lived and died in this remote place. There were dark patches that marked old Neanderthal camp fires and the charcoal told us which plants had been growing outside the cave, and when they had done so. There were limpet and mussel shells that had been collected from the beach by Neanderthals. And I knew that there were also, invisible to the eye, millions of pollen grains that spoke to us about the vegetation of a distant past.

Imagine my excitement all those years ago when I found my first Crag Martin bones in a context that associated them with 40,000-year-old Neanderthals in Gorham's Cave, the same cave where I had been ringing them for years. The martins had been there all the time. We have now identified the remains of 145 bird species in a small group of caves and rock shelters on the eastern side of Gibraltar. Put into some kind of context, that represents 26% of the Western Palearctic breeding species that we will be looking at in this book in a mere 6-kilometre stretch of coast. But it is a stretch that contains a unique 300,000-year-old archive. Here we had a wealth of information about the birds of today and what they had done yesterday, and that got me straying more and more, while I discussed the Neanderthals with my colleagues in the cave, into the world of the Pleistocene. This had also been Moreau's world, as he had clearly seen and recognised that we ignored past climate and vegetation change at our peril.

Having introduced the man, it is now time to bring in the bird. It is the Iberian Azure-winged Magpie *Cyanopica cooki*. When I started off in science, this bird was considered conspecific with the Eastern Azure-winged Magpie *C. cyanus,* and it was one of those birds with an unusually disjunct distribution. It occurs in south-western Iberia (but curiously not in Gibraltar – the significance of this parenthesis will become apparent in a moment) and in the Far East, in parts of China, Japan and Korea. It seemed an impossible exemplar of natural range fragmentation, and the popular view for years was that these colourful birds had been brought back from China by Portuguese mariners in the late medieval period and then escaped, going feral, in south-western Iberia (Finlayson 2007). Azure-winged Magpies were in some distant way part of Marco Polo's legacy.

Then, during our excavations, the remains of these magpies were found in Gorham's and Vanguard Caves, in contexts dating back well into the Pleistocene, to 40,000 years ago and beyond (Cooper 2000). This discovery altered everything. The magpies were native to the south-west of Europe and had not been introduced by humans. That simple observation changed the way in which I viewed the Palearctic and its birds. My first reaction was to try and understand, at the local level, how it was that Azure-winged Magpies had lived in Gibraltar and did not do so now (my parenthesis above should now become relevant).

That question took a little while to answer and it was resolved in collaboration with my wife Geraldine, who was studying the past landscapes outside Gorham's Cave, and my good friend Pepe Carrión from the University of Murcia who was looking at the pollen samples. The answer went something like this. The deep blue Mediterranean that I would stare at from the cave was today in an unusually high position, lapping the beach immediately below the cave. For the greater part of the Late Pleistocene, global temperatures had been lower that today and, as water became trapped as ice at the poles, sea levels dropped. The submerged shelf outside our cave was shallow and we calculated that at times the coast had been as much as 4.5 kilometres away. In fact, for most of the past 100,000 years sea level had averaged around 80 metres below present levels, and it had been as much as 120 metres lower (Finlayson, 2009). Here we had a vivid image of the scale of past natural climate change and its impact on the landscape. I sometimes take people to the cave by boat and we sail out until we are 4.5 kilometres from the cave, at which point I surprise my visitors by telling them that the whole stretch of sea that we have sailed across was once the hunting ground of Neanderthals, Spotted Hyenas and Leopards.

It was indeed a rich landscape between 50,000 and 28,000 years ago, dominated by herds of grazing animals, especially Red Deer *Cervus elaphus*, wild horse *Equus ferus* and aurochs (wild cattle) *Bos primigenius*. It had a suite of predators, and also scavengers; Bearded Vultures *Gypaetus barbatus* nested on Gibraltar's cliffs, which were practically at sea level (a lesson for those who see the relictual few as having some kind of montane fixation), and there were lots of Red Kites *Milvus milvus,* Egyptian Vultures *Neophron percnopterus* and Griffon Vultures *Gyps fulvus,* with a few Black Vultures *Aegypius monachus* thrown in. It was a Mediterranean Serengeti, a vast complex of savannas and wetlands on sand. And the tree that grew preferentially on this sand, as it does today in the coastal National Park of Doñana in south-west Spain, was the Stone Pine *Pinus pinea*. If there was to be a third character in my story it would be this tree, because of its close association with the Azure-winged Magpie.

Go to Doñana today, a mere 125 kilometres from Gibraltar as the magpie flies, and you will find lots of Azure-winged Magpies on Stone Pine savannas. The rising sea over the last 10,000 years flooded the pine savannas off Gibraltar; there are no magpies there now. But go back to the Late Pleistocene and you will find a land of stone pines and magpies, right outside Gorham's Cave. By contrast, at this time there is nothing at Doñana but open ocean – the sand spit that would close the coast off and create the dunes for the pines and the magpies would only form some time after the Romans had arrived in Iberia, with Doñana's marsh the product of silting caused by deforestation upriver as men cut trees to make invincible armadas. Spain's emblematic national park has a recent origin, and has been largely shaped by the hand of humanity.

So this answered the first question. Azure-winged Magpies, Stone Pines, dunes and coasts had been in constant flux within the local context of south-western Iberia. There is another point that I want to make before moving up a scale with our magpies. When I started all this work another prevalent misconception was that Stone Pines were somehow unnatural. In most contexts they were planted and we had somehow helped them along. But now we know that this is not true. We have planted many forests of Stone Pine, but these trees have been growing on the coastal dunes of this part of the world for a very long time, as the evidence from Gorham's Cave has shown.

There seems to have been an obsession with regarding the potential natural vegetation of the Iberian Peninsula as somehow dominated by oak forests, which we have later cut down. But it has once again been Pepe Carrión who has bravely begun to take on long-established and firmly entrenched positions. He has ably demonstrated, using a wealth of pollen data from across the Iberian Peninsula, that the natural vegetation over much of this land was dominated by various species of pine (Carrión and Fernández 2009). Results such as this one also have a major impact on how we understand the history of the birds of the Palearctic.

I started to delve deeper into the Azure-winged Magpie's message to us as far back as 2000. I realised then that there must have been, at some point in the remote past and very likely long before the Pleistocene glaciations, suitable habitat for Azure-winged Magpies all the way from Portugal to Japan. I had been trying to understand the wider picture of Neanderthal distribution, and I was convinced that the belt of mountains that stretched from Iberia to China had provided a continuum of topographically varied landscapes that had been the mainstay of the Neanderthal economy. I called this belt the heterogeneous mid-latitudes (Finlayson *et al.* 2000) and later (Finlayson 2004) the mid-latitude belt (Figure 2.3, p. 28). I also introduced the southern ranges, from the Arabian Peninsula south down the Rift Valley to South Africa, as an extension of this belt that crossed latitude lines. When I started to look at birds I realised that many followed a similar pattern of distribution – Bearded Vulture, wheatears, rock thrushes and choughs, for example – but these were not limited to birds of rocky habitats. I found similar patterns for wetland birds and for birds of savannas (by which I mean the gamut of habitats with trees at densities low enough not to be forest). The Azure-winged Magpie was the smoking gun. I presented my preliminary results at a conference that I hosted in Gibraltar in 2007; Jacques Blondel, another of those legendary ornithologists whose writings influenced me unmeasurably, listened to what I had to say and encouraged me to publish. He was, incidentally, one of the few people who had tackled the history of the avifauna of the Western Palearctic (Blondel and Mourer-Chauviré 1998) and had a clear vision of the role of history in its formation (Blondel and Vigne 1993). I decided, after much deliberation, that a short paper would not do justice to the results. I needed to verify my data, add to them and then think of publishing them as a book. And this is it!

There are two epilogues to this introduction and they both matter because they had a major influence on my thinking. The first is that, following the publication of the discovery of the Azure-winged Magpie fossils, interest in these birds gathered momentum and two papers were published that compared the DNA of eastern and western magpies (Fok *et al.* 2002; Kryukov *et al.* 2004). Most interesting were the estimates for the split between the two populations, admittedly broad given uncertainties about mitochondrial clock calibration – anything between 3.35 and 1.04 million years ago. This placed the split either in the Early Pleistocene or, more likely in my opinion, the Pliocene.

The second epilogue comes from the world of human evolution that I had also kept up with and had been writing about. A paper appeared in the journal *Nature* at the end of 2005 in which archaeologists Robin Dennell and Wil Roebroeks argued that a corridor of savannas, across which our early ancestors had dispersed, had stretched from Portugal to China during the Pliocene. They called it 'savannahstan' (Dennell and Roebroeks 2005; Dennell 2009). It seemed that we were independently coming to similar conclusions. When, in 2009, I presented my improved dataset at the Calpe 2009 Conference, Robin Dennell was in the audience. His lecture at the conference added to his earlier paper and we shared one of those indescribably stimulating moments of intellectual convergence.

Acknowledgements

Many people have helped me with this book, too many to list, but I owe my deepest gratitude to my wife, Geraldine, who has been as patient with me as always while I immersed my mind in this work. She has been the person that has been closest to me and has given me the best professional advice. We have shared many of the ideas in this book, over many years spent studying birds in the field. I am also grateful to my son, Stewart, for his support and companionship in the field. I would not have embarked into the world of the past without the friendship and support of colleagues who opened my eyes to the world of the Quaternary and, later, the Tertiary. Joaquín Rodríguez Vidal and Francisco Giles Pacheco have been my mentors.

Three other friends have been pivotal when the time has come to discussing the ideas put forward here – José S. Carrión, Darren Fa and Juan José Negro. Steve and Julie Holliday have been wonderful companions and friends who have shared many moments with us in the field during the preparation of this book and who have helped me get to remote haunts. I have also discussed many of the topics of this book with Steve, whose ornithological knowledge has been invaluable. My good friend Mario Mosquera has been a colleague and field companion over many years. I am grateful to Bryan Rains for giving up his time to show me eagles on Mull.

I would like to thank Nigel Redman and, especially, Jim Martin at T & AD Poyser for their encouragement, support and patience throughout. The manuscript benefited greatly from their comments and advice. I also thank the five guest photographers who have graced this book with their brilliant images – Paul Bannick, Stephen Daly, Ian Fisher, Peter Jones, Stefan McElwee, and Roberto Ragno, and my son Stewart.

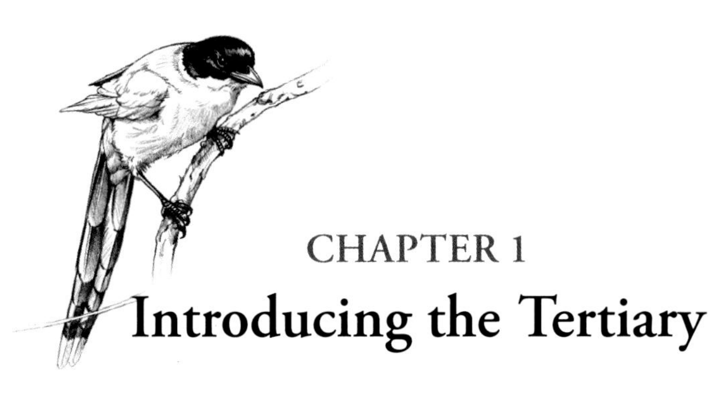

CHAPTER 1
Introducing the Tertiary

Sometimes, on rare occasions, a piece of information is revealed that opens up a window of opportunity for a new understanding of data that we may have been unimaginatively staring at for years. This revelation may come through personal inspiration, perhaps while in the field, or it may be an insight gleaned from a new publication or even by looking afresh at an old one. A paper by John Klicka and Bob Zink of the Bell Museum of Natural History of the University of Minnesota, published in 1997 in the journal *Science,* was one such eye-opener. The paper's title said it all – *The importance of recent Ice Ages in speciation: A failed paradigm.*

This paper challenged the long-established view that the Pleistocene, particularly the Late Pleistocene during which time much of the Palearctic was engulfed in ice sheets, had been a period of speciation for vertebrates in general (Mayr, 1942, 1963; Simpson, 1944) and birds in particular (Mengel, 1964; Bermingham *et al.*, 1992). Klicka and Zink argued that most North American songbird lineages had instead had a protracted history of speciation over the past 5 million years (mya). These results have been contested by others (Johnson and Cicero, 2004; Weir and Schluter, 2004; Cicero and Johnson, 2006) and reinforced by the authors (Zink *et al.*, 2004), which would seem to leave the debate open to further discussion. Lovette (2005) has provided a balanced analysis of the situation; the most closely related sister taxa and lineages seem to have split during the Pleistocene, and there seems a bias in favour of such diversification among boreal over low-latitude species (following Weir and Schluter, 2004); but this is only the tail-end of a long history of lineage-branching; the majority of birds have had long tenures as independent lineages that predate the Pleistocene. Although the debate and analyses have focused on North American songbirds, the general conclusions are pertinent to the Palearctic.

A separate but equally significant debate concerns the origins of the main avian lineages, once thought to be the result of diversification after the K/T Event (the massive asteroid impact together with major volcanic activity and sea-level changes that was responsible for the mass extinction of dinosaurs and other animals at the end of the Cretaceous) of 65 mya. Clear evidence of a pre-impact diversification of mammals, going back as far as 170 mya, has emerged from fossil sites in China, Madagascar and Portugal (Hu *et al.*; 2005; Weil, 2005; Ji *et al.*, 2006), and it now seems that many lineages of present-day birds were already present in the Mid- to Late Cretaceous, and subsequently survived the K/T Event (Cooper and Penny, 1997; Cracraft, 2000; Clarke *et al.*, 2005; van Tuinen *et al.*, 2006; Brown *et al.*, 2008; Pratt *et al.*, 2009). Not everyone agrees and some still situate the avian radiation after the K/T event (Feduccia, 2003; Chubb, 2004; Poe and Chubb, 2004; Ericson *et al.*, 2006).

What is important to us here is that modern bird lineages have been around for a very long time, and their origins are ancient. Whether before or after 65 mya, these lineages have had a long history and have been exposed to the massive climatic and ecological changes of the Tertiary (65–2.6 mya; traditionally, the boundary between the Tertiary and the Quaternary was considered to lie at 1.8 mya, but a recent revision has placed it at 2.6 mya, within the older bracket of the Pliocene, the last epoch of the Tertiary) and Quaternary (which encompasses the Pleistocene and the Holocene, the latter representing the last 10,000 years of global warming). This means that the Tertiary represents, in terms of time taken, the equivalent of 24 Quaternary periods. The relatively short length and climatic variability of the Quaternary has not generated large-scale speciation, and has not seen the evolution of new genera of birds.

We have already seen (see p. 10) how the azure-winged magpie split pre-dated the glaciations of the

Middle and Late Pleistocene and had probably occurred during the Pliocene. This is a marker of a wider picture of events that may have been taking place throughout the Tertiary, and which may have gathered momentum during the Miocene (after 10 mya until 5.33 mya) and Pliocene (5.3–2.6 mya), for reasons that I will explain in this chapter. It is noteworthy that this was also the time during which significant evolution and radiation of the human lineage took place (Finlayson, 2009). Moreau's (1972) chapter on the fluctuating ecology of the Palearctic was limited to the Pleistocene. This is hardly surprising given the established view at the time regarding the importance of the Pleistocene and the lack of detailed information on earlier geological periods. But it is also an indication of his insight that he limited the discussion to the distribution and re-distribution of species and not to the formation of new species.

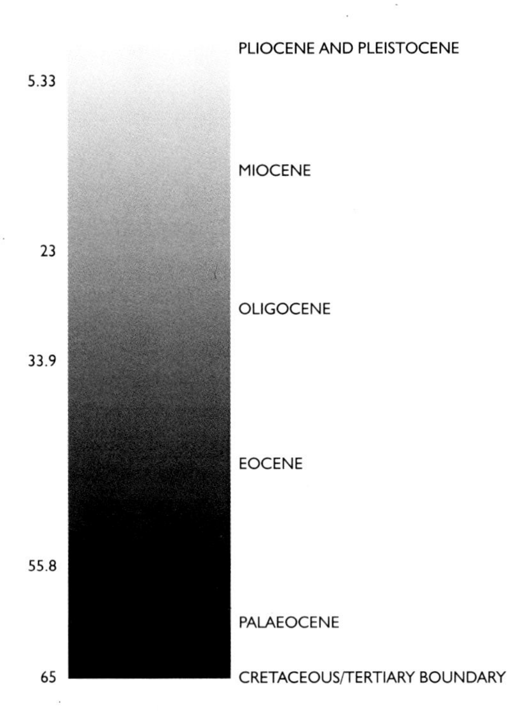

Figure 1.1 *Approximate time-line for the main geological periods described in this book, starting at the K/T boundary 65 million years ago. Numbers on the left of the column are millions of years. The major periods of the Tertiary and Quaternary are to the right of the column.*

There are two possible biological responses of species to climate change: (a) their geographical ranges can shift, contract or expand, which is what Moreau (1972) was looking at; or (b) they can evolve by adapting to changing environments without moving and they can also, when isolated, branch off in different evolutionary directions and become different species. If the above fail then extinction will follow (Bennett, 1997). My basic argument in this book is that species have frequently responded to climate change by shifting geographical position to keep to their preferred ecological conditions. When these conditions have disappeared altogether, species have invariably become extinct. But when ecological change has been in a single direction and has been slow, gradual or has persisted for long periods, then species have had the chance to change and evolve with the new conditions. The climatic changes of the Pleistocene have been so severe and rapid away from the tropics that there has been little room for evolution and speciation, and redistribution has dominated. There is little evidence, as we will see, of large-scale avian extinctions in the Pleistocene either, which suggests that the species that made it to the Pleistocene were the 'survivors' that had the ability to withstand the oscillations of the climate of the last two million years. We will test this prediction in the chapters that follow. For the rest of this chapter I will summarise the main climatic and ecological changes of the Tertiary and Quaternary.

FROM THE PALAEOCENE TO THE END OF THE OLIGOCENE (65–23 MYA)

This long span of time covering 42 million years is the backdrop to the latter part of the Tertiary that will be of greater interest to us. It marks the zenith of a 'hothouse' world that progressively drifted towards the 'icehouse' world in which we live. A rise in global temperatures at the start of this period (the Palaeocene, 65–55 mya) culminated in one of the warmest episodes in the Earth's history. Global tectonic events increased hydrothermal activity in the sea floor, flooding the atmosphere with carbon dioxide (CO_2) (Rea *et al.*, 1990; McElwain, 1998). Deep oceanic water temperatures were between 9° and 12°C warmer than today, and the Antarctic Ocean's surface temperatures were of the order of 21°C (Kennett and Stott, 1991). The warming was rapid; global sea-surface temperatures rose by 8°C in 10,000 years.

During this time there was an intermittent connection between North America and Europe via Greenland, and as evergreen forests stretched across high latitudes of the northern hemisphere, the world witnessed the expansion of early tree-dwelling primates (Beard, 2008). The position of the continents influenced the situation. Africa and India were separated from Eurasia by the Tethys Ocean, and the Mediterranean Sea had not formed. Australia was still joined to Antarctica, which was connected via a series of islands to South America. Warm water was carried towards high latitudes, keeping a low-temperature latitudinal gradient and the poles ice-free.

The vegetation belts of Eurasia, Greenland and North America (or 'Holarctica') followed latitudinal bands with a limited amount of polar broadleaved deciduous forest in the extreme north, a broad belt of subtropical (broadleaf evergreen) woodland to its south (across what would today be Canada and northern Siberia) and paratropical (seasonally dry) forest occupying a band to the south (including much of today's United States, south-central Siberia and temperate Europe). To the south of this lay a broad belt of tropical forest covering the whole of Africa, Madagascar, India, southern Asia and southern Europe. There was also a block of ancient woody savanna on the early high ground of a very young Tibetan massif (Janis, 1993).

This Eocene thermal optimum would come to an end, leading to a steady downward trend in global temperatures to the present day; there have been many oscillations and temperature reversals along the way, of course, and not just during the Pleistocene (Bennett, 1997). But the world never saw temperatures of this magnitude again, and present-day predictions of human-generated global warming are minuscule in comparison. The main culprits of the downward temperature spiral (for there were several) were the Earth's land masses, which were gradually drifting toward their positions of today. In the process they altered the flow of ocean currents and air masses.

The most significant of these movements was probably the first major phase of uplift of the Tibetan Plateau, as India collided into Eurasia around 54 mya. Over the following 45 or so million years the plateau, an area roughly half that of the United States, was thrust five kilometres into the air, influencing atmospheric patterns, deflecting jet streams and intensifying the monsoon (Ruddiman and Kutzbach, 1991). As rainfall increased, larger amounts of atmospheric CO_2 were dissolved. The resulting reduction in atmospheric CO_2 reduced the greenhouse effect, and global temperatures subsequently decreased. Other geological events intensified the effect: there was massive volcanic activity on the North Atlantic seabed, two Antarctic marine gateways opened up (Drake Passage between Antarctica and South America and the Tasmanian Passage between Antarctica and Australia), the Andes and Rockies were elevated, and the Central American Seaway between North and South America began to close (Figure 1.2). All these factors contributed to the decrease in global temperatures and, because these events were irreversible, the trends continued in one direction – towards cooling.

The world of the Eocene hothouse is relevant because its birds included many kinds that would disappear from the Eurasian landmass as the tropical forests disappeared and climatic conditions cooled. There was undoubted global extinction of species and families, but there was also a significant degree of loss at the regional level. Many of these birds of tropical forests left lineages that are today only represented among the birds of the Neotropical, Afrotropical and Indo-Malayan regions. Among the fossils recovered from the Eocene and Oligocene of Eurasia (from sites such as the Messel Shales in Germany) are trogons (Mayr, 2005a, 2009), barbets (Mayr, 2005b), turacos (Musophagiformes), motmots (Momotoidea), hummingbirds (Trochilidae), mousebirds (Coliiformes; Mayr and Peters, 1998; Mayr, 2000; Mayr and Mourer-Chauviré, 2004), hornbill-types (Bucerotes; Mayr, 2006a), New World Vultures (Cathartidae), secretary birds (Sagittariidae) and parrots (Psittaciformes) (Mayr, 2005c). There are also taxa that became globally extinct, for example the swifts of the family Jungornithidae (Mayr, 2003) or the galliformes of the Gallinuloididae (Mayr, 2006b). This early contraction of range and loss of bird taxa was the first step towards the present configuration of the Palearctic avifauna.

Many of these groups, which are now missing from the Palearctic avifauna, are typified by an absence of migratory behaviour, while many extant Palearctic groups which have tropical representatives include

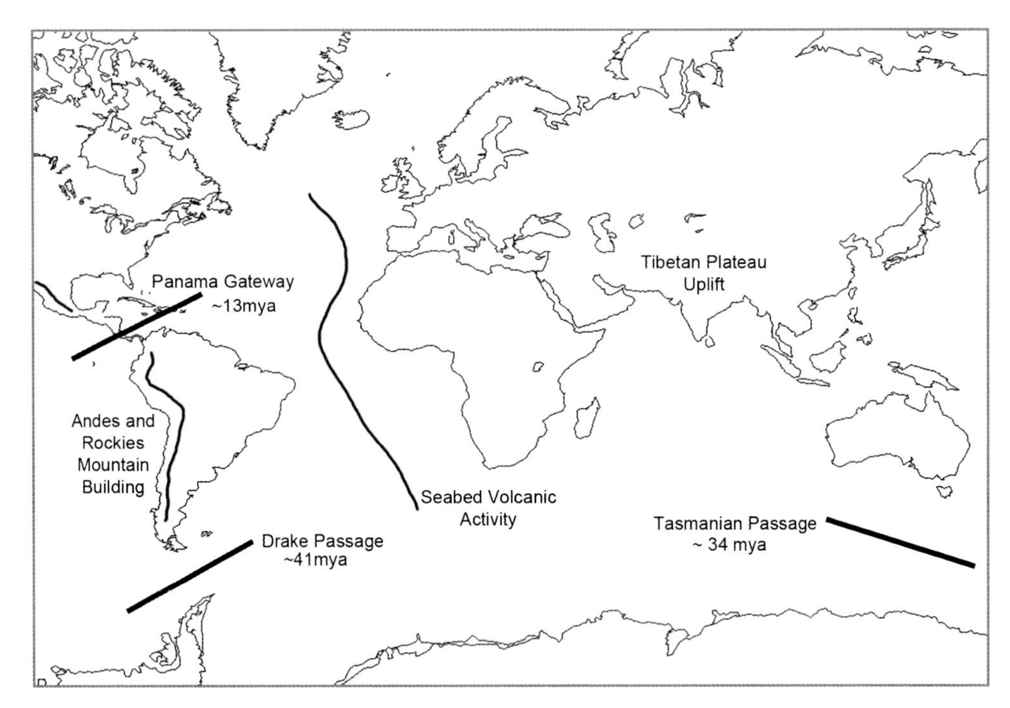

Figure 1.2 *Major tectonic events affecting Tertiary climate change superimposed on a present-day map. Dates for the opening of Tasmanian and Drake passages and closing of Panama are approximate and indicate the start of protracted processes. Mountain building similarly was protracted, with rates of uplift varying at different times during the Tertiary. For example, the uplift of Tibet started by 54 million years ago and activity continued to at least 7 million years ago.*

migratory species. It is impossible to disentangle whether migratory behaviour is missing from those absent from the Palearctic today simply because they live in situations that do not require migration or whether it is because they are *unable* to migrate. In the hot world of the Eocene, migration in the northern hemisphere's tropical forests may have been unnecessary; but there must have been limited, short-range migrations among birds of high latitudes that experienced annual cycles of long and short days, and those of the paratropical forests which had a dry winter season, just as there are intra-tropical, rainfall related migrations today (Moreau, 1966; Curry-Lindahl, 1981 a, b). These would have been the precursors of the long-distance migrations that characterise many Palearctic birds today.

Into the Oligocene

The Oligocene world, still considerably warmer than today's, was cooler than that of the preceding Eocene. The Eocene-Oligocene (*c.* 34 mya) boundary was marked by an unprecedented and abrupt cooling (Janis, 1993), much more pronounced than the gradual lowering of temperatures during the Eocene. This seems to have been caused by a combination of factors that included the opening of the Antarctic gateways and lowered atmospheric CO_2 levels, at a time when the Earth's orbit of the sun was at its most distant (therefore minimising solar insolation). Continental temperatures dropped by an average of 8.2°C in 400,000 years, and the first Antarctic ice sheets formed, but there was no corresponding increase in aridity (Zanazzi *et al.*, 2007).

After this, the continuing rise of the Tibetan massif, along with the elevation of the Rockies, Pyrenees, Carpathians, Zagros and other mountain ranges, during the Oligocene and into the Miocene contributed towards a global tendency towards aridity and the formation of the great deserts of today (Carrión, 2003). By restricting the flow of humid air beyond the mountain ranges, continental interiors dried up and more water was retained in the new polar ice caps.

Epoch	Fossil genera	Genera that became extinct	Genera extinction %
Eocene	77	54	70.13
Oligocene	44	15	34.09
Miocene	128	39	30.47
Pliocene	143	2	1.4
Pleistocene	187	0	0

Table 1.1 *Tertiary and Quaternary European fossil bird genera, and inter-epoch rates of extinction.*

The vegetation belts across Oligocene Holarctica were very different from those of the Eocene. The polar broadleaved forest disappeared and the belt of subtropical broadleaved evergreen woodland, which had occupied a high-latitude position, was displaced further south across much of the area taken up today by the Sahara, Arabia and across southern Asia to China. Instead the most northerly latitudes were taken over by a broad belt of temperate mixed (coniferous and deciduous) woodland, and to the south, across present-day temperate Eurasia and Northwest Africa, was a belt of temperate broadleaved deciduous woodland. Woody savannas were localised on the growing Tibetan Plateau. Tropical forest was limited to equatorial areas, with belts of seasonal paratropical forests to the north and south (Janis, 1993).

These changes are dramatically reflected in the number of avian genera that were lost between the Eocene and Oligocene. Following Mlíkovský's (2002) list of Cenozoic bird genera in Europe, I have calculated that more than 70% of the Eocene genera were extinct by the Oligocene, while just under 35% disappeared between the Oligocene and the Miocene (Table 1.1). Twenty-one new genera made their appearance in the Oligocene. There are no Palaeocene, and only two Late Eocene, genera represented among the Palearctic birds of today (Table 1.2). These are the *Recurvirostra* avocets and *Coturnix* quails. The number

Sub-epoch	Genera making first appearance	Genera still present today	Genera present today %
Late Palaeocene	2	0	0
Early Eocene	19	0	0
Mid-Eocene	20	0	0
Late Eocene	38	2	5.26
Early Oligocene	4	0	0
Mid-Oligocene	10	1	10
Late Oligocene	7	2	28.57
Early Miocene	46	20	43.48
Mid-Miocene	26	21	80.77
Late Miocene	27	21	77.78
Early Pliocene	16	14	87.5
Late Pliocene	38	38	100
Early Pleistocene	41	41	100
Mid-Pleistocene	2	2	100
Late Pleistocene	3	3	100
Holocene	2	2	100

Table 1.2 *Genera of European birds by first appearance and representation today.*

of Oligocene genera remaining in the present – *Puffinus* shearwaters, *Phoenicopterus* flamingoes and *Alcedo* kingfishers – is proportionately higher, though few in number. The Eocene–Oligocene transition reflects a major turnover of bird genera, which appears concomitant with the significant climatic and ecological changes that took place around 34 mya.

THE MIOCENE (23–5.33 MYA)

By the start of the Miocene the world had been transformed. The world's land masses were practically in the positions of today, and polar ice caps had formed and grown (albeit non-permanently). As a result sea-levels had dropped significantly. Ecosystems and faunas were radically altered, the broadleaved forests of the poles vanished, and tropical forests were restricted to low latitudes. Among the mammals, herbivores increased in diversity and abundance and the first apes appeared on the scene (Finlayson, 2009). But the Miocene commenced with a period of global warming that lasted from 23 to 15 mya, during which time tropical and subtropical forests expanded once more within Africa and into Eurasia as far as Siberia and Kamchatka (Janis, 1993). A huge territory, from the Iberian Peninsula to China and from Kenya to Namibia, was a mass of tropical and subtropical forest across which a diversity of apes thrived (Begun, 2003), giving a clear indication of the climatic and ecological conditions across this huge area given the tropical requirements of apes today, confined as they are (bar humans, of course) to forested regions of Africa and South-east Asia.

Around 19 mya Africa and Arabia collided into Eurasia, severing the ancient Tethys Ocean and creating the Mediterranean Sea. The connection between Eurasia and Africa was intermittent until 14 mya. After that, and right up to today, Africa and Eurasia have formed a single land mass (Finlayson, 2009). The interlude of global warming at the start of the Miocene came to an end and the general downward trend in global temperatures was renewed around this time. Coincidental was a period of intense uplift of the Tibetan Plateau from 13 mya (Clark *et al.*, 2005), which may have continued regionally until after 7 mya (Wang *et al.*, 2006), the permanent establishment of the Antarctic ice cap, and the start of glacial climates in the Arctic (Carrión, 2003). Once more the tropical forests retreated and the northern hemisphere latitude vegetation belts restored, with notable innovations. The north was dominated once more by temperate woodland (of mixed coniferous and deciduous trees) with a wide band of temperate broadleaved deciduous woodland to its south. These forests resembled the modern taiga in character. The pattern resembled that of the end of the Oligocene, except the temperate broadleaved deciduous woodland did not extend south to the new Mediterranean Sea. There the vegetation across both shores of the sea, and south to the tropical zones, had a distinctly seasonal character, with winter rainfall and a summer dry season. The vegetation was dominated by forests of pines and oaks, with other typical components including cedars, hemlocks and *Arbutus* strawberry trees (Carrión, 2003).

To the east of the Mediterranean the paratropical forest, with summer rainfall, occupied a low latitude band across Arabia to China. This was dominated by beeches, laurels, pines and spruces. To its north was a band of subtropical woodland and woodland savanna, which included mangroves and swamps. This woodland included cypresses, maples, poplars, willows and oaks. South of the Mediterranean and paratropical belts lay the tropical forest, in Africa and across southern Asia (Janis, 1993). But the lifting of the Tibetan Plateau interrupted the mid-latitude belts, which were no longer continuous from east to west. Grasslands and steppe appeared for the first time, covering a large area of the arid continental interior of the Eastern and Central Palearctic. The wooded steppes were open and dominated by junipers and *Celtis* nettle trees. Temperate woodland, similar to that in the far north, appeared around the great mountain mass as it was elevated.

Grasses

A major and significant ecological novelty arose during the Miocene, one that would have a major impact on the world and from which many bird species would benefit. This was the emergence of the C4 grasses.

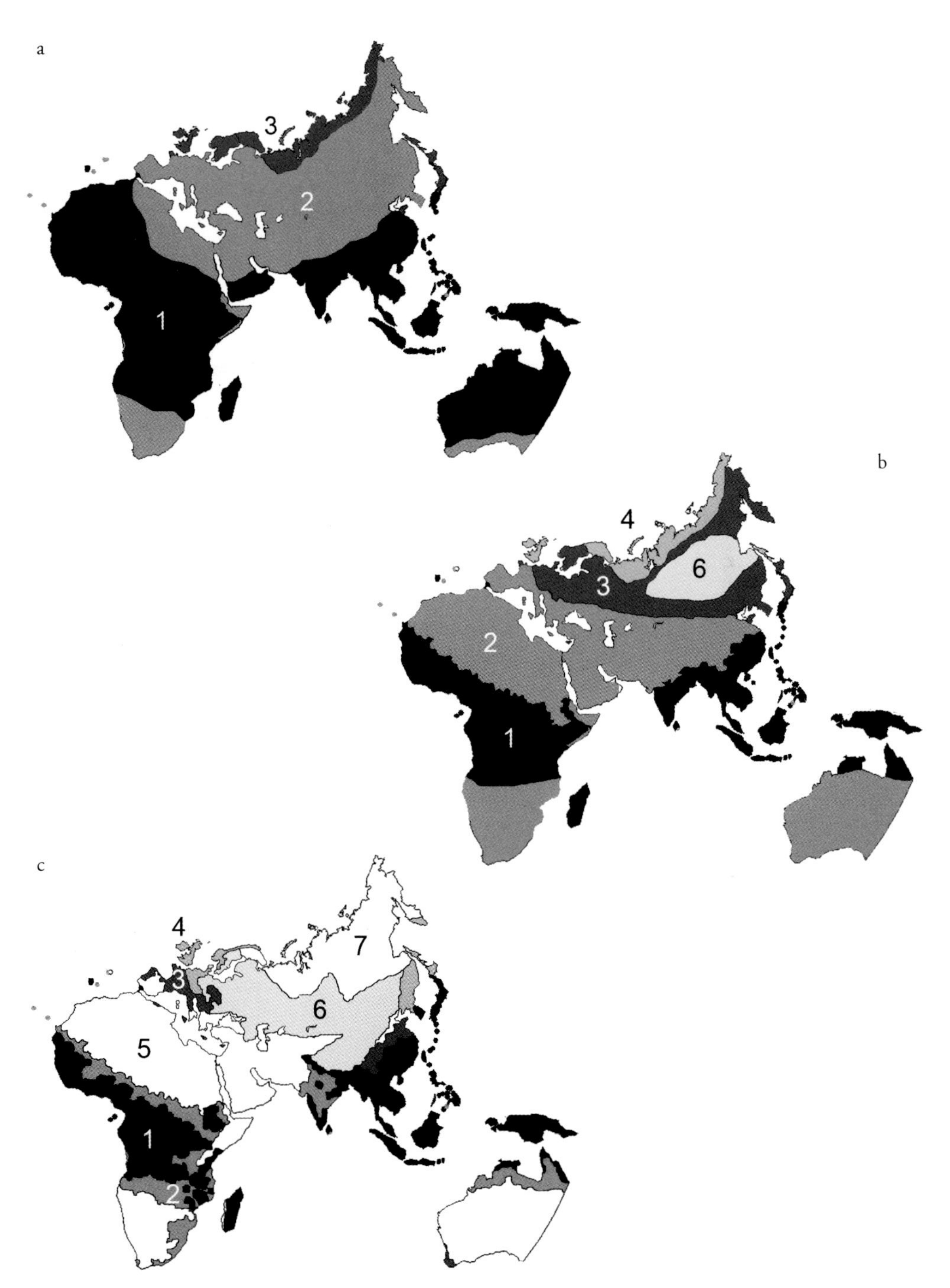

Figure 1.3 *Idealised representation of (a) Eocene and (b) Miocene bioclimates of the Old World compared to the present day (c). The Old World goes from domination by warm and wet tropical climates to the temperate and dry climates of today, with humid climates restricted to the tropics and oceanic seaboards. Key: 1 warm/wet (annual temperature T > 20°C, annual rainfall R > 1200mm); 2 warm/humid (T > 20°C, R 600-1200mm); 3 temperate/wet-humid (T 10–20°C, R > 600mm); 4 cool/ wet-humid (T 0–10°C; R > 600mm); 5 warm/dry (T > 20°C; R < 600mm); 6 temperate/dry (T 10–20°C; R < 600mm); 7 cold/dry (T < 0°C; R < 600mm). Based on sources cited in the text.*

Lineage	Eocene	Oligocene	Miocene	Pliocene	Pleistocene
1(a) (i) Shrikes, corvids, orioles			(1)/2	4	
1(a) (ii) Hirundines, warblers, larks			6	3	6
1 (a) (iii) Tits				2	
1 (a) (iv) Chats and thrushes			2	8	4
1 (a) (v) Sparrows, finches, buntings & pipits			4	7	4
1 (c) Falcons			1		
1 (d) Terrestrial non-passerines	(17)	(3)/1	(9)/3	1	2
1 (e) Owls	(6)	(1)	(4)/3	5	
1 (f) Diurnal Raptors	(1)	(2)	6	(1)/1	6
2 Shorebirds, gulls, terns and auks	1	(4)	(1)/7	6	11
3 Water birds	(10)	(3)/1	(9)/10	1	4
4 Cranes, rails, bustards and cuckoos	(12)	(1)	(7)/4	(1)/5	1
5 Swifts and nightjars	(12)		2		2
6 Pigeons, flamingoes and grebes	(2)	(1)/1	(2)/1	4	3
7 (a) Gamebirds	(5)/1		(2)/1	4	3
7 (b) Wildfowl	(5)	(3)	(3)/8	2	5
8 Ratites	(2)		1		

Table 1.3 *First presence of genera by epochs. Figures in table are number of genera. Numbers in brackets are extinct genera.*

Grasses were able to use a photosynthetic pathway different to that of established (C3) plants, and they did so most efficiently in warm climates with low CO_2 concentrations. They first appeared around 20–25 mya but had relatively little impact (Pagani *et al.*, 2005) until 6–8 mya, when decreases in atmospheric CO_2, coupled with enhanced aridity, gave these plants the edge in warm environments and they started to expand geographically (Cerling *et al.*, 1997; Pagani *et al.*,1999). This represented the start of the new world of grasslands and savannas. The expansion started in Africa around 8 mya in the equator and reached the cooler south of Africa by 5 mya (Ségalen *et al.*, 2007), but the process of expansion was protracted and there were important gains at 1.8 and 1 mya, by which time grassland had come to dominate large areas of tropical Africa.

The Miocene was a time of upheaval during which we detect the clearest signals of the modern avifauna of the Palearctic. We have seen that there was considerable loss of Oligocene bird genera, though not as spectacular as during the Eocene-Oligocene transition, and more disappeared during the Miocene, which represents the last epoch with major generic extinction (Table 1.3). The analysis presented in Table 1.3 is based on genera present in the European fossil record. It shows the Miocene as a key period in which loss of genera is compensated by a greater number of genera that persisted to the present day. Before the Miocene, extinction greatly exceeded persistence, and after this the pattern was reversed. The modern genera of Palearctic birds are therefore the product of accumulation since the Miocene.

When did migration commence?

To try to answer this important question I show in Table 1.4 the European fossil genera by epoch, and I have indicated those genera which today include migratory and trans-Saharan migratory species. This is an underestimate because we do not know how many other, extinct, genera might have also been migratory, and

Epoch	Genera Present	Migratory Genera	Trans-Saharan Genera
Eocene	76	2 (2.63%)	1 (1.32%)
Oligocene	43	4 (9.3%)	1 (1.32%)
Miocene	105	44 (41.91%)	22 (51.16%)
Pliocene	120	68 (56.67%)	37 (30.83%)
Pleistocene	164	106 (64.63%)	57 (34.76%)

Table 1.4 *Distribution of fossil migrant genera by epochs. Migratory genera refer to genera that today have migratory species.*

it is an approximation because it does not follow that these genera were migratory in the past. But, at least, we do know that they had the *potential* to become migratory. The results show a clear increase of migratory potential in the Miocene. The increase is not only absolute but is also proportional – there are more migratory genera that first appeared in the Miocene per total genera in the Miocene than before or after. There is also a sustained first appearance of more migratory genera after the Miocene suggesting a progressive build-up of migrants. If these results are correct then we can suggest that migratory behaviour in the Palearctic intensified during the Miocene and continued thereafter. Given what we have seen regarding the increased seasonality and reduction in tropical habitats from the mid-Miocene, this is not altogether surprising.

THE PLIOCENE (5.33–2.6 MYA)

The Pliocene had an important prelude at the end of the Miocene. It was a protracted geological event with ecological implications as great as the uplift of the Tibetan Plateau. It commenced around 8.5 mya, well within the Miocene, with the restriction of water circulation from the Atlantic into the Mediterranean, as Africa closed into Europe. By then the eastern end of the Mediterranean Sea had closed as the Arabian Peninsula connected with the Eurasian landmass. Unlike today, Atlantic water entered the Mediterranean along two channels, but these became increasingly constrained as new land was uplifted by the pressure of the African plate as it pushed northwards. As insufficient water replenished evaporation, the Mediterranean started to become a series of saline lakes (Finlayson, 2009).

At 5.59 mya, the connection between the Mediterranean and the Atlantic was severed. This was a world of extreme aridity. But not all was dry and saline. One consequence of the desiccation of the Mediterranean was that a summer low pressure system developed over the hot and dry basin. With the south-westerly monsoons not fully fledged as Tibet continued to rise, the Mediterranean summer lows drew in moist air from the Indian Ocean, creating a south-easterly monsoon. Today's south-westerly monsoon draws moist air up to the Himalayas. The water is then discharged as rainfall, which feeds the major rivers that empty into the Bay of Bengal. But the south-easterly monsoons of the Miocene-Pliocene took this moist air towards north-east Africa and much of the eastern half of the present-day Sahara. We can see the remnants of this climatic period in Lake Chad and the Nile – this is what is left of what became a land of mega lakes. At its height, water collected in four huge basins within the Sahara. These became massive inland freshwater seas that drained northwards into the eastern Mediterranean. There the water dropped from great heights down massive cataracts into a saline lake (Lake Cyrenaica). The four lakes drained an area of 6.2 million square kilometres, an area eleven times the size of France (Griffin, 2002; Finlayson, 2009).

We get an inkling of what the environment around 'Lake Mega-Chad' was like in the Late Miocene and Pliocene from palaeontological work carried out in the area, which has produced some of the earliest known hominid fossils (*Sahelanthropus tchadensis*). These riverine and lake environments covered huge areas. Seasonally inundated lands and gallery forest interfaced directly with savanna and desert, generating a high diversity of life including freshwater fish, soft-shelled turtles, tortoises, pythons and lizards, mammals from hippos to sabre-toothed cats to hyaenas, and a diverse community of water birds including cormorants, darters, ducks, storks and herons (Vignaud *et al.*, 2002; Louchart *et al.*, 2004). At this stage at least trans-Saharan migration would not have presented the kinds of problems that it would for birds in later times.

Filling the Mediterranean

The Miocene prelude was brought to a dramatic conclusion at the very start of the Pliocene, 5.33 mya. A river had been eroding land in the west of the Mediterranean Basin, cutting back slowly towards the Atlantic Ocean. When it reached the level of the Atlantic, which was 3000 metres above the dry basin to its east, oceanic water started trickling in. It was a slow trickle for a few decades (dated with extraordinary precision to 26 years; see Loget *et al.*, 2005 and Blanc, 2006), during which time it began to cut a deep channel. But once it opened up sufficiently a huge cataract of water fell down the abyss. The western basin of the Mediterranean filled within ten years, and water spilled over to the eastern basin, filling it in a year (Blanc, 2002; Loget and van den Driessche, 2006). The new sea altered the climate of Europe and North Africa almost overnight. The increasing trend towards the south-westerly monsoon, as Tibet's influence magnified and the Mediterranean summer low disappeared, generated aridity, and modern types of desert, semi-desert and arid grasslands began to spread. The world's climatic and vegetation zones were beginning to increasingly resemble those of today. The warm, wet and forested planet was gone. Rainforests were shrinking in extent and woodlands had started to break up; this was a significant time in the evolution of the early human lineage (Finlayson, 2009).

In parallel with these events, Central Asia started to become increasingly arid in the Early Miocene (*c.* 22 mya). The trend towards a windier and drier climate intensified in the Late Miocene (there were peaks of dust accumulation at 13–15 and 7–8 mya) and especially in the Pliocene and Pleistocene, after 3.5 mya (Guo *et al.*, 2002). These events were related to the periodic expansion of sea ice in the Arctic Ocean and the appearance of ice sheets in northern Eurasia. We should recall here that the separation of azure-winged magpie populations coincided with the Pliocene intensification of Central Asian aridity, and this phenomenon has been implicated in the isolation of other birds across Eurasia, with consequent speciation that can be linked to periods of aridity when deserts isolated populations of forest and other birds (Voelker, 2010). Significant changes seem to have affected Palearctic mammals during this time, too, with a notable increase in hypsodont species (with high-crowned teeth and enamel allowing for greater wear) and grazers, reflecting the shift towards more open biomes (Fortelius *et al.*, 2006).

Birds of arid and montane habitats

Given this ecological scenario it is not surprising to find bird genera typical of arid habitats as well as those of rocky and montane habitats appearing for the first time in the Pliocene fossil record – the *Tetrax* bustards, *Charadrius* plovers, *Gyps* vultures, *Athene* owls, *Streptopelia* doves, *Falco* falcons, *Alauda* and *Eremophila* larks, *Pyrrhocorax* choughs, *Oenanthe* wheatears, *Monticola* rock thrushes, *Prunella* accentors and *Emberiza* buntings. The rate of genera extinction between Miocene and Pliocene was comparable to that between Oligocene and Miocene, not surprising perhaps given the ecological changes described above, but it was on a scale lower than at the Eocene-Oligocene transition (Table 1.1). If we attempt to provide an estimate of new genera per sub-epoch, taking the length of each into account, we find that there have been three peaks of genera formation: (a) Late Eocene; (b) Early Miocene; and (c) Early Pliocene-Early Pleistocene. We have to treat these results with some caution as there is likely to be a bias towards the present (since more recent fossils stand a better chance of preservation), and the first fossil dates need not represent first appearance. The Pliocene to Early Pleistocene stands out as a point at which many new genera make their first appearance. The Early Miocene is its prelude. In contrast, the Middle-Late Pleistocene seems insignificant in comparison.

CHAPTER 2

The changing ecology of the Palearctic in the Pleistocene

I showed in the previous chapter that the Pleistocene was relatively insignificant in terms of species formation and extinction, and that it was range shifts, contractions and expansions that instead characterised this period. The reasons for the difference with the preceding Miocene and Pliocene, when range shifts presumably also played some part, are the shortness of this period and the increasing small-scale climatic oscillations in the coldest and driest world since the K/T Event. I attribute the poverty of new species and genera to lack of time and directional climatic trends. The absence of significant extinction is, in my view, attributable to the fact that weeding had occurred in the Miocene and to a lesser degree Pliocene, so that the species that reached the start of the Pleistocene, and ultimately made it to today, were the survivors.

This message is reinforced in Figure 2.1, which shows the first recorded presence of present-day species in the European fossil record (Mlíkovský, 2002). The picture is clear: most species appear during the Pliocene and Early Pleistocene and very few do so after this (after *c.* 780,000 years ago). Bearing in mind first fossil records probably mean that the species were present before that, and seeing the clear peak in species during the Late Pliocene and Early Pleistocene, we can conclude that the emergence of modern Palearctic species (in most cases evolving from ancestral species present since the Miocene) happened around 3.5 mya and after that. This date coincides with the main period of Central Asian desert formation, and comes after the opening of the Strait of Gibraltar and the flooding of the Mediterranean Sea, with the consequent

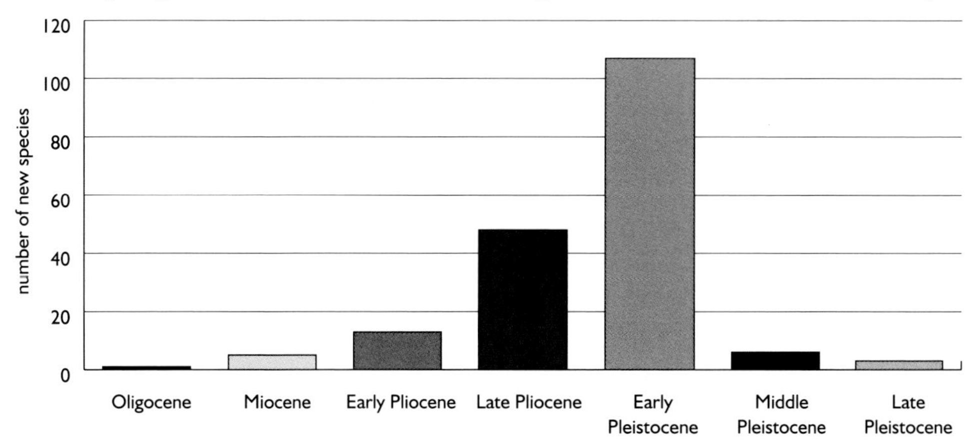

Figure 2.1 *First presence of Palearctic bird species in the fossil record. The importance of the Late Pliocene and Early Pleistocene in generating novelty at species level is clear, and contrasts with the comparative insignificance of the Middle and Late Pleistocene. Early Pliocene (5.33–3.6 mya), Late Pliocene (3.6–1.8 mya), Early Pleistocene (1.8–0.78 mya), Middle Pleistocene (0.78–0.125 mya), Late Pleistocene (0.125–0.01 mya). Note the boundary of the Pliocene-Pleistocene has recently been revised to 2.588 mya but I have kept to the earlier boundary as fossil sites employed in this analysis have used the earlier definition. For the purposes of this analysis, the importance of the Late Pliocene–Early Pleistocene is not affected.*

aridification of the climate of the Palearctic. The emergence of modern species therefore has much more to do with habitat fragmentation caused by aridity than with the waves of cold that engulfed the Palearctic, especially from the Middle Pleistocene. This result is supported by Voelker's (2010) genetic study of a range of Palearctic songbirds that became isolated and split into different species precisely at this time, and as a result of the expansion of the deserts of Central Asia.

THE CLIMATE OF THE PLEISTOCENE

During the Pleistocene, the world came under the increasing grip of cold periods (or glacials), usually though not always accompanied by aridity, and climatic instability. The harsh conditions intensified at the start of the Middle Pleistocene (*c.* 780,000 years ago), when the Earth's climate came under the grip of 100,000-year cycles of cold and warm (caused by regular and periodic changes in the Earth's orbit, tilt and wobble), which replaced the lower amplitude climatic oscillations (in 41,000-year cycles) of the Early Pleistocene (Ruddiman *et al.*, 1986). The world of the Middle Pleistocene was a colder one than that which preceded it in the Late Pliocene and Early Pleistocene, especially after 400,000 years ago. The glacials were preceded and succeeded by long periods when temperatures were returning from or heading towards warm maxima; these interglacials, which at times saw temperatures higher than those of today, were short, lasted around 10,000 years (Burroughs, 2005) and took up no more than 10% of the Pleistocene (Lambeck *et al.*, 2002 a, b). For much of the Pleistocene the Palearctic was a colder place than it is today.

Climatic variations

At millennial scales the dominant climate cycles are known as Dansgaard-Oeschger (DO) temperature oscillations. These were linked with the alternation between warm periods and glacials and stadials (shorter episodes of ice expansion during warm interglacials) in the North Atlantic (Dansgaard *et al.*, 1993; Alley *et al.*, 1999). Less frequent, and of shorter duration, were moments of intense cold which were the result of massive discharges of ice as the Laurentide Ice sheet surged through the Hudson Strait. These events are known as Heinrich Events (Heinrich, 1988; Bond and Lotti, 1995: Alley *et al.*, 1999). There were also brief periods of warming in the midst of glacials (interstadials), all of which added to the general climatic instability (A. Voelker, 2002). Typical of lower latitudes were alternating phases of rainfall (pluvials) and aridity (inter-pluvials) which would have had a particular bearing on Palearctic birds that migrated to these latitudes in the winter. These oscillations caused repeated advances and contractions of forest and open habitats in Europe and semi-desert in the Mediterranean and North Africa.

A characteristic feature of these cycles of cold and warm, wet and dry, was the speed of change. Global warming at the start of the last interglacial was estimated at 5.2°C per thousand years (Ruddiman and McIntyre, 1977) and temperatures rose by 7°C in 50 years at the end of the last glacial maximum (LGM) (Dansgaard *et al.*, 1989) while Stuiver and Grootes (2000) recorded 13 cold-warm transitions between 60,000 and 10,000 years ago, each of which took 50 years to complete. An early study revealed that the temperate forests of northern France were replaced by pine, spruce and birch taiga at the end of the last interglacial in the space of between 75 and 225 years (Woillard, 1979), and in Italy changes in vegetation were rapid over the last 102,000 years, with forest and wooded steppe biomes replacing steppe and back, at an average interval of 142 years (Allen *et al.*, 1999). These rapid changes in vegetation would ultimately have affected many terrestrial vertebrates, including birds, in the Palearctic (Finlayson and Carrión, 2007).

Mammals responded to the onset of the 100,000-year cycles at the start of the Middle Pleistocene by contracting or shifting geographical range, by immigration from northerly latitudes and by adaptation to the new conditions. There was extinction too, all of which contributed to a drastic reorganisation of the Palearctic mammalian fauna (Finlayson, 2009). In contrast, as we have seen, there was relatively little speciation or extinction among the Palearctic's birds, which instead seem to have responded by adjusting their geographical ranges. Specific examples will be discussed in the species accounts in the following chapters.

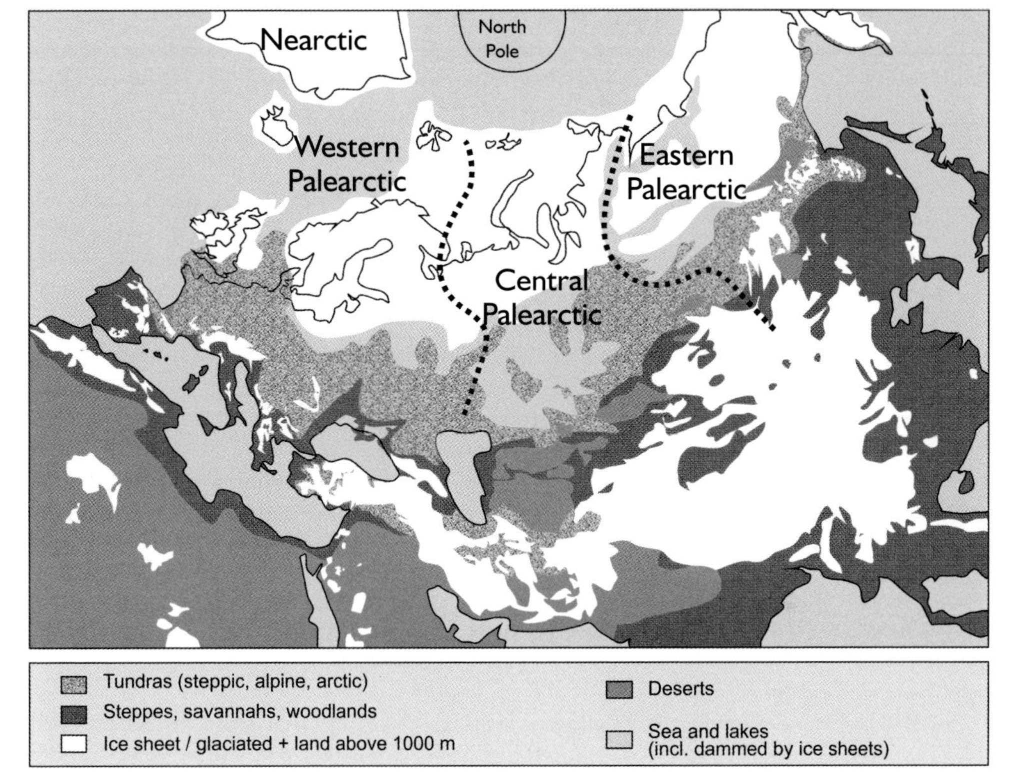

Figure 2.2 *The Palearctic at the last glacial maximum. Note the huge areas covered by ice, ice-dammed lakes, and tundra, all of which would have had a significant impact on the distribution of Palearctic breeding birds. This map shows the boundaries of the Western, Central and Eastern Palearctic adopted throughout this book. Map courtesy of Professor J. S. Carrión, University of Murcia.*

Sea levels

During the cold glacials, sea-level dropped between 90 and 130m below present levels (Shackleton and Opdyke, 1976; Rohling *et al.*, 1998, Lambeck *et al.*, 2002a), creating large expanses of coastal habitat in many areas that are now submerged. Sea-level high-stands (sometimes higher than present levels) were associated with interglacials. In general, sea-level rise appears to have been rapid. During the deglaciation leading to the last interglacial (known to geologists as the Marine Isotope Stage (or MIS) 6), sea-level rise was of the order of 20m per 1,000 years, and sea-levels reached between 2 and 12m above the present level (van Andel and Tzedakis, 1996). The subsequent drop towards the LGM was of the order of between -118 and -135m below present levels (Clark and Mix, 2002).

The barriers that were formed by the glaciations, and the land bridges that reopened as sea levels dropped with the cold, did cause isolation and reorganisation of geographical ranges, as the Central Asian desert formation had done previously. There seems to have been some level of lineage-splitting among some birds (Voelker, 2010) but at a smaller scale than in the Pliocene. It seems likely that the rapid shifts in climate did not allow sufficient time for much more, and what we do observe is part of a protracted process of lineage differentiation which, in the Pleistocene, occurred largely at the level of populations, subspecies or very closely related species. This differentiation seems to have taken place largely in refugia (places where animals and plants survived during the worst moments of the glaciations), particularly in the Iberian Peninsula, Italy, the Balkans, the Caucasus, and parts of southern Siberia. This has been amply documented for a variety of animals and plants, and it is in such refugia that a wide range of species, including humans, survived

(Hewitt, 1996, 2000; Willis, 1996; Taberlet *et al.,* 1998; Willis *et al.*, 2000; Taberlet and Cheddadi, 2002; Finlayson *et al.*, 2006). There is also increasing evidence that some species survived further north in what have been termed 'cryptic' refugia (Stewart and Lister, 2001), all of which adds up to a picture of survival of the most adaptable species in a range of geographical locations, from which they re-colonised other parts of the Palearctic when climatic conditions improved.

The last glacial cycle

The toughest period climactically of all was that from 125,000 to 10,000 years ago, the last glacial cycle, which covered the Late Pleistocene. Some of the remaining warm-climate mammals from the Palearctic (Straight-tusked Elephant *Elephas antiquus*, Narrow-nosed Rhinoceros *Stephanorhinus hemitoechus*, Barbary Macaque *Macaca sylvanus*, Hippopotamus *Hippopotamus amphibius*, European Water Buffalo *Bubalus murrensis*) did not make it and were gone early in this period, which came to be dominated by cold steppe-tundra mammals (Woolly Mammoth *Mammuthus primigenius*, Woolly Rhinoceros *Coelodonta antiquitatis*, Reindeer *Rangifer tarandus*, Musk Ox *Ovibos moschatus*, Irish Elk *Megaloceros giganteus,* Arctic Fox *Alopex lagopus*) of the Palearctic, with periodic westward range extensions of arid Central Asian mammals (particularly Saiga Antelope *Saiga tatarica*) as far as the Iberian Peninsula.

The severity of the last glacial cycle illustrates the radical changes that the birds of the Palearctic were subjected to. It was the culmination of the progressive climatic deterioration and increasing instability that marked the Pleistocene as a result of the progressive glaciation of the northern hemisphere (Finlayson, 2004). It marked the final contraction and extinction of all tropical and subtropical woodland from southern Europe, and the rise and geographical expansion of xeric (dry-adapted) plants during the LGM, around 20,000 years ago (Carrión *et al.*, 2000).

The last glacial cycle started with the significant warming and melting of ice around 130,000 years ago, which led to a brief interglacial (known as MIS 5e) when global temperatures were warmer than today. This was followed by a rapid climatic deterioration towards a glacial at 75,000 years ago (MIS 4), then a long period of ice withdrawal before the next, and most severe, advance towards the LGM (*c.* 20,000 years ago – MIS 2), when the Atlantic polar front reached a latitude of 38°N (the position of Lisbon) at the LGM (Calvo *et al.*, 2001). Between the two glacials, between 60,000 and 25,000 years ago, temperatures were lower than in the interglacials but higher than during the glacials. This was a long period, covering 35,000 years, which was typified by the high variability of the climate at the scale of millennia, centuries and even decades (van Andel and Tzedakis, 1996). Intense cold, aridity and climatic instability were the hallmark of the last glacial cycle that the Earth has experienced to date.

CHANGES IN REGIONAL ECOLOGY

The Eurasian Plain

This is the vast region of lowlands that spans the Palearctic, from the British Isles to the Pacific east. The way in which trees colonised this vast area during interglacials was heavily dependent on where they started from (*i.e.* the location of the glacial refugia) and the speed and duration of warm conditions. Some interglacials were more humid (or oceanic) than others, and this also added to the variability of the vegetation's response (Zagwijn, 1992). Much of Europe was covered by temperate mixed oak forest during the last interglacial (Vandenberghe *et al.*, 1998; Turner, 2002; Kukla *et al.,* 2002). This was an oceanic interglacial, a relatively uncommon type that only took up 12% of the whole of the last 500,000 years, and typified by a continuous sequence of geographical expansion of elm *Ulmus,* oak *Quercus,* hazel *Corylus,* yew *Taxus,* and hornbeam *Carpinus*, and then by fir *Abies,* spruce *Picea* and pine *Pinus.* The response of trees to warming during the continental interglacials was significantly slower.

At the other end of the scale, at the height of the LGM, two-thirds of the British Isles was under glaciers, as was much of Scandinavia, the Baltic and the Alps. The rest of Europe north of the Mediterranean region was under permafrost (CLIMAP, 1976; Maarveld, 1976). In ice-free areas bare ground and a strange flora predominated, usually low in height and poor in cover (Birks, 1986; Zagwijn, 1992; Finlayson and Carrión, 2007). In general, the Eurasian plain during this glaciation was desolate, with tundra, pockets of sparse vegetation and scattered and isolated refugia of trees (Figure 2.2). The severity of the climate was made worse by abruptness of the changes (Allen *et al.*, 1999; Barron *et al.*, 2004; Tzedakis, 2005). This severity is reflected by temperature estimates based on the presence of beetles that are sensitive to such changes (Coope, 2002); during the build-up to the LGM, between 40,000 and 25,000 years ago, the mean July temperature in the British Isles (an oceanic temperate region) was close to 10°C and winter temperatures were around -20°C. Summer temperatures were close to the lower limit below which tree growth could not take place.

In summary, it seems that the vegetation of the Eurasian plain, in the west at least, oscillated between conifer woodland, with shrub tundra to the north, when it was warm, and a tundra-cold steppe mosaic with polar deserts to the north during the coldest periods (van Andel and Tzedakis, 1996; Finlayson and Carrión, 2007).

To the east, on the Russian plain, an open and harsh steppe replaced the dense temperate forests of the interglacials (Rousseau *et al.,* 2001) as the climate became increasingly arid and continental prior to the LGM (Soffer, 1985). Large tracts of land suffered extremely low temperatures, high aridity and a covering of permafrost. The cold steppe was a unique combination of tundra, steppe and grassland, punctuated by sparse arboreal vegetation, even during the coldest periods, of pine, beech and oak along the rivers valleys. These may have provided refugia for some species of birds.

During a brief respite, around 33,000–24,000 years ago (known as the Briansk Interstadial), a mix of tundra, forest-tundra and tundra-steppe covered the Russian plain. The southern limit of Arctic plants was nonetheless 1200 km further south than it is today, and it reached a further 600 km further south during the LGM. The southern Russian plain was milder, especially around the Black Sea, where steppe and forest-steppe predominated. The Crimean Peninsula, with its topographic variability that contrasted with the stark flatness to the north, allowed a greater diversity of vegetation to survive (Markova *et al.*, 2002). These temporary climatic improvements were also noticeable to the north. In northern Siberia, open larch forest with alder *Alnus glutinosa* and dwarf birch *Betula nana* developed in the Taimyr Peninsula between 48,000 and 25,000 years ago (Andreev *et al.*, 2002).

With the passing of the LGM there was a noticeable thermal improvement by 14,700 years ago, which lasted 2,000 years, and boreal woodland started to replace steppe and tundra in north-western Europe. Spruce forest and birch-conifer woodland followed (Huntley and Birks, 1983). This was the time when humans are thought to have entered North America. A brief reversal, known as the Younger Dryas, between 12,900 and 11,600 years ago (Burroughs, 2005) brought the tundra back as far south as France and the British Isles, but rapid climatic amelioration leading to the conditions of the present day followed after 11,000 years ago, and today's climatic world had taken shape by 10,000 years ago. Woodland took over as steppe retreated to Central Asia and tundra to the Arctic.

One dramatic aspect of the Last Glacial Cycle, of huge potential importance to populations of Palearctic waterbirds, was the emergence of gigantic inland freshwater seas across Siberia during this time (Figure 2.2). These were created as ice sheets covered the Russian Arctic, preventing the outflow of water from the Siberian rivers northwards. The corresponding wetland habitats that would have emerged on the margins must have attracted vast numbers of waterbirds and the present, fragmented, distribution of a number of these species (which we will see in the species accounts) may be a reflection of the loss of this former waterbird paradise. When the climate warmed the ice dams became unstable and large areas suffered cataclysmic superfloods (Rudoy, 2002; Finlayson and Carrión, 2007).

I end this section with a note of caution. For too long we have been given the image, partly reflected above, of a Eurasian plain that was either forested or denuded of vegetation. But we have seen already that glacials and interglacials only covered a small fraction of the time, most of which was taken up toing and froing between the two extremes. This means that for large periods vegetation was also shifting between these extremes, creating regional mosaics of habitat, many of which were neither dense forest nor treeless plains. It seems that even during the warm interglacials patches of open vegetation persisted within forests, along floodplains, on areas with infertile soils, and in continental and sub-Mediterranean areas (Svenning, 2002). This observation has significant implications for the survival of Palearctic birds during glacial cycles, and has been hitherto unappreciated.

The mid-latitude belt

One major change that I have not yet commented on is the emergence of the mountain chains that resulted from the process of collision of Africa, Arabia and India into Eurasia. We have seen how the uplift of the Tibetan massif had profound implications for the climates, biomes and animals of the Palearctic. But the change that I am referring to here was the emergence of a belt of varied topography, contrasting with the flat expanses of the Eurasian plain to the north, which ran from the Himalayas west to the Iberian Peninsula. I have referred to this mid-latitude belt (MLB) earlier in this book (see p. 10). The important point to note now is that it created, from the Miocene onwards, a discontinuity in the arrangement of the vegetation belts that had previously followed a latitudinal pattern. On the Eurasian plain, arctic, boreal and temperate belts continued to respect latitude, but the MLB became a source of fragmentation and discontinuity that would have significant implications for speciation and geographical range in many species. In addition, the potent combination of high mountains, deserts (to which we can include the Sahara) and the Mediterranean Sea created an effective barrier that severed the connection between the Palearctic and the tropics, except in the Far East.

The best-studied part of the MLB is the west, where the Mediterranean and proximity to the Atlantic generated unusual oceanic conditions that contrasted with the much more arid conditions of the central and eastern MLB, where the highest mountains also happened to be located. The climatic conditions of the MLB would have varied in a similar manner to those of the Eurasian plain, except that the regions closest to the high mountains would have suffered severely from the local formation and expansion of mountain glaciers and the loss of water supply to the continental areas, which consequently became massive deserts. Geography and local climatic peculiarities controlled vegetation much more closely than on the Eurasian plain (Suc *et al.*, 1994).

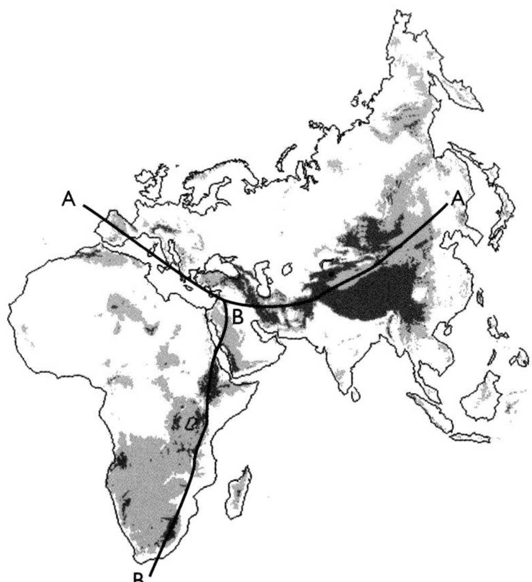

Figure 2.3. *The mid-latitude belt of heterogeneous topography. Land between 750m and 1500m in light grey; land over 1500m darker grey. Line A–A indicates the approximate mid-line of the belt. Line B–B indicates the southward extension to South Africa.*

The width of the Mediterranean and the major mountain masses of the MLB became barriers for plant and animal movement. One feature of the MLB was the extension of high ground well south, and this feature permitted the corresponding intrusion, and later survival, of temperate and boreal vegetation (Rivas-Martínez, 1981, 1987; Zagwijn, 1992). The effect was not limited to plants; we will discover a number of birds with geographical ranges that reflect these southward intrusions during glacials.

A major difference from the Eurasian plain across much of the MLB was the importance of moisture as a critical bioclimatic variable, with temperature playing a supporting role (Tzedakis, 1994; Willis, 1996). The influence of wet and dry cycles, which resembled those of tropical latitudes, would have varied at different points of the glacial cycles (Narcisi, 2001). Many areas of the MLB would have experienced harsh climatic conditions during glacials and stadials (Rose *et al.*, 1999) but there would have been areas of relative stability (Prokopenko *et al.*, 2002) and localised refugia which, even during the coldest periods, would have retained a warm and wet climate. The most notable of these were at the two extremes of the MLB, the south-western Iberian Peninsula and south-eastern China (Finlayson and Carrión, 2007). Significant patches of Mediterranean vegetation survived even the coldest and driest moments, and these areas provided major refugia for many bird species.

Shifts from forest cover to open vegetation in the MLB were even more abrupt, given the influence of altitude bands in close proximity to each other (Peteet, 2000), than on the Eurasian plain. Tzedakis (1994) reported that between 70 and 80% of each glacial-interglacial cycle was taken up by intermediate conditions between forest and treeless habitats. This observation recalls that made for the Eurasian plain, but in the MLB, with its topographic heterogeneity at small scales, the vegetation mosaics and patterns were more diverse and continuously changing.

In Greece, the intermediate periods were characterised by steppe-forest, forest-steppe and steppe vegetation, with the extremes being desert-steppe and forest (Tzedakis, 1994). In Italy open and arid habitats were also typical of glacials with less open, humid, environments during interglacials (Montuire and Marcolini, 2002). In Iberia, habitats ranged from open steppe in the central areas during the coldest moments through conifer steppe during cold and arid phases on the coast to Mediterranean woodland in the warmest times. In the south-western Iberian refugium conditions varied little from warm savannas and woodlands at any stage, with only the intrusion of montane pines in the coldest moments being indicative of change. These southerly latitudes never received the cold steppe-tundra mammal fauna, which did not stray much further south than the level of Madrid, but Arctic seabirds did reach the latitude of Gibraltar, suggesting increased summer-winter seasonality (Finlayson and Carrión, 2007).

One final note of interest concerns the differences between the various glacial refugia in terms of the nature of the vegetation that each harboured. The western Balkans, and to a lesser degree the Alps and the Italian mountains, seem to have been the major European refugia for broadleaved trees during the last glaciation (Bennett *et al.*, 1991; Zagwijn, 1992; Willis, 1996; Tzedakis *et al.*, 2002). The Iberian Peninsula was, instead, a refugium of sclerophyllous (dry scrub) Mediterranean vegetation (Carrión *et al.*, 2000; Finlayson, 2006). The Near East and south-western Asia were arid and not as important as refugia for temperate plants. In North Africa, the southward migration of the dry subtropical high-pressure zone during glacials generated aridity with a reduction of Mediterranean woodland at the expense of steppe and semi-desert (Hooghiemstra *et al.*, 1992; Dupont, 1993). These vegetational differences had implications for birds also reliant on these refugia.

The Sahara

The Sahara, the northern half of which marks the southern boundary of the Western Palearctic, is a major feature of direct and critical importance to Palearctic birds, especially those that perform long migrations over it. This boundary between the Palearctic and the Afrotropical zoogeographical regions has not always occupied this position (around 30°N) but has instead shifted significantly since the Miocene (starting 23 mya), reaching at times a northerly position at 50°N (Pickford and Morales, 1994). During these moments, the African influence on the Palearctic was high and is reflected, for example, in the presence of crocodiles in the Iberian Peninsula at 23–22, 18–16, 9 and 7.5–5.0 mya. When in this position there was climatic continuity between west and east from Iberia to China, which facilitated faunal interchange. The last time this occurred was in the Pliocene, around 3 mya, a time that we have already identified as critical for Palearctic birds, with significant disruption of geographical ranges.

A major shift towards increased cooling and aridity of the African climate started at the end of the Pliocene, around 2.8 mya, and it marked the wet and dry oscillations that have characterised it since that time (deMenocal, 1995). In north-east Africa the climate became progressively drier for long periods that were interspersed by short pluvial episodes (Crombie *et al.*, 1997). During the Middle Pleistocene a notable increase of aridity occurred after 200,000 years ago (Jahns *et al.*, 1998).

The north-west African climate has been very variable over the past 100,000 years, with a succession of wet and dry phases lasting between 1,000 and 10,000 years. The Sahel was largely vegetated for much of the last glacial cycle; the dominant vegetation was grassland that showed sudden and sharp peaks of expansion and contraction, with major expansions around 105,000, 80,000 and 10,000 years ago in the Sahara itself, at times when trees became important in the Sahel (Tjallingii *et al.*, 2008). The most recent wet phase lasted from 10,000 to 5,000 years ago, once the Holocene global warming had started; at this time the Sahara was covered by tropical grasslands, with forests and large permanent lakes (deMenocal, 2008). This was the last time that 'Lake Mega-Chad', which had its origins in the south-easterly monsoonal regimes of the Late Miocene (see p. 21), extended over a huge area (more than 400,000 km^2 – larger than the Caspian Sea, and up to 173m deep) (Drake and Bristow, 2006; Schuster *et al.*, 2005; Sepulchre *et al.*, 2008).

The ecology of this vast area, comparable in size to the United States, was rich and beautifully represented in the rock art painted by people who lived in a lush environment where desert now reigns (Roberts, 1998). This also means that as recently as 5,000 years ago, Palearctic-African migratory birds had an easy passage across the Sahara, which must also have been a hugely important breeding area for many species of wetland and grassland birds. But the abrupt loss of this ecologically rich region to desert that followed was part of a series of millennial dry-wet oscillations that had affected this region since the Middle Pleistocene, and embedded in the deeper trend toward aridity that has affected this region of Africa since the end of the Pliocene.

Tropical wintering areas of Palearctic migrants

A description of the tropical areas of Africa and southern Asia is beyond the scope of this book, but comment is warranted before concluding this chapter. The cold-warm cycles that affected the northern hemisphere also were largely replaced by wet-dry oscillations in the mid-latitude belt and Sahara and were also the dominant force in the tropics (deMenocal, 1995); this was reflected in expansion and contraction of tropical rainforest and mangrove at the expense of savanna and grassland in Africa (Lezine *et al.*, 1995). These broad changes would have affected differentially the wintering areas of Palearctic migrants which would, overall, have benefitted during the driest phases when rainforest extent was reduced and savanna and grassland extent was greatest, given that few Palearctic migrants winter in tropical rainforest (Moreau, 1972). South-east Asia remained wetter than Africa but was not exempt from the effects of aridity. There, too, rainforest gave way to savanna at times that coincided with significant lowering of sea-levels, which exposed large areas of land (e.g. Sundaland, a vast area including the islands of modern Malaysia and Indonesia east to the Wallace Line) (Bird *et al.*, 2005). In between, Arabia and southern Asia resembled North Africa more than South-east Asia in the response of vegetation to climate change, with huge areas taken over by desert during dry phases, and dry forest and savanna occupying tropical areas of south Asia (Field and Lahr, 2005).

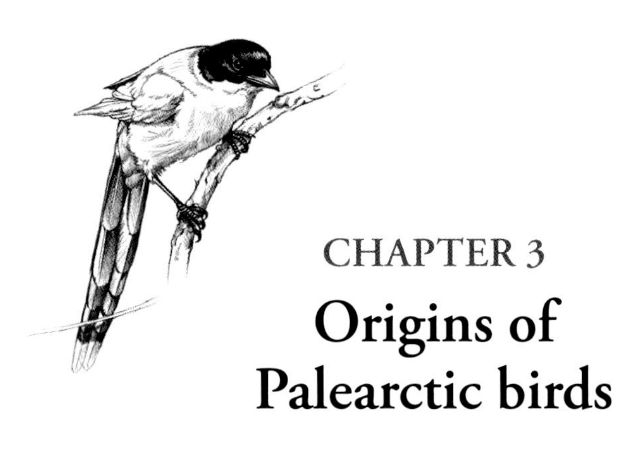

CHAPTER 3
Origins of Palearctic birds

The preceding chapters have set the climatic and ecological scene that led to the origins of the Palearctic avifauna. In the species chapters that follow we will see how these origins appear to have been *in situ* within the large area of tropical and subtropical bioclimates that constituted the Eurasian-African land mass, together with an element of immigration that was overwhelmingly from South-east Asia, with a smaller fraction from the Nearctic and a relatively tiny contribution from Africa. The Palearctic avifauna's lineages are ancient, in most cases dating back to the Cretaceous (and therefore predating the K/T event that led to the extinction of the dinosaurs). The climatic events of the Eocene and Oligocene seem to have weeded out many genera that were unsuited to the cooling and drying world that was developing, while others were becoming confined to the tropical forests that were being restricted to low latitudes. Many of these genera still persist in the tropics of America, Africa and southern Asia and are effectively imprisoned within these warm and wet, mainly forested, environments.

Groups that remained within the Palearctic entered a phase in which new body plans, suited to a world of cool and dry bioclimates, emerged during the Miocene, a point that saw the end of the last lineages that had done well in the warm world of the early Tertiary and the beginning of a radiation of new forms that were able to handle the new ecological conditions. The culprit of this ecological change seems to have been the uplift of the Tibetan plateau, assisted by the re-organisation of ocean circulation resulting from the rearrangement of the continents. Ultimately, the extant Palearctic avifauna is the product of plate tectonics.

In the overall picture of climate change and the origins of new genera and species of Palearctic birds, aridity seems to have played a greater role than temperature. Aridity in northern Africa and Arabia replaced a land of mega-lakes at the start of the Pliocene. Wet and dry oscillations became typical of the African climate after 2.8 mya, and the Sahara experienced repeated phases of wet and dry that continued to the present day, the last wet period having ended as recently as 5,000 years ago. Similarly, the Central Asian deserts had been forming since the Miocene. By the end of the Pliocene a wide band of desert and semi-desert, stretching from north-west Africa, across the Sahara and Arabia and across Central Asia, dominated the middle and lower latitudes of the Palearctic.

Added to this new phenomenon were the expansion of the grasses and the emergence of grasslands and savannas across this belt. A huge area, once the domain of tropical forests, was now taken over by largely treeless habitats that fluctuated, depending on climate, between savannas, grasslands and wetlands on the one hand and deserts and semi-deserts on the other. A new habitat of huge extent, and a constant in this changing world had also arrived. This was the world of rocky habitats that provided a continuous habitat chain from the Himalayas to the Atlantic margin of the Palearctic.

So this was the new world the Palearctic avifauna faced at the start of the Pleistocene. These groups had run the gauntlet of climate deterioration and seemed built to last. They were the survivors. So when the glaciations hit the Palearctic, these birds had enough in them to allow them to survive everything that was thrown at them. The climatic changes of the Pleistocene were too many, too short and too fast to allow novelty and, instead, birds moved around the continental land mass, some faring better than others, as ecological conditions changed. Those that did well in wetlands received the bonus of the formation of large inland freshwater seas in Siberia.

The species accounts

Having introduced the story of the climate and the birds of the Palearctic, it is now time to take a closer look at the species. The data on which the analyses are based are presented in Appendix 1 (see p. 242), which includes 862 species, 556 of which breed in the Western Palearctic. This book concentrates solely on species that breed in the Palearctic. I do not include vagrants or migratory species that pass through the region but do not breed there. My analysis of information is based exclusively on climatic conditions, habitat and ecology in the breeding areas.

When I plan a long trip in the car, I take out a map with a scale that allows me to plan the entire route. These maps summarise the route and sacrifice a great deal of detail that would get in my way. The essentials are there, the main roads, main towns and distances that will allow me to get to the desired location. Once there a more detailed map allows me to navigate streets and roads to find my final destination. That detailed map would have been useless at the start of the journey as too much detail would have prevented me from understanding the route. In the same way, my aim here is to provide a road map, which is the history of the birds of the Palearctic, from which we can then look at species in greater detail. But we cannot begin to understand the detail without the big picture. This is why I never fully understood my swifts (see p. 8) – I was looking at them from too close. But that proximity makes sense now that I understand their bigger picture. Like the road map, to be able to make sense of the mass of data that emerge from looking at so many species, I have had to simplify the information. So I look at forests as a structural habitat in which trees occur at high density but I dispense with distinguishing between types of forest, not because these details are not important but because it would provide unnecessary clutter and prevent us from getting to our destination.

The dataset has the following headings:

(a) Bioclimatic tolerance

Each species is given a rank based on an examination of the breeding distribution of each species globally and compared with a bioclimatic map of the world (see Figure 1.3c, p. 19). Tolerance is a measure of how many of the bioclimates on the map the species occupies:

A = specialist (occupies between 1 and 20%)
B = semi-specialist (21–40%)
C = moderate (41-60%)
D = semi-generalist (61–80%);
E = generalist (81–100%)

Bioclimatic tolerance is a reflection of a species' ability to live in a wide range of ecological contexts in a range of climates. It is not a measure of ability to withstand extremes of climate, but rather to survive in a range of climate-modulated environments, hence bioclimates.

(b) Latitude

Each species is allocated a latitude band on which its geographical range is centred.

A = arctic (70°N)
B = boreal (60°N)
C = temperate (50°N)
D = mid-latitude belt (warm species) (40°N)
E = subtropical (30°N)
F= multilatitude – species occupies several latitude bands.

(c) Temperature

From the global maps used to calculate tolerance, each species is given a position on a temperature gradient that goes from cold (1%) to hot (100%). Species at the cold end (between 1% and 20%) are A; this is followed by B (21–40%), C (41–60%), D (61–80%) and the warmest, E (81–100%).

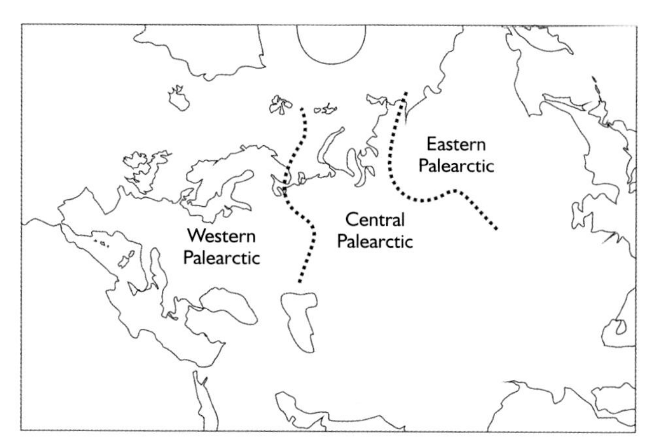

Figure 3.1 *The Western, Central and Eastern Palearctic. The southern boundary has shifted between 30° and 50°N on a number of occasions in the last twenty-five million years (Pickford and Morales, 1994). The current boundary is close to the 30°N southern limit.*

(d) Humidity

As with temperature, I allocate a category for each species, from A (at the dry end of the gradient, 1–20%) to E (at the wet end, 81–100%).

(e) Montane

I record species that are montane. In cases where they are strictly montane I do not allocate a temperature or humidity rank, which would be impossible to allocate from the large-scale map.

(f) Palearctic occurrence

Three columns record the presence (+) or absence (-) of each species in the Western, Central and Eastern Palearctic. Symbols without parenthesis indicate a widespread distribution. Symbols in parentheses indicate fragmented or marginal distribution. The boundaries of the Western, Central and Eastern Palearctic are shown in Figure 3.1.

(g) Habitat

Two columns record feeding and nesting habitat (as they are not always the same). Letters indicate the broad habitat used:

F = forest, habitats with a high density of trees
O = open, habitats without trees
M = mixed, habitats between F and O, including savanna, shrubland and mosaics
R = rocky
W = wetland, including all kinds of water habitats, including coastal ones, except marine
Ma = marine
A = aerial.

(h) Diet

A column indicating the diet category:

O = omnivore – includes plant and animal matter
M = mixed-strategy carnivore – includes a range of animal matter in the diet
E = endothermic carnivore – only consumes warm-blooded prey
I = insectivore – only consumes insects and other arthropods
H = herbivore – only consumes plant matter;
F = fish
N = necrophyte – eats carrion.

(i) Migratory Behaviour

A column recording whether a species is fully migratory (M), partially migratory (P, with only some populations migratory) or sedentary (S).

A note on the species order

The arrangement of the chapters follows the recent genetic analysis by Hackett *et al.* (2008). Species and genus-level taxonomy follows a combination of Sibley and Monroe (1990), Clements (2007) and various papers, cited in each section.

Given that recent analyses based on genetic relationships have significantly rearranged the relationships of major bird taxa, traditional ways of listing species are no longer tenable. For example, Hackett *et al.*'s results show that cuckoos form part of a cluster of groups alongside cranes, rails and bustards (and other examples will become apparent throughout this book). Therefore, I have chosen to represent species according to their phylogentic relationships, and I start with the groups that have separated out most recently, working backwards towards the most ancient. Passerines are therefore at the beginning, reflecting their numerical importance as well as their significance as an major group that expanded relatively recently in the Palearctic.

I have supplemented Hackett *et al.*'s arrangement with Treplin *et al.*'s (2008) detailed classification of the passerines. I divide the species into eight groups which correspond to Hackett *et al.*'s major nodes and I subdivide the main node (which is the largest and contains the passerines, parrots, falcons, owls, raptors and terrestrial non-passerines) to make it more manageable.

A series of numbers appears after each of the lineages in the list below. For Group 1 (Passerines and allies) this reads 510/3296 – 15.47%. The first number is the number of species in the Palearctic, the second is the world total, with a percentage of one to the other.

Group 1: Passerines and allies (510/3296 – 15.47%)
- (a i) Corvoidea – crows and allies (37/265 – 13.96%)
- (a ii) Sylvioidea – warblers and allies (127/941 – 13.5%)
- (a iii) Paroidea – tits (18/71 – 25.35%)
- (a iv) Muscicapoidea – flycatchers and allies (106/698 -15.19%)
- (a v) Passeroidea – sparrows and allies (116/411 – 28.22%)
- (b) Psittaciformes (parrots – non-Palearctic group)
- (c) Falconidae – falcons (13/63 – 20.63%)
- (d) Coraciiformes etc. – terrestrial non-passerines (26/348 – 7.47%)
- (e) Strigformes – owls (23/250 – 9.2%)
- (f) Accipitridae and Pandionidae – diurnal raptors (44/249 – 17.67%)

Group 2: Charadriiformes – shorebirds and allies (145/352 – 41.19%)

Group 3: Waterbirds, including pelicans (Pelecaniformes), storks and allies (Ciconiiformes), shearwaters and petrels (Procellariiformes) and divers (Gaviiformes) (60/282 – 21.28%)

Group 4: Mixed land-water bird group, including Gruiformes (cranes), bustards (Otididae) and cuckoos (Cuculiformes) (29/262 – 11.07%)

Group 5: Apodiformes and Caprimulgiformes, swifts and nightjars (13/167 – 7.78%)

Group 6: Pigeons and allies (Columbiformes), sandgrouse (Pteroclididae) tropicbirds (Phethontidae), flamingos (Phoenicopteridae) and grebes (Podicipediformes) (29/354 – 8.19%)

Group 7: Gamebirds and waterfowl (76/341 – 22.29%)
- (a) Galliformes – game birds (24/215 – 11.16%)
- (b) Anseriformes – waterfowl (52/126 – 41.27%)

Group 8: Ratites – of limited relevance to the Palearctic and not considered further in this book (1/10 – 10%)

The world number of species requires explanation. This figure is based on an additional database that was compiled for this book and includes only those families that have at least one species represented in the Palearctic. Families like the New World Warblers (Parulidae), penguins (Spheniscidae) of the parrots (Psittacidae) were therefore left out of the database. The database was compiled from Sibley and Monroe (1990) with appropriate updating from later studies referred to in the texts of the chapters that follow. The categorisation applied to the Palearctic database was also done for this second database (see Appendix 1), so bioclimatic tolerance, habitat and migratory behaviour was recorded for each species. This information allowed for comparisons to be made between Palearctic and non-Palearctic species within lineages. A total of 5,064 species is included in the database, which is roughly half of the world's species. It means that the Palearctic birds are derived from lineages that represent half of the birds of the world, the others having no representation today although we have already seen that a number did so in the remote past. Of these 5,064 species, 863 (17%) breed in the Palearctic and 557 (64.5% of the Palearctic species) do so in the Western Palearctic. These species will be examined in the following chapters.

An introduction to the species accounts

Chapters 4 to 18 will focus on Western Palearctic birds at the level of the individual species, to help us understand why we observe many species of one particular family or genus while in others we only observe a few. Can we say anything about the success of particular lineages of birds and the limitations of others? I will adopt a multi-scale taxonomic approach, looking at the species first by the major lineages that we have already established (the separations of these lineages is ancient and transcends Palearctic considerations). Next I will look at families, which also represent ancient separations of particular groups of birds. In some cases (e.g. warblers Sylviidae) I may add a further category – the subfamily or the tribe – which helps to separate out particular ways of life within the family. Finally, I will look at the genus (and the species within) which is, in my view, the taxonomic category that reflects particular adaptive strategies. These are also ancient, but less so than the higher taxonomic classes, and the existing Palearctic genera were around before the Pleistocene; as we have seen, most were there in the Miocene and Pliocene.

Each chapter will feature an introduction to the families, including their history and relationships (the treatment varying depending on the complexity of each group and the degree to which their genetics have been studied); a section on climate; a section on habitat, which will situate the Palearctic species in the context of the group globally; a description of the fossils known from the Palearctic and particularly Europe, which is the best-studied area; and finally a genus-by-genus treatment of the Palearctic species themselves.

CHAPTER 4

Corvoidea – shrikes, crows and orioles

Group 1, subgroup a (i)

This is the first of five chapters covering the passerine superfamilies that are represented in the Western Palearctic. They are all members of the oscine suborder Passeri; the suborder Tyranni (the suboscines) is not represented (Treplin *et al.,* 2008). In Sibley and Monroe's (1990) classification, the Tyranni, a largely Neotropical group but with representatives in the Old World tropics, comprise 291 genera and 1,151 species. By contrast, the Passeri comprise 870 genera and 4,561 species. If the identification of a broadbill (currently African and Asian Eurylaimidae) from the Early Miocene of Germany is correct (Mlíkovský, 2002), it points to the distribution of the Tyranni once being much wider, and including the Western Palearctic.

Group 1 in our classification is huge, with more than half (59.2%) of all the species in the world. It includes the passerines, falcons, terrestrial non-passerines (rollers, bee-eaters, hoopoes, etc.), owls and diurnal raptors. The only subgroup that either never made it or vanished from it very early on was the parrots. Group 1 is a very diverse group that represents an ancient radiation of species; it would be unmanageable, and probably not very informative in the context of this book, to deal with all of them together. So I will tackle the various subgroups, which we have already identified in the previous chapter, in order, following the genetic work of Hackett *et al.* (2008). In this chapter I cover the superfamily Corvoidea, which comprises the shrikes (Laniidae), corvids (Corvidae) and orioles (Oriolidae). Sibley and Monroe (1990) allocate the orioles to the Corvidae but here I follow Clements (2007) in separating them into their own family.

The Corvidae and Oriolidae are represented by 23 Palearctic species. These represent 15.44% of the world species belonging to these families. The corvids, with 21 species, are better represented (17.5% of the world's species) than the orioles (2 species, 6.9%). The other family in this group, the shrikes (Laniidae), has a much stronger proportional Palearctic component with 14 species, representing 38.89% of the world's shrike species. This indicates that the shrike model has done particularly well, in terms of species, in the Palearctic but the oriole one has been very restricted. The corvids have produced the most Palearctic species but fewer, in proportion to the global number of species, than the shrikes. The distribution of the Corvoidea is even across the Palearctic with 22 species in the Western, and 20 each in the Central and Eastern. The corvids are marginally under-represented in the Western (12 species against 14 in Central and Eastern) but the shrikes are better represented (10 species against 5 in Central and Eastern). Each sub-region of the Palearctic is home to a single species of oriole.

The corvids have been the most successful at achieving high levels of bioclimatic tolerance, with four generalists and three semi-generalists among them (see Appendix 1). By contrast, the shrikes appear to be much more specific to particular bioclimatic conditions, and there are no generalists or semi-generalists among them. This difference may partly explain the high number of shrike species, since bioclimatic specialisation leads to range fragmentation and the conditions for speciation. The high number of shrike species in the Palearctic may therefore represent high levels of range fragmentation. This is borne out by the distributional data of the groups: 8 of 21 corvids (38.1%) have pan-Palearctic ranges but only 1 of 14 shrikes (7.14%) does; 9 of 14 shrikes (64.29%) occur in only one Palearctic sub-region while only 10 (47.62%) of the corvids do. The orioles, though few, are closer to the corvid than the shrike pattern (Appendix 1).

LANIIDAE

The shrikes are an Old World family with two representatives in the New World (Loggerhead Shrike *Lanius ludovicianus* and Northern *L. invictus*). Of the three genera in the family, two (*Corvinella* and *Eurocephalus,* with two species in each) are wholly African, and the third (*Lanius* with 32 species) includes Asian and Palearctic as well as African species. There are no species in Australasia. The diversification of the *Lanius* shrikes appears to date to the Miocene/Pliocene boundary around 6 mya (Gonzalez and Wink, 2008), which would be consistent with the break-up of forests at the expense of savannas and grasslands that would have provided ample opportunities for open-vegetation predators.

Three Palearctic species are closely related and separated off at an early stage from the remaining shrikes: Red-backed *L. collurio,* Isabelline *L. isabellinus* and Bull-headed *L. bucephalus* (Gonzalez and Wink, 2008). With slight or no overlap, they appear to be west-east geographical counterparts across the Palearctic. To this group we should probably add the south Asian Bay-backed Shrike *L. vittatus* and the East Asian Brown *L. cristatus* and Tiger *L. tigrinus* Shrikes. The Palearctic species are highly migratory, especially the two Western Palearctic species (Red-backed and Isabelline) and the Brown and Tiger Shrikes. The winter quarters are also geographically separate, Red-backed in southern Africa, Isabelline in east and north-east Africa and south Asia (generally north-west of the largely resident Bay-backed Shrike), and Brown, Tiger and Bull-headed in south-east Asia.

A second group includes the Lesser Grey Shrike *L. minor* and the Common Fiscal *L. collaris* of sub-Saharan Africa (Gonzalez and Wink, 2008). The south and south-east Asian Long-tailed Shrike *L. schach* is probably part of this group (Mundy and Helbig, 2004). The Woodchat Shrike *L. senator* is more distantly related than the Lesser Grey and Common Fiscal are to each other but nevertheless fits within this shrike grouping (Gonzalez and Wink, 2008). These species are therefore geographical counterparts: Woodchat and Lesser Grey west and east within the Western and Central Palearctic, Common Fiscal (and presumably other Fiscals) in sub-Saharan Africa, and Long-tailed Shrike in southern Asia.

The Masked Shrike *L. nubicus* of the southern Palearctic appears unique, taking up a position between the previous group and the larger shrikes of the 'Great Grey' complex (Gonzalez and Wink, 2008). The 'Great Grey' complex is not well understood and requires further study. The birds seem to share a common ancestor from the beginning of the Pleistocene, *c.* 1.8 mya, and the radiation of the species in this group may, in part at least, be a response to habitat rearrangements during the glaciations (Klassert *et al.,* 2008). The North American Loggerhead Shrike *L. ludovicianus* is the basal member of the group that diverged first from the rest. Next to separate out was the Chinese Grey Shrike *L. sphenocercus.* This species, intriguingly, forms a cluster with the Southern Grey Shrike *L. meridionalis* of the Iberian Peninsula and the North American species *L. invictus* (usually regarded, incorrectly, as a subspecies of Great Grey Shrike *L. excubitor*). A second, distinctive, cluster links the Middle Eastern Levantine Grey Shrike *L. aucheri* with the Central Asian Steppe Grey Shrike *L. pallidirostris* and the Great Grey Shrike *L. excubitor* of the northern Palearctic. This cluster is associated with the North African Algerian Grey Shrike *L. algeriensis* and the Canary Islands Grey Shrike *L. koenigi* (Klassert *et al.,* 2008).

The situation thus requires revision but it would seem that there have been two lineages within this group: (1) the North American Loggerhead Shrike and the temperate-humid Chinese, Iberian (*L. meridionalis*) and North American Northern Shrike; and (2) a group inhabiting drier environments across the northern Palearctic, Central Asia and North Africa. A possible interpretation is as follows: during the relatively mild and wet early part of the Pleistocene a population of grey shrikes became distributed widely across the Palearctic from Iberia to China and into northern North America. The lineage that lead to the Loggerhead Shrike of North America separated around 1.2 mya, probably during a cold period that isolated them from Old World populations. Subsequently, a second entry into North America from this stock led to the Northern Shrike. The Chinese Grey Shrike *L. sphenocercus* had separated slightly earlier from this same lineage which was probably widespread across Eurasia. With the cooling and drying associated with the Pleistocene, the North American population adapted, but Palearctic populations living at the temperate humid extremes (Iberia and China) became isolated and eventually became two distinct species (which we should call the

Southern Grey Shrike *L. meridionalis* and the Chinese Grey Shrike *L. sphenocercus*). While this may at first seem counter-intuitive, the North American population should therefore be considered either a subspecies of the Southern Grey Shrike (thus *L. m. invictus* not *L. e. invictus*) or, perhaps more appropriately, a species in its own right (Northern Shrike *L. invictus*).

The shrikes of the second lineage, living in continental areas of the Palearctic, adapted to the new conditions and diversified within dry or arid contexts. They should probably be given specific status as part of a Great Grey Shrike superspecies: northern (Great Grey *L. excubitor*), Central Asian arid (Steppe Grey *L. pallidirostris*), Middle Eastern arid (Levantine Grey *L. aucheri*), North African (Algerian Grey *L. algeriensis*) and Canary Islands (Canary Island Grey *L. koenigi*). They appear to have emerged around 0.9 mya, although the North African/Canary Islands shrikes may have diverged from the common stock earlier than this.

Climate

Shrikes are birds of warm climates and occupy a wide range of the humidity gradient. A small number of species have colonised temperate and cold climates, and these include some of the Western Palearctic species. Shrikes are bioclimatic specialists and the Palearctic species are among the most tolerant of all the shrikes (Figure 4.1). The Palearctic shrikes occupy a broad spectrum of temperature and humidity regimes and are particularly well-represented within arid bioclimates (Figure 4.2).

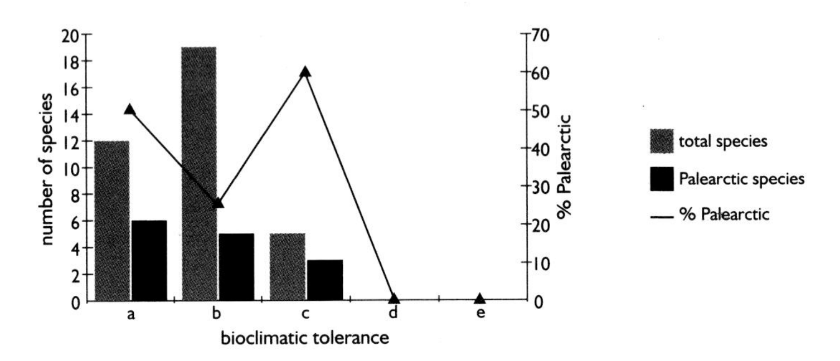

Figure 4.1 *Bioclimatic tolerance of shrikes. Total species refers to all the world's species. Percentage Palearctic is the proportion of Palearctic species of the world total in each category. Bioclimatic tolerance levels: (a) Specialist – occupies between 1 and 20% of the world's bioclimates; (b) Semi-specialist (21–40%); (c) Moderate (41–60%); (d) Semi-generalist (61–80%); (e) Generalist (81–100%).*

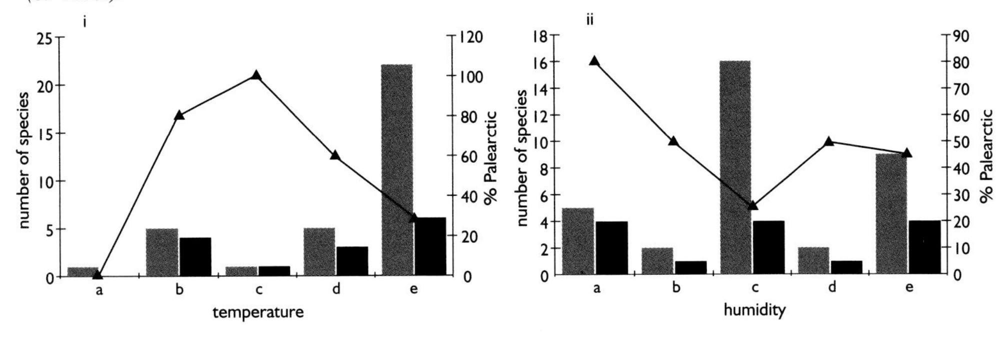

Figure 4.2 *Temperature (i) and humidity (ii) tolerance of shrikes. Total species refers to all the world's species. Percentage Palaearctic is the proportion of Palaearctic species of the world total in each category. For (i): temperature tolerance levels: (a) occupies the coldest portion of the temperature range, between 1 and 20% of the gradient which has 1% as coldest and 100% as warmest; (b) 21–40% of temperature range; (c) 41–60% of temperature range; (d) 61–80% of temperature range; (e) 81–100% of temperature range (warmest portion of the gradient). For (ii), (a) occupies the driest portion of the humidity range, between 1 and 20% of the gradient which has 1% as driest and 100% as wettest; (b) 21–40% of humidity range; (c) 41–60% of humidity range; (d) 61–80% of humidity range; (e) 81-100% of humidity range (wettest portion of the gradient). Key as in Figure 4.1.*

Habitat

Shrikes are birds of open country, being particularly at home in savanna and shrubland and in open habitats (Figure 4.3). This predilection for open habitats, which includes almost treeless situations in some species, coupled with an ability to occupy warm and temperate dry climates, has undoubtedly been a key factor behind the success of the shrikes in the Palearctic.

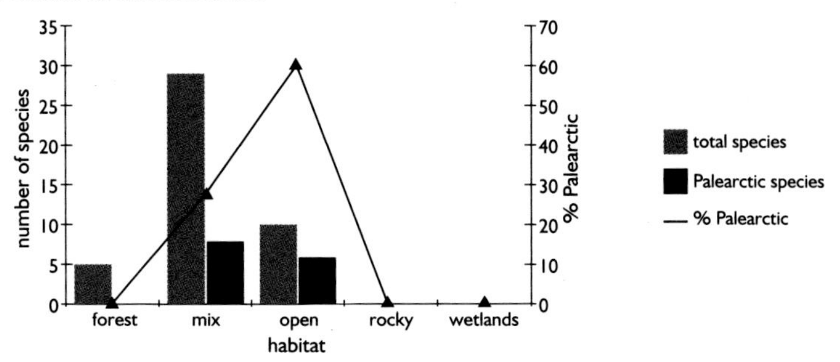

Figure 4.3 *Habitat occupation by shrikes. Total species refers to all the world's species. Percentage Palearctic is the proportion of Palearctic species of the world total in each category. Further details of habitat definitions are given on page 33.*

Migratory behaviour

Almost half (41.67%) of the shrikes are migratory or have migratory populations (Table 4.1), an added feature that would have made them successful in the cooling continental climates of the Plio-Pleistocene. The proportion of Palearctic migratory shrikes is higher (64.29%) than the global figure, reflecting an ability to cope with seasonality.

Fossil shrikes

Lanius shrikes are first recorded from the Late Miocene of Hungary (Mlíkovský, 2002). Shrikes are found in different parts of Europe from the Early Pleistocene and include all the typical species (Appendix 2). The misidentification of some species as Great Grey, when they may include other grey shrikes, cannot be discarded. Shrikes are uncommon in the fossil record, a feature typical of many small passerines, and are found in situations that would not be unusual today. They provide little by way of evidence of changing distribution during the Pleistocene. Their ability to live in open habitats and migrate south to tropical regions in the winter would have made them successful in the steppe-tundra and marginal parkland habitats of glacial Europe.

Shrikes	Global	Palearctic	Global %	Palearctic %	Palearctic of global %
Migratory	15	9	41.67	64.29	60
Sedentary	25	6	69.44	42.86	24
Total Species	36	14			

Corvids	Global	Palearctic	Global %	Palearctic %	Palearctic of global %
Migratory	4	4	3.39	19.05	100
Sedentary	118	20	100	95.24	16.95
Total Species	118	21			

Table 4.1 *Migratory behaviour in global and Palearctic shrikes and corvids.*

Taxa in the Palearctic

Lanius

Thirteen (40.6%) of the species in the genus *Lanius* occur in the Palearctic. The majority of world species are inhabitants of warm and dry climates but there is a good representation in the warm and humid climates of tropical Africa and, to a lesser degree, south-east Asia. Thus the break-up of the tropical forests of the Tertiary and their replacement by open and mixed habitats across much of the southern Palearctic favoured the shrike way of life. This is borne out by the large number of species with distributions centred around 40°N (Latitude Category D) where, as we have seen, the break-up of forest into dry and arid mixed and open habitats was most pronounced: 10 (76.92%) of the 13 species have distributions centred on this line of latitude. Their limited bioclimatic tolerance, as we have seen, severely restricted the breadth of their geographical ranges which must have promoted the formation of new species in isolation. Only the Great Grey Shrike has a pan-Palearctic distribution (which is continuous and is centred on 60°N, Latitude Category B) and, along with Red-backed and Lesser Grey, they are the most bioclimatically tolerant at moderate (C) status. They contrast strongly with the corvids in this regard. The remaining species are semi-specialists (B, Woodchat, Southern Grey, Chinese Grey, Tiger and Bull-headed) or specialists (A, Masked, Levantine Grey, Algerian Grey, Canary Islands Grey and Isabelline).

In contrast with the omnivorous corvids, the shrikes are all mixed strategy carnivores and they invariably inhabit open (six species) or mixed (seven) habitats. Eight species are fully migratory, and one partially. Only the southern 'grey shrike' endemics are sedentary; they resemble the African and south-east Asian shrikes, which are also resident.

Success across much of the Palearctic in the genus *Lanius* has been linked with the ability to occupy highly seasonal environments in the northern spring and summer, followed by migration south for the winter. We can infer that a range of species inhabited the Palearctic, Africa and south-east Asia during the Tertiary. These became increasingly successful (measured by number of species) as conditions in the Palearctic became drier and forested habitats opened up. But their dietary conservatism precluded residency, and these northernmost shrikes moved south after breeding. As deserts, particularly the Sahara, made many areas to the south inhospitable, trans-Saharan migration evolved to different degrees in a number of species – Red-backed, Lesser Grey, Woodchat, Masked and Isabelline. This example is a powerful contrast to the omnivorous corvids that were able to achieve higher bioclimatic tolerances and a higher degree of residency (none are trans-Saharan migrants) than the shrikes. This example illustrates what we have seen and will see again in the following chapters: that migratory behaviour evolved from within the Palearctic of the Tertiary as conditions became increasingly colder, drier and seasonal, but it was tempered by the abilities of lineages to achieve high bioclimatic tolerances (reflected in diet and habitats occupied). The occupation of particular latitudinal bands was also significant, those living around 40°N suffering the effects of wet-dry seasonality and those around 60°N and 70°N suffering those of cold and short winter days.

Tchagra

This genus is largely extralimital. A single species – Black-crowned Tchagra *Tchagra senegallus* – is widely distributed across arid and semi-arid regions of Africa and reaches the Maghreb (race *cucullatus*), which lies within the limits of the Western Palearctic. This species is one of five in this African genus; they belong to a distinct family – Malaconotidae (bush-shrikes and allies) – which has 46 species, all of them African. In my interpretation the bush-shrikes remained on the southern margins of the deserts that split the Palearctic from Africa, with one species managing to penetrate the southern portion. In broad habits they resemble the *Lanius* shrikes except that they are sedentary; it is difficult to know if residency reflects the areas occupied by the bush-shrikes or whether, instead, it reveals an inability to develop migratory behaviours, which would explain the range limitations of the genus and the family as a whole. It is tempting to speculate on the latter. The *Lanius* shrikes, living in comparatively high latitudes, would have been exposed to degrees of daylight seasonality even during the Tertiary and so may have had a migratory predisposition; the bush-shrikes, on the other hand, were exposed far less to such seasonal regimes of day-length, which may have precluded any kind of adaptive response to seasonality. If this interpretation is correct it may advance us some way towards understanding the nature of the origins and evolution of migration. A corollary and consequential prediction is that we should not expect bush-shrikes to be particularly good candidates for colonisation of the Palearctic in scenarios of global warming. That there is no fossil record of *Tchagra* in the Palearctic may simply be due to scarcity of individuals, but it is equally tempting to suggest that it reflects an inability to colonise seasonal environments, even during the powerful global warming events of some Pleistocene interglacials.

CORVIDAE

The corvids form part of an early branch of passerine birds related to the ancestor of all oscines, which evolved within the Papuan-Australian area in the early part of the Tertiary. The corvids were part of a dispersal event into Asia at a time when the Papuan-Australian tectonic plate came close to it (Ericson *et al.*, 2002, 2003). It therefore comes as no surprise to find that the origins of the four major corvid lineages are south-east Asian (Ericson *et al.*, 2005). The oldest branches are made up almost wholly of south-east Asian forest birds (genera *Temnurus, Crypsirina, Dendrocitta, Platysmurus, Cissa, Urocissa*), with the choughs *Pyrrhocorax* being exceptional (Ericson *et al.*, 2005). The choughs represent a successful divergence, limited to two species (Red-billed *P. pyrrhocorax* and Alpine *P. graculus*). These ancient corvids, having adapted to open rocky terrain presumably in the Himalayas, managed to spread and colonise all the mountainous areas across the mid-latitude belt to Iberia and Morocco (Cramp and Perrins, 1994). This is a recurring pattern among Palearctic birds. Both species are present in the Western Palearctic and are therefore the representatives of the oldest corvid lineage in the region.

The later corvid lineages appear associated with occupation of drier habitats, retaining species within south-east Asia, but spreading into large areas of the Palearctic, Africa and the New World. One group includes two genera, *Cyanopica* and *Perisoreus*, which are represented in the Western Palearctic by the Iberian Azure-winged Magpie *C. cooki* and the Siberian Jay *P. infaustus*. *Perisoreus* jays are birds of coniferous forest, which spread across northern parts of the Holarctic, retaining a species within the Himalayas. The *Cyanopica* magpies, which would have formerly had a wider distribution to the south of *Perisoreus*, became isolated during the Plio-Pleistocene in Iberian (*C. cooki*) and east Asian (*C. cyanus*) refugia (Cooper, 2000; Fok *et al.*, 2002; Kryukov *et al.*, 2004). The *Cyanopica* magpies have curiously retained a predilection for coniferous woods (more open than in *Perisoreus*), but also occupy broadleaved and mixed open savanna-type woodland. All the species of *Perisoreus* and *Cyanopica* are geographical counterparts, within and between genera. In the case of the Siberian Jay there is an east-west separation at subspecies level as with other corvids (Haring *et al.*, 2007): nominate *infaustus* in north-eastern Europe, *ruthens* in northern Russia, *yakutensis* in north-east Siberia, *sibericus* in east-central Siberia and *maritimus* in east Asia.

Perisoreus is an example of what I will define as a 'launch-pad model', which appears frequently among Palearctic birds. These are species that appear to have originated in south-east Asia in tropical forests. With the uplift of the Tibetan massif, populations of these species adapted to altitude bioclimatic belts that had latitudinal equivalents to the north. In the Himalayan altitudinal zones their ranges were severely restricted. On occasions when climatic cooling pushed the appropriate latitude belt south, and the corresponding altitude belt down, the two merged and allowed these species to disperse into the latitudinal habitat. Once in it they were able to spread widely and occupy large areas of the Palearctic. In some cases they were able to colonise the Nearctic too. With subsequent warming, Himalayan and Palearctic populations became isolated from each other and diverged. This model applies to species at different altitude-latitude bioclimatic positions and dispersal-contraction would have affected them differently.

Another, related, group includes Central Asian and African genera that have become adapted to very arid conditions: the *Podoces* ground Jays (four species) in Central Asia, and Stresemann's Bush Crow (*Zavattariornis stresemanni*) and Piapiac (*Ptilostomus afer*) in the dry tropical African grasslands. More distantly related, but within the same group, are the *Pica* magpies. Together, the Black-billed *P. pica* and Yellow-billed *P. nuttalli* Magpies have occupied large areas of open habitats across the Holarctic. The Black-billed Magpie is separable into subspecies on a west-east gradient (Haring *et al.*, 2007): *melanotos* in Iberia, *pica* in western, central and parts of eastern Europe, *bactriana* east of the Urals, *hemileucoptera* in Central Asia, *leucoptera* in south-central Siberia, *camtschatica* in north-east Siberia, *jankowskii* in east Asia and *sericea* in Japan and Korea. The corvids in this group complement the previous group geographically and occupy the drier and more open parts of the habitat gradient. They are geographical counterparts, within and between genera. Put together we can propose an ancient geographical spread of corvid lineages across the Palearctic, the Nearctic and Africa of species that have occupied a gradient from coniferous forest, especially in the north (*Perisoreus*), through open woodland-savanna (*Cyanopica*) to open grasslands with scattered trees (*Pica*), dry grasslands (*Zavattariornis, Ptilostomus*) and desert (*Podoces*).

The American jays (*Aphelcoma, Calocitta, Psilorhinus, Cyanocorax, Gymnorhinus, Cyanocitta, Cyanolyca*) appear to have become established following a dispersal from south-east Asia via Beringia shortly after the corvid's Asian radiation (Ericson *et al.*, 2005). In the absence of competitors a number became inhabitants of open country. By contrast, the Old World jays (*Garrulus*) stayed in forest habitats. The Eurasian Jay *G. glandarius* spread across forests in south-east Asia and across the northern and Central Palearctic, including the Western Palearctic. Two other species occupied forests in the Himalayas (Lanceolated *G. lanceolatus*) and tropical Japan (Lidth's *G. lidthi*). Five subspecies of the Eurasian Jay are recognised along a west-east gradient (Akimova *et al.*, 2007): *glandarius* in Europe, *brandtii* in Asia, *krynicki* in the Caucasus and Turkey, *iphigenia* in the Crimea and *japonicus* in Japan. The jays conform with the launch-pad model.

The Old World jays group alongside the nutcrackers (*Nucifraga*) and the crows (*Corvus*). The nutcrackers are a further example of the launch-pad model. The Spotted Nutcracker *N. caryocatactes* has a similar distribution to the Eurasian Jay, including parts of south-east Asia south of the Himalayas and across a wide area of the Palearctic, although not as far west into Europe as the Jay. As with other corvids eastern (*macrorhynchos*) and western (*caryocatactes*) populations are separable at the subspecies level. The Spotted Nutcracker and its North American counterpart, Clark's Nutcracker *N. columbiana*, are specialists of high-latitude and altitude pine forests. The Old World jays and the nutcrackers therefore represent a separate occupation of Palearctic forests from that of the *Perisoreus* jays. The species within the genera *Garrulus* and *Nucifraga* are geographical counterparts.

The crows of the genus *Corvus* represent a radiation into open habitats across the Palearctic and the New World as well as within southern Asia, Australasia and the Pacific. The two jackdaws, Eurasian *C. monedula* in the Western and Central Palearctic and Daurian *C. dauricus* in the east, are geographical counterparts. There is further separation at the subspecies level within the Eurasian Jackdaw: *spermologus* in western and central Europe; *soemmeringi* in the Balkans and *monedula* in north-east Europe including Russia. Similar east-west differentiation occurs, at the subspecies level, within the Carrion Crow *C. corone* (*corone* western Europe; *cornix* eastern and northern Europe; *orientalis* Central and Eastern Palearctic; *sardonius* Near East; *capellanus* Iraq) and the Rook *C. frugilegus* (*frugilegus* Western and Central Palearctic; *pastinator* Eastern Palearctic) (Haring *et al.*, 2007). There is no such east-west differentiation in the Raven *C. corax* but the Canary Islands population is separable at the subspecies level (*C. c. tingitanus*, Baker and Omland, 2006).

Climate

Like the shrikes, corvids are specialists of warm climates but, unlike them, they are preferentially at the humid end of the gradient (Figure 4.4 and 4.5). Curiously, penetration into temperate climates has been via the arid climate route, suggesting that adaptation to dry conditions has enabled subsequent occupation of the less humid parts of the range. A smaller number of species have managed to occupy the colder bio-climates and a significant number of species have adapted to montane conditions. Like the shrikes, corvids are bioclimatic specialists and Palearctic species are among the most tolerant of all the corvids (Figure 4.4). The Palearctic corvids occupy a broad spectrum of temperature and humidity regimes, especially in temperate and fairly wet climates, but are absent from the coldest environments (Figure 4.5).

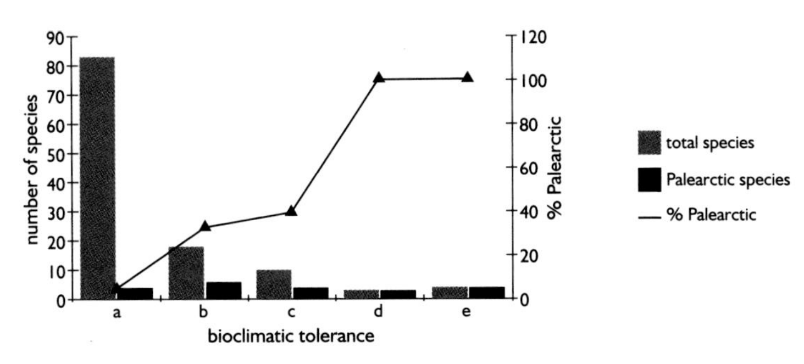

Figure 4.4 *Bioclimatic tolerance of corvids. Definitions as in Figure 4.1 (p. 38).*

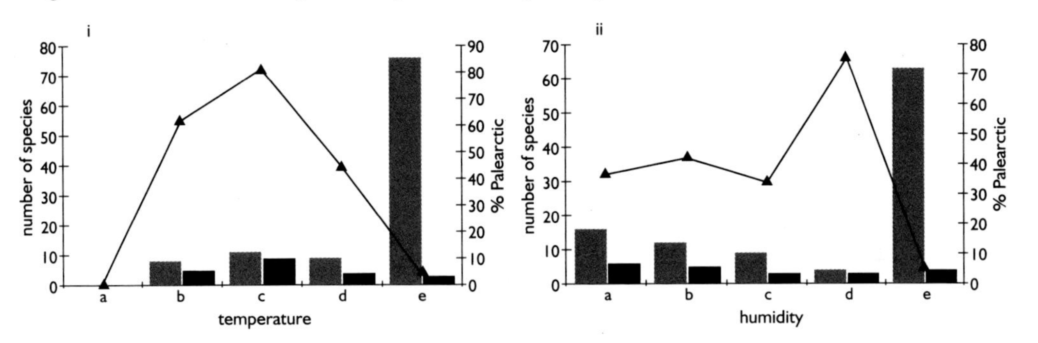

Figure 4.5 *Temperature (i) and humidity (ii) tolerance of corvids. Definitions as in Figure 4.2 , key as in Figure 4.1 (p. 38).*

Habitat

Corvids are essentially forest-dwellers with a high proportion of species having occupied open woodland, savanna and scrub habitats (Figure 4.6). The Palearctic corvids are derived from a subset of open-country dwellers, including treeless, especially rocky, habitats and savannas. Nesting on rocky surfaces represents a departure from tree-nesting and appears to have permitted a few species to exploit the vast expanses of treeless habitats of the Palearctic. These species would have benefitted from the opening up of forests during cooler phases of the Pleistocene. This seems to have been a common trend among Palearctic birds from diverse lineages.

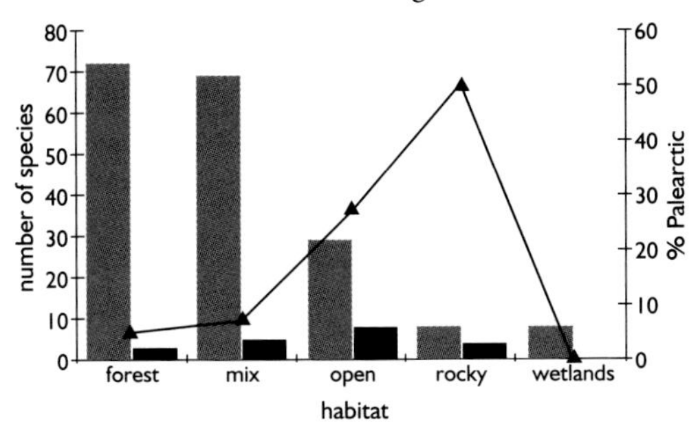

Figure 4.6 *Habitat occupation by corvids. Definitions and key as in Figure 4.3 (see p. 39).*

Migratory behaviour

Corvids are sedentary birds (Table 4.1). Only four species, all Palearctic – Eurasian Jackdaw, Daurian Jackdaw, Rook and Carrion Crow, have migratory populations. The plasticity of corvid diet, being broadly omnivorous, would appear to have been the solution of the few species that managed to establish themselves in the Western Palearctic.

Fossil corvids

Corvids are recorded in the European fossil record from the Middle Miocene, in the form of the extinct genus *Miocorvus* (Mlíkovský, 2002). Three size-classes of modern *Corvus* are evident by the Late Pliocene: a small group, represented by the Jackdaw; an intermediate-sized group (present from the Late Miocene) represented by the Carrion Crow complex and the Rook; and a large group represented by the Raven, in which I include related forms that are probably part of a single lineage (e.g. *C. antecorax*). Magpie and Alpine Chough are present from the Early Pliocene, and Eurasian Jay, Spotted Nutcracker and Red-billed Chough from the Late Pliocene. Thus all modern corvids were about before the start of the Pleistocene.

Corvids become increasingly abundant in the fossil record from the Early to the Late Pleistocene (Appendix 2). As with other birds this reflects preservation of more recent sites and fossil material. The genus *Corvus* is well represented, the Raven complex being the most numerous species, followed by Jackdaw, Carrion Crow and Rook. The best represented genus is *Pyrrhocorax,* with the Alpine Chough more numerous than the Red-billed. The abundance of the choughs, species typical of open ground, montane and rocky habitats, reflects the situation in Europe for large parts of the Pleistocene. The Alpine Chough, today a bird of the high mountains, appears in many lowland localities, often alongside the Red-billed Chough, as at Gibraltar, which is at sea-level. Alpine Choughs are not just numerous but they also appear in places in which they are absent today, such as lowland areas of the Iberian Peninsula, France and Germany (Tyrberg, 1998, 2008). It would seem that conditions in Europe during the Pleistocene favoured the spread of the Alpine Chough into lowland areas; at the same time the Red-billed Chough does not seem to have abandoned its sites in the lowlands. Thus the independent response of the two species to the climatic conditions of the Pleistocene created associations of the two species that have no present-day analogue.

Two boreal woodland corvids that appear to have extended, or shifted, their ranges in response to cooling and drying conditions. The Spotted Nutcracker appears in localities in which it is not present today: Germany, Italy, France, and north-east Spain including Mallorca; the Siberian Jay is similarly found in Hungary, Germany and France (Tyrberg, 1998; 2008).

Of particular interest is the presence of the Iberian Azure-winged Magpie *C. cooki* in the fossil record from Gibraltar (Cooper, 2000) indicating, as discussed above, a once-widespread distribution across the Palearctic that was severed by climate-driven habitat change.

Taxa in the Palearctic

Garrulus

There are three species of *Garrulus*, two with highly localised distributions on islands (in Japan) and mountains (the Himalayas), but the dominant species is the Eurasian Jay, which is distributed across the Palearctic and also on the southern and eastern foothills of the Himalayas. It is a semi-generalist with a broad latitudinal range (Latitude Category F). This bird represents a single and highly successful Palearctic lineage of sedentary, forest-dwelling omnivores. Its broad bioclimatic tolerance has permitted a wide and continuous distribution without apparent room for other species within the genus.

Perisoreus

There are three species in this Holarctic genus. One is North American and a second is localised in the mountains of western China. The third species, the Siberian Jay is a bioclimatic semi-specialist with a pan-palearctic distribution centred on 60°N (Latitude Category B). Like the Eurasian Jay it is a sedentary, forest omnivore. Only one species has dominated the Palearctic, despite its rather restricted bioclimatic tolerance, and this is probably because of its northern distribution that has allowed it to thrive across the homogeneous boreal forest belt. It is only in the Western Palearctic that its range is localised although it was wider during the colder periods of the Pleistocene.

Podoces

The four species of ground jay have distributions centred on 40°N (Latitude Category D) but none are found in the Western Palearctic. These are sedentary omnivorous species of open, arid, habitats (all have arid – A - humidity mid-points). Two are bioclimatic semi-specialists (B) and two are specialists (A). Consequently it is not unexpected to find them restricted to single regions of the Palearctic (three Central and one Eastern). There is only a single, doubted, record of *Podoces* in the fossil record of the Western Palearctic: a Late Pleistocene specimen from the German site of Baumannshöhle bei Rübeland in Sachsen-Anhalt (Tyrberg, 1998). Given the well-established westerward spread of steppe mammals (and some birds) during the arid phases of the Pleistocene, the possibility that ground jays expanded westwards during the Late Pleistocene (as suggested by Moreau, 1954) should not be discarded. The relationship between the ground jays and Stresemann's Bush Crow and Piapiac of the dry African tropics certainly indicates a broader distribution of the common ancestor of these birds.

Cyanopica

Here I separate the Iberian Azure-winged Magpie *C. cooki* from the Asian Azure-winged Magpie *C. cyanus* on the basis of genetic evidence. These two species are situated at either end of the Palearctic and their common ancestor would have had a pan-palearctic distribution. Both are sedentary omnivores that live in open woodland or savanna (mixed habitat class) and have distributions centred on 40°N (Latitude Category D). The estimated timing of the separation of the two populations (estimates vary between 1.2 and more than 3 million years ago; Fok *et al.*, 2002; Kryukov *et al.*, 2004) indicates that the breakdown of the *Cyanopica* range was an early process that probably anticipated the Pleistocene turmoil but may have continued during this period. Their distribution may be seen as complementary with the ground jays along 40°N, the jays fitting into the arid, treeless, central parts of the 40th parallel Palearctic where azure-winged magpies must once have lived when moister climates permitted the expansion of wooded habitats.

Urocissa

The Red-billed Blue Magpie *U. erythrorhyncha* of the Eastern Palearctic is one of five species that are Indo-Malayan and Himalayan in origin and distribution. This species, a bioclimatic specialist of warm and wet environments, penetrates north into the Eastern Palearctic where it is a resident omnivore of forests. Its range has been limited by its bioclimatic and habitat requirements.

Pica

The Black-billed (or Eurasian according to Clements 2007, who distinguishes three species) Magpie is, like the Siberian Jay and the Spotted Nutcracker, part of a Holarctic species complex. It differs from the jay and the nutcracker in that it is an inhabitant of open habitats rather than forest. It is a mixed strategy carnivore (arguably it could be defined also as an omnivore) with a pan-palearctic distribution, and is largely sedentary. Like the nutcracker, the magpie is a bioclimatic generalist that must have shifted its range in accordance with changes in its habitat during the Pleistocene. Its success and failure would, in a way, have been the opposite to that of the nutcracker: gaining when forests withdrew and losing when forests expanded. Along with Raven, Rook and Carrion Crow it must have formed part of a complex of open habitat, bioclimatically tolerant omnivores and mixed-strategy carnivores that were able to remain sedentary throughout by the nature of their adaptable diets.

Nucifraga

There is a similarity between this genus and aspects of *Garrulus* and *Perisoreus*. There are two species of nutcracker that jointly have a Holarctic distribution, but whereas the *Perisoreus* jays are fairly specialised bioclimatically, the Spotted Nutcracker is a generalist, being one of only four corvids with this status. It is a sedentary omnivore of forest with a pan-palearctic distribution and a temperate-centred range (50°N, Latitude Category C), but with populations that reach even the southern slopes of the Himalayas. It therefore has a more southerly-centred range that the Siberian Jay but does not have the broad latitudinal span of the Eurasian Jay. Its range in the Western Palearctic, like the Siberian Jay's, is fragmented, but it is another species that had a wider distribution in the west during the cooler phases of the glaciations when the temperate and boreal forests shifted range southwards and westwards.

Pyrrhocorax
The choughs are unique. There are only two species and both are Palearctic birds that have adapted to life in the mountains of the mid-latitude belt, centred on 40°N (Latitude Category D), with the Red-billed Chough penetrating south into the Ethiopian Highlands. The Red-billed Chough is a bioclimatic generalist, while the Alpine Chough is a semi-specialist. They are largely sedentary omnivorous inhabitants of rocky habitats, from sea level to the highest mountains, and they are separated by altitude with some overlap. They have achieved pan-palearctic distributions but these tend to be fragmented in the Western and Central Palearctic. During the glaciations the range of the Alpine Chough was broader than today and included lower altitudes. During these times both species appear to have overlapped, being found in the same fossil localities, although it is not possible to determine whether the two species occupied the same sites but at different times of year.

Corvus
Eight of 43 species in this globally distributed genus occupy the Palearctic. In terms of bioclimatic tolerance they range from generalists (Raven) through semi-generalists (Carrion Crow, Rook), moderates (Eurasian Jackdaw, Daurian Jackdaw, Large-billed Crow *C. macrorhynchos*), semi-specialists (Collared Crow *C. torquatus*) to specialists (Brown-necked Raven *C. ruficollis*). They exemplify a rule that seems to be generalised – that there is usually only one bioclimatic generalist (usually widespread) in a genus.

Corvus crows are species of open and mixed habitats and two of them – Raven and Eurasian Jackdaw – typically breed in rocky habitats. The generalists and semi-generalists have pan-palearctic ranges (but the distribution is discontinuous in the Western Palearctic in respect of the Raven). The combined distribution of the closely related jackdaws is pan-palearctic, suggesting the recent split of a pan-palearctic bioclimatically-moderate species into two. In a similar way, the Brown-necked Raven appears as a 'desert version' of the Raven and the two may represent the split of a more widely distributed species with the aridification of areas of the Palearctic in the Plio-Pleistocene. The Collared and Large-billed Crows are Eastern Palearctic species.

Most are sedentary but Carrion Crow, Rook and Eurasian Jackdaw have some migratory populations, and Daurian Jackdaw is fully migratory. Of these species five (62.5%) are found in the Western Palearctic – Raven, Carrion Crow, Rook, Eurasian Jackdaw and Brown-necked Raven. The same five are found in the Central Palearctic but only the first three remain in the Eastern Palearctic, which additionally has Large-billed Crow, Daurian Jackdaw and Collared Crow in its avifauna.

ORIOLIDAE

The orioles are an Old World family closely related to the Corvidae (Barker *et al.*, 2004; Treplin *et al.*, 2008). They are predominantly Indo-Malayan in distribution, which belies their south-east Asian origin. Only two species occupy parts of the Palearctic: the Golden Oriole *Oriolus oriolus* in the western and Central Palearctic and the Black-naped Oriole *O. chinensis* in east Asia. They both belong to a predominantly south-east Asian genus that includes Afrotropical and New Guinean representatives.

Climate

Orioles are restricted to warm climates, mainly humid ones. They contrast with the corvids in being climatic specialists with limited tolerance. Of all the orioles, the Golden Oriole is the most bioclimatically tolerant (Appendix 1). The dependence on warm and humid climates is much stronger than in the corvids, which must have been a major factor limiting the spread of the group outside the tropics.

Habitat

Orioles are birds of forest and mixed habitats. Their complete absence from open habitats and their inability to adapt to nesting on rock surfaces (contrasting with the corvids) may have been additional factors limiting their distribution. Nevertheless, it is surprising that savanna and shrub-dwellers might not have succeeded in the Palearctic. That they did not further strengthens the argument that climatic specialisation has been the limiting factor in the distribution of the orioles.

Migratory behaviour

Only four of 27 oriole species (14.8%) have migratory populations, among which are the two Palearctic orioles. Curiously, both species have resident populations in the southernmost parts of their ranges, in south and south-east Asia. The lack of migratory species may not represent an inability to develop migratory behaviour but may instead be a consequence of the few species that have adapted to highly seasonal climates.

Fossil orioles

The orioles are sparsely represented in the fossil record (Appendix 2), not surprising as they are forest birds with limited chances of finding their way into caves, where fossils are best preserved.

Taxa in the Palearctic

Oriolus

Only two species occupy the Palearctic. The pair resembles the jackdaw pair in that, between them, they have a pan-palearctic distribution – Golden Oriole in the Western and Central Palearctic and Black-naped in the Eastern. Golden is a semi-generalist and Black-naped a moderate; they are migratory omnivores of forest and mixed habitat. The orioles have been more limited than the corvids in their success within the Palearctic. The two fairly bioclimatically tolerant species have monopolised the Palearctic to the exclusion of all other orioles.

CONCLUSIONS

The Corvoidea are highly illustrative of the strategies that have succeeded in the Palearctic. These include:

(a) **Bioclimatic generalisation**. A number of highly tolerant species have achieved broad distributions but this appears to have been lineage-dependent. Corvids have achieved this, but orioles only to a limited degree and shrikes have been unable to do so. Generalist and semi-generalist strategies appear to have been most successful among species with ranges centred on 50°N (Latitude Class C: Spotted Nutcracker, Black-billed Magpie, Rook) or with broad latitudinal ranges (Latitude Class F: Raven, Carrion Crow, Eurasian Jay, Golden Oriole). The exception is the Red-billed Chough, a generalist with a 40°N (Latitude Class D) distribution, which may be understood by the altitudinal range of the species that may be equivalent to the latitudinal one of the other tolerant species. They are all species that have been able to maintain broad ranges by shifting position as habitat belts moved. This was not possible, except for the rock-dwelling Red-billed Chough, to the south, where mountains fragmented habitats and ranges.

Bioclimatic tolerance (Category E – generalist) appears only once in each genus. Only in *Corvus* do we have generalists (Raven) and semi-generalists (Category D – Rook and Carrion Crow), and size differences may have been important in enabling more than one broad-tolerance species to persist.

(b) **High mid-latitude belt representation.** 21 of the 36 species (22 if we add the Black-crowned Tchagra) have ranges centred on 40°N. Only the Red-billed Chough is a generalist. These birds are highly specialised bioclimatically: 7 specialists (Category A – 33.33%), 9 semi-specialists (Category B – 42.86%) and 4 moderates (Category C – 19.05%). It is not surprising that they all, with the exception of the chough which proves the rule, have fragmented ranges. Eight are exclusive to the Eastern Palearctic, three to the Central and six to the Western Palearctic.

There is only one forest-dweller among them, the Red-billed Blue Magpie of the Eastern Palearctic, which must have retained connections with tropical populations to the south. The rest are birds of open (10 species) and mixed (10 species) habitat. There seems to have been no room for forest-dwellers in this group.

It is not surprising that the Iberian and Asian Azure-winged Magpies became isolated. They are bioclimatically specialised corvids, living in mixed habitats that include trees, distributed across the mid-latitude belt whose ranges have been severely decimated by repeated aridification.

(c) ***Corvus* versus *Lanius*.**These two genera provide us with a useful contrast that show us the extremes of ways to survive in the Palearctic. The crows belong to a lineage that has produced the complete range of bioclimatic tolerances in the Palearctic and they have become widespread and successful, largely sedentary omnivores of mixed and open habitats. The shrikes, on the other hand, have remained bioclimatically specialised; they are all mixed-strategy carnivores occupying similar habitats to the crows, but they are highly migratory.

(d) **Representation in the Western Palearctic.** 23 of the 37 (62.16%) Palearctic Corvoidea are found in the Western Palearctic. Only two genera – the arid Central Palearctic *Podoces* and the humid forest Eastern Palearctic *Urocissa* – are absent. It is possible that *Podoces* reached the Western Palearctic during the glaciations of the Pleistocene (though this remains to be confirmed).

(e) **Multi-species and single species genera.** There are only two genera – *Corvus* (8) and *Lanius* (13) – that are species-rich, and they have succeeded with different strategies (see (c) above). Curiously, both genera have been absent from forest habitats. A number of single-species genera have dominated large areas of the Palearctic, including the west, by becoming bioclimatically tolerant and/or as inhabitants of temperate and boreal forests – Spotted Nutcracker, Eurasian Jay, Siberian Jay – or open/rocky habitats – Red-billed Chough and Black-billed Magpie. In all cases the genus has been dominated by a single species; where more than one species has been present the remainder have limited ranges (latitudinally and longitudinally in *Garrulus* and altitudinally in *Pyrrhocorax).* Its seems that bioclimatic generalisation in the Palearctic is largely the domain of single species within each genus; when achieved congeners must adopt more specialised strategies.

CHAPTER 5

Sylvioidea – hirundines, warblers and larks

Group 1, subgroup a (ii)

The Sylvioidea form a large and complex group. I will treat the history of the various families and genera individually, as our knowledge of each is not of the same detail and lumping all the information, from families as ecologically diverse as swallows, warblers and larks, in an introductory section would be awkward. A few general comments are necessary at this stage. The arrangement of genera in this chapter follows Treplin *et al.* (2008) but I have had regard of the earlier paper by Alström *et al.* (2006) that adds detail of some genera not covered in the later paper. I have, for example, included the long-tailed tits (Aegithalidae) here, as they appear to be a lineage that is part of a large cluster of genera within the Sylvioidea that includes *Phylloscopus* and *Cettia*.

The basic arrangement of the families identifies four (perhaps five) early lineages that split from the main tree at an early stage, first the larks (Alaudidae), followed by the *Locustella* warbler lineage, the *Acrocephalus–Hippolais* line and then the hirundines (Hirundinidae) (Treplin *et al.*, 2008). The split of the Aegithalidae probably occurred around the time of that of the hirundines. Following these early splits came the divergence of the *Phylloscopus* lineage after which there is a branching into two groups, one containing the babblers (Pycnonotidae) and the other containing the *Sylvia* warblers, white-eyes (Zosteropidae) and *Garrulax*. The *Sylvia* warblers are set aside from the latter two groups. This broad characterisation gives us sufficient information to delineate basic lineages but the details are complicated and not fully understood.

When did this diversification take place? At present we cannot give a clear answer to this question but there is sufficient information to allow some speculation. As we have seen, the Miocene was a significant time when many modern genera made their first appearance and it is very likely that it was during this epoch that the radiation of the Sylvioidea took off. This would have been followed by substantial splitting within individual lineages in the superfamily. Fossils offer some glimpses and show that a number of present-day species were present in the Early Pleistocene (Appendix 1), with some as far back as the Early Pliocene. This must mean that the genera were present in the Pliocene and probably earlier. Some molecular clock-estimates of divergences support this view: in the Miocene, Gray's Grasshopper Warbler *Locustella fasciolata* is estimated to have branched off around 9.3 mya, Savi's Warbler *L. luscinioides* at 8.5 mya and Lanceolated Warbler *L. lanceolata* at 8.1 mya; the Grasshopper Warbler *L. naevia* split later, in the Pliocene, around 3.3 mya (Voelker, 2010). This is a clear indication that the genus was well-established in the Miocene. An earlier set of estimates for *Sylvia* warblers (Blondel *et al.*, 1996) put the timing of the split between this genus and *Acrocephalus* at 19.2–20.8 mya, which is very early in the Miocene. Branching within the genus *Sylvia* seems to have occurred, in a protracted manner, throughout the Miocene and Pliocene, with exceptional cases in the Pleistocene. Put together, this evidence allows us to conclude that the initial radiation of the Sylvioidea took place early in the Miocene (and perhaps started in the Late Oligocene), took off during the Miocene and proceeded during the Pliocene. Given the nature of many of the genera, adapted to open and arid environments, this is not altogether surprising.

Seven sylvioid families have Palearctic representatives. They are the Hirundinidae (eight species, 8.99% of the world's species), Pycnonotidae (four species, 2.92%), Hypocoliidae (one species, 100%), Cisticolidae (four species, 3.4%), Sylviidae (83 species, 16.71%), Aegithaliidae (one species, 12.5%) and Alaudidae (25 species, 27.77%). In terms of Palearctic representation the larks (Alaudidae) are the species that have done proportionately best from this lineage. They are followed by the sylviid warblers, which resemble the Corvidae of the previous chapter. The swallows are poorly represented (in proportion comparable to the orioles of the previous chapter) while the bulbuls (Pycnonotidae) and the cisticolas (Cisticolidae) have hardly had an impact in terms of species. Unlike the Corvoidea, the Sylvioidea are unevenly distributed across the Palearctic, with a clear bias in favour of the Western Palearctic (83 species) with 67 in the Central Palearctic and only 42 in the Eastern Palearctic. This result is due in large measure to the paucity of *Sylvia* and *Hippolais* warblers and larks in the east. We will explore the characteristics of the groups in this chapter to see if we can establish why this eastern deficit should have occurred.

HIRUNDINIDAE

The history of the swallows follows a sequence of three radiations. The original swallows were nest-site excavators which, among the Palearctic species, include the sand martins (Sheldon *et al.*, 2005). The sand martins form part of a core group (which includes many New World genera) that became widespread across Africa, southern Asia and into Australia. The Palearctic marks the northern limit of the group's range in the Old World and is solely represented by the genus *Riparia.*

The second radiation, apparently centred in Africa, involved the mud-nesting swallows that in the Palearctic are represented by the genera *Ptyonoprogne, Hirundo, Cecropis* and *Delichon* (Sheldon *et al.*, 2005). The success of this radiation seems to have depended on the change of nesting habit, from nest excavation to nest construction, which widened breeding location opportunities. The crag martins *Ptyonoprogne* are the basal group among the mud-nesting cluster that includes *Hirundo,* while the other two genera are part of a separate cluster that also includes the cliff swallows (*Petrochelidon*). The third radiation involved the New World endemic swallows.

The genus *Hirundo* has been studied in some detail (Dor *et al.*, 2010) and the 14 species within it form four clusters, the two basal ones being restricted to Africa. A Pacific group is ancestral to the fourth cluster, which includes the Barn Swallow *H. rustica.* Only four species are found outside Africa, which seems to be the region where this genus originated. This is therefore a pattern unusual among Palearctic birds, many of which evolved *in situ* or are derived from immigration from south-east Asia. The expansion of the Barn Swallow is estimated to have been a recent phenomenon, within the last 100,000 years (that is, in the Late Pleistocene) (Zink *et al.*, 2006; Dor *et al.*, 2006), following a split from an African ancestral population. We are therefore observing a very recent phenomenon of geographical expansion, one that even involved a secondary re-colonisation of the Eastern Palearctic from North America, and one that stands out as rare among the birds of the Palearctic. This phenomenon must, however, represent the latest dispersal which must have followed from range contraction during earlier glaciations, as the species is first recorded in the Palearctic fossil record of the Early Pliocene (Appendix 2).

Climate

Hirundines are bioclimatic specialists, and the species that have been successful in the Palearctic are atypical in being highly tolerant. All of the world's generalist hirundines breed in the Palearctic (Figure 5.1). They are birds of warm and wet climates, and the Palearctic species are less restricted, occupying intermediate climatic positions (Figure 5.2). In combination, these unique features among hirundines have allowed survival in the Palearctic.

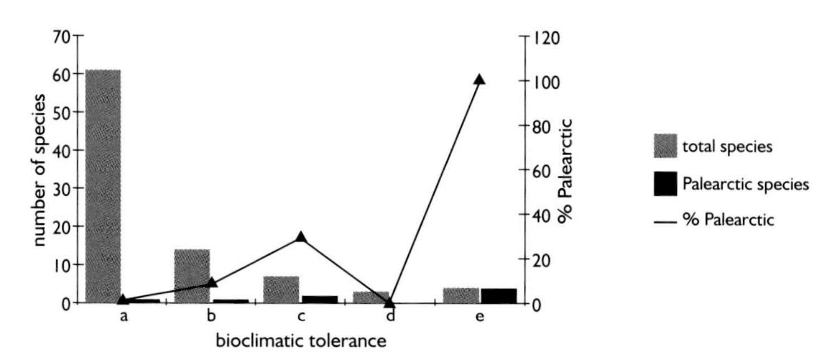

Figure 5.1 *Bioclimatic tolerance of hirundines. Definitions as on Figure 4.1 (see p. 38).*

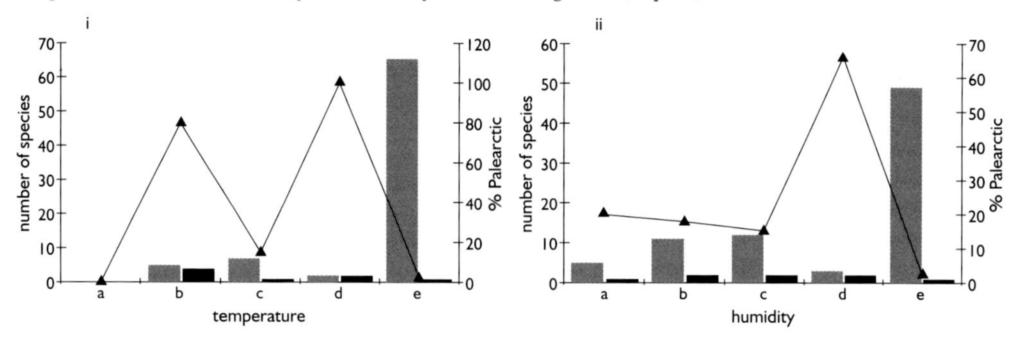

Figure 5.2 *Temperature (i) and humidity (ii) tolerance of hirundines. Definitions as in Figure 4.2 (see p. 38). Key as in Figure 5.1.*

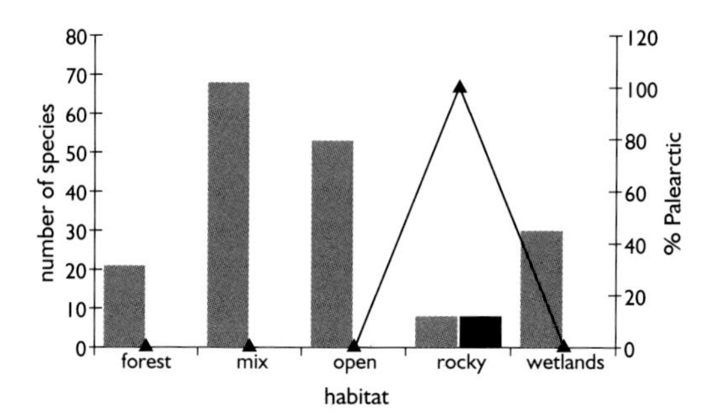

Figure 5.3 *Habitat occupation by hirundines. Definitions as on Figure 4.3 (see p. 39). Key as in Figure 5.1.*

Habitat

Hirundines are predominantly birds that exploit the aerial environment, where they catch insects. They are particularly good at flying over areas that do not restrict their movement, so they are comparatively scarce in forest and common in mixed and open habitats (Figure 5.3). A significant number of species are also birds of wetland habitats, which are rich in insects. The ability to nest in rocky situations has also opened up opportunities for birds across the mid-latitude belt and, later, widely in association with human dwellings. These birds were therefore predisposed for life in the seasonal and drying Palearctic.

Migratory behaviour

Migratory behaviour is frequent among hirundines and contrasts with the predominantly sedentary behaviour of most families of the Sylvioidea (Table 5.1). The migratory habit has been especially important

Hirundines	Global	Palearctic	Global %	Palearctic %	Palearctic of global %
Migratory	28	7	31.46	87.5	25
Sedentary	71	2	79.78	79.78	2.82
Total Species	89	8			

Acrocephalines	Global	Palearctic	Global %	Palearctic %	Palearctic of global %
Migratory	77	58	32.63	95.08	75.32
Sedentary	182	7	77.12	77.12	3.85
Total Species	236	61			

Sylviines	Global	Palearctic	Global %	Palearctic %	Palearctic of global %
Migratory	29	18	11.07	81.82	62.07
Sedentary	249	9	95.04	40.91	3.62
Total Species	262	22			

Larks	Global	Palearctic	Global %	Palearctic %	Palearctic of global %
Migratory	14	10	15.56	40	71.43
Sedentary	86	23	95.56	92	26.74
Total Species	90	25			

Table 5.1 *Migratory behaviour in sylvioids.*

among these birds, which are wholly dependent on aerial insects, and has been the key, along with bioclimatic tolerance and ability to nest in rocky habitats, to the success of the Palearctic hirundines; they are highly migratory with the exception of the Rock Martin *Ptyonoprogne fuligula*, which occupies a marginal range in the south, and the Eurasian Crag Martin *P. rupestris*, which is partially migratory and is less widespread that the other Palearctic species.

Fossil hirundines

Hirundines are recorded from the Pliocene and Early Pleistocene and are well represented (as far as passerines, which tend to appear sporadically, go). Sand Martin *Riparia riparia* is, as we would expect, the least represented while the rock dwellers appear frequently in cave sites (Appendix 2) There is little indication from the fossil sites of presence in geographical areas not occupied today.

Taxa in the Palearctic

Hirundo

The Barn Swallow is the sole representative of the genus in the Palearctic, and has a Holarctic range. The Wire-tailed Swallow *H. smithii* might have been added to the list but it barely, if at all, penetrates the southernmost Central Palearctic and has been left out. The Barn Swallow is one of 14 species, most of which are tropical African – warm, wet climate specialist and semi-specialist sedentary species. It is a generalist with a broad latitudinal range (Latitude Category F) and has a continuous pan-palearctic distribution. It seems to represent another example of a generalist taking up the available space, at genus level, throughout the region. It is an aerial insectivore, and therefore highly migratory, nesting naturally in rocky situations and hunting in mid-air over a variety of situations.

Cecropis

The Red-rumped Swallow *C. daurica* is, like the Barn Swallow, the only Palearctic species of its genus. It is one of seven species that are tropical African and Indo-Malayan – they are warm climate specialists and semi-specialists, and largely sedentary. The Red-rumped Swallow is another bioclimatic generalist with a pan-palearctic range centred on 40°N (Latitude Category D); it is discontinuous within the Western Palearctic. An aerial insectivore, it resembles the Barn Swallow in nesting, feeding and migratory habits and is another generalist that has monopolised the Palearctic at genus level.

Riparia

The Sand Martin is a generalist with a discontinuous Holarctic range and a broad latitudinal distribution (Latitude Category F) but it is not the only Palearctic representative of the genus as the Pale Sand Martin *R. diluta,* a bioclimatic specialist with a range centred on 50°N (Category C) also breeds in the Central Palearctic; it is probably a recent split from the Sand Martin. They are two of five species in the genus, the others being specialists and semi-specialist residents of warm climates in Africa and Indo-Malaya. The two Palearctic sand martins nest in river and other banks and feed aerially on insects. They are highly migratory. The Sand Martin would seem to bear out the single-species monopoly within a genus, in the sense that the second species is highly specialised bird of dry steppe environments.

Ptyonoprogne

The Eurasian Crag Martin is a bioclimatically moderate species with a distribution centred on 40°N (Latitude Category D). To its south and west lives a second species, the Rock Martin, which is a bioclimatic semi-specialist with a range centred on 30°N (Latitude Category E). These two species have complementary geographical ranges across the mid-latitude belt and to the south in the Sahara. They are two of three species in the genus, the third species occupying a range in south Asia outside the Palearctic. All nest in rocky habitats and feed on insects caught in mid-air. The two southern species, and southern and western populations of the Crag Martin, are sedentary but those living in the northernmost regions, the most continental ones in the centre and those at high altitude, are migratory.

This small genus represents a variant of the 'launch-pad' model, in which rocky dwellers spread across the mid-latitude belt from source areas in the Himalayas. Three species have emerged from a Palearctic-African-south Asian common ancestor that probably lived in the Miocene or Pliocene. Living in mountain regions, they spread across the mid-latitude belt and adjacent mountain ranges, but they would have been prone to physical separation of populations, resulting in speciation. Those in the south remained closest to the common ancestor – a sedentary, rock-dwelling martin – but the one that managed to survive to the north developed migratory behaviour. Because its habitat was not severed by deserts to the extent that other habitats were affected, the Crag Martin populations became short-distance migrants, simply moving south into suitable wintering habitat, a few reaching the other side of the great desert (Moreau, 1972). A comparable genus is *Pyrrhocorax* (see p. 46), the Crag Martin even having a pan-palearctic distribution, which is uncommon among Latitude Category D species, except that the omnivorous nature of the choughs minimized the need to migrate whereas it became a necessity in the case of an aerial insectivore whose food ran out in the winter.

Delichon

The distribution of the three house martins recalls that of the *Garrulus* jays: a widely distributed species, and two others with limited ranges around the Himalayas. The House Martin *D. urbicum* supports the view regarding the exclusivity of the bioclimate generalist strategy to a single species in a genus. It is a species with a broad latitudinal distribution (Latitude Category F), which is continuous across the entire Palearctic. A second species, the Asian House Martin *D. dasypus,* occurs in the Eastern Palearctic and does not overlap in range with the House Martin; it is bioclimatically moderate with a range centred on 40°N (Latitude Category D). The third species, Nepal House Martin *D. nipalensis*, is extralimital, breeding south of the Himalayas. These species are also illustrative, as the crag martins were, in that they must represent a split from a common ancestor with a focus on the eastern side of Asia (there are no breeding house martins in Africa). The Nepal House Martin, the southern populations of the Asian and the populations of House Martin that breed on the southern slopes of the Himalayas are sedentary; but Palearctic populations of the Asian House Martin and House Martin are highly migratory. Success among this group of aerial insectivores has, once again, depended on the ability to develop migratory strategies.

PYCNONOTIDAE

This is a marginal family in the Palearctic, which is part of a sister cluster to *Sylvia.*

Taxa in the Palearctic

Pycnonotus

The Palearctic bulbuls belong to one of two large clusters, in this case a group of Asian species that includes two African species, one of which is the Common Bulbul *P. barbatus* (Moyle and Marks, 2006). The bulbuls are specialists of warm and wet climates, being predominantly species of south and south-east Asia. They are resident omnivores of forest and mixed habitats. They have hardly made an impact, with four of the 40 species (10%) barely penetrating the south of the region: Chinese *P. sinensis,* White-cheeked *P. leucogenys,* White-spectacled *P. xanthopygus* and Common Bulbuls. Three have fragmented ranges centred on 30°N (Latitude Category E) in the Western Palearctic and another in the Eastern Palearctic. Being highly sedentary bioclimatic specialists seems to have impaired the success of the bulbuls in the Palearctic.

HYPOCOLIIDAE

Hypocolius

This is an unusual family that has a single species, the Grey Hypocolius *H. ampelinus.* It is a bioclimatic semi-specialist living on the southern fringes of the extreme east of the Western Palearctic and west of the Central Palearctic. It is a short-distance migrant and an omnivore that lives in mixed habitats, usually in semi-arid contexts. It seems to be a unique case of a single species that has remained restricted to a very specific part of the Palearctic.

CISTICOLIDAE

This is a large family of African species with four genera that also occur in Asia and the Palearctic (*Cisticola, Rhopophilus, Scotocerca* and *Prinia*). Much of the evolutionary history of the family seems to have taken place in Africa (Nguembock *et al.*, 2007), which might explain the paucity of species in the Palearctic. The two largest genera among those present in the Palearctic – *Cisticola* and *Prinia* – are part of the same cluster of three that seem to compose this family. The Cisticolidae are bioclimatic specialists of warm, usually wet, bioclimates although some have moved towards the dry end of the spectrum. They occupy a range of habitats and are predominantly sedentary.

Taxa in the Palearctic

Cisticola

This is a predominantly African genus with many species, of which one (2%), Zitting Cisticola *C. juncidis,* occurs in the Western Palearctic. It is bioclimatically moderate, the only one of a genus of specialists and semi-specialists, and some Palearctic populations are short-distance migrants (Finlayson, 1979); this is of particular interest as all other species are sedentary. Presumably, this ability to move has enabled it to survive as an insectivore of open habitats (grasslands) with a range centred on 40°N (Latitude Category D). It contrasts with the bulbuls, which have not managed to expand this far into the Palearctic.

Rhopophilus

The Chinese Hill (or White-browed Chinese) Warbler *R. pekinensis* is the only species in its genus. It is a bioclimatic specialist (Latitude Category D) of mixed habitats with an omnivorous diet and sedentary habits. Given this profile it is not surprising that it should have a restricted distribution within the Palearctic.

Scotocerca

The Streaked Scrub Warbler *S. inquieta* is also in a genus of its own. It is a bioclimatic specialist with a fragmented range along the southern fringes of the easternmost Western and the Central Palearctic (Latitude Category E). It is a sedentary omnivore that occupies mixed habitats. This profile has clearly restricted its geographical expansion.

Prinia

The Graceful Prinia *P. gracilis* is the only Palearctic member of this large genus of 26 largely African and Indo-Malayan species. It is one of five semi-specialists in a genus otherwise dominated by bioclimatic specialists of warm, mostly humid, environments. The Graceful Prinia occupies a limited range on the southern fringe of the eastern end of the Western Palearctic. It is a sedentary omnivore of mixed habitat which seems to survive on the northernmost limit of the range of the prinias. This case resembles that of the bulbuls.

SYLVIIDAE

Climate

The warblers are birds with low bioclimatic tolerances, a distinction that applies to the two large subfamilies Acrocephalinae and Sylviinae (Figure 5.4). The Palearctic species are the most tolerant in the family. The acrocephaline warblers include a greater proportion of semi-specialists within the Palearctic than the sylviines, which are mostly concentrated around the moderate and semi-generalist categories.

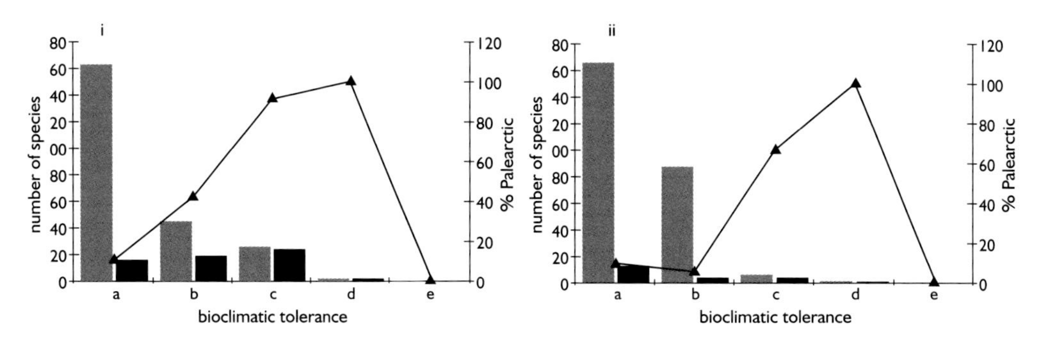

Figure 5.4 *Bioclimatic tolerance of acrocephaline (i) and sylviine (ii) warblers. Definitions as in Figure 4.1 (p. 38). Key as in Figure 5.1.*

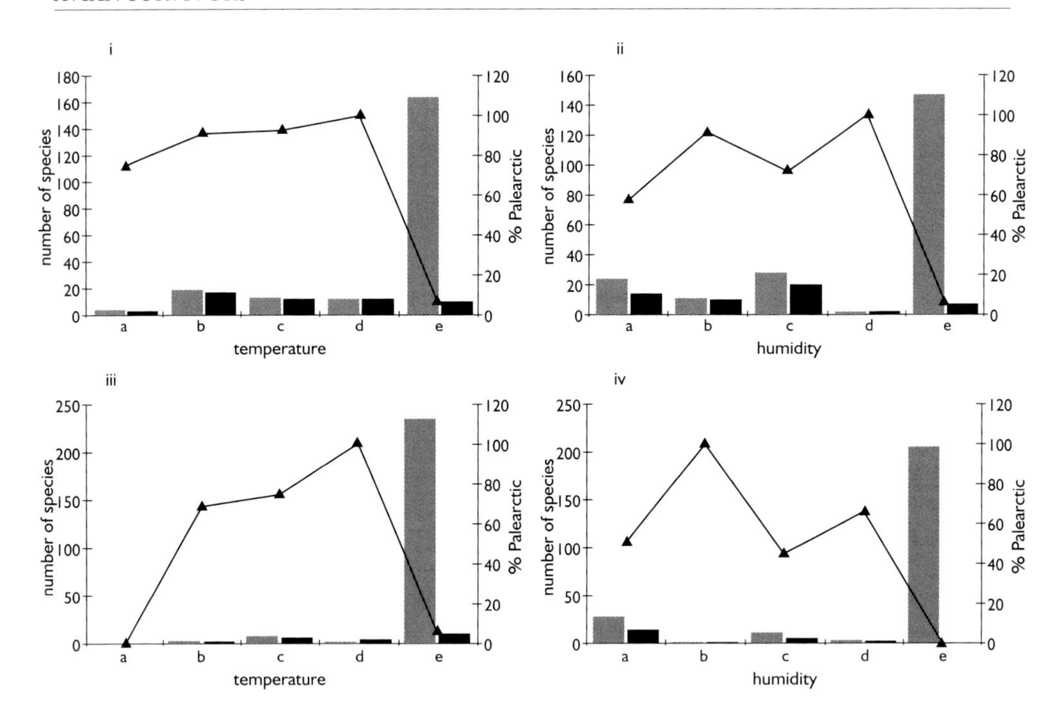

Figure 5.5 *Temperature (i) and humidity (ii) tolerance of acrocephalines; temperature (iii) and humidity (iv) tolerance of sylviines. For definitions see Figure 4.2 (p. 38). Key as in Figure 5.1.*

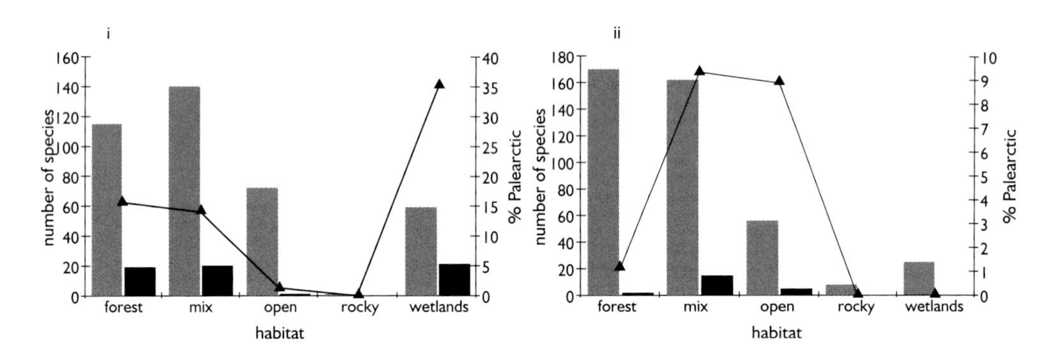

Figure 5.6 *Habitat occupation by acrocephalines (i) and sylviines (ii). Definitions as in Figure 4.3 (p. 39). Key as in Figure 5.1.*

These are species of warm (Figure 5.5 i and iii) and wet (Figure 5.5 ii and iv) climates. The Palearctic species move towards the cooler end of the spectrum but never reach the coldest climates. The acrocephalines are fairly tolerant of the cooler climates and occupy a wide range of positions around the warm, temperate and cool parts of the spectrum. The sylviines, on the other hand, range away from the warmest climates but are restricted from the cooler ones.

The Palearctic warblers show a clear tendency towards the dry end of the humidity range, atypical in a family that globally is concentrated at the wet end of the spectrum. Both acrocephalines and sylviines reveal a bimodal pattern with peaks in the humid and arid parts of the spectrum, the sylviines apparently being more focused on the drier end than the acrocephalines.

Habitat

The warblers are birds of forest and mixed habitats, with a high proportion also in open habitats and wetlands. The Palearctic warblers are a subset: the acrocephalines are birds of wetlands (21 species), forest (19 species) and mixed habitats (20 species), while the sylviines are birds of mixed (15 species) and open (five species) habitats. The forest acrocephalines are mainly *Phylloscopus* with *Hippolais, Acrocephalus* and *Locustella* in mixed and wetland habitats.

Migratory behaviour

Migratory behaviour is not uncommon among warblers and it is especially pronounced among Palearctic species (Table 5.1). Migratory behaviour has therefore been an important component in the success of warblers in the Palearctic, especially as they do not appear to have been particularly successful at bioclimatic tolerance.

Fossil warblers

The only comment which we can make about the fossil warblers, given their extreme rarity (Appendix 2), is that the main genera appear to have been present since the beginning of the Pleistocene, as we would have expected.

Taxa in the Palearctic

Urosphena

This genus of three south-east Asian species occurs only in the Eastern Palearctic. It is represented by the Asian Stubtail *U. squamiceps,* a bioclimatic semi-specialist with a range centred on 40°N (Latitude Category D). The Asian Stubtail is highly migratory, wintering in south-east Asia while its congeners to the south are sedentary. This genus reflects a pattern that is repeated in other genera of tropical forest, bioclimatically specialised insectivores and omnivores that have maintained links (at genus and species levels) with areas of the Eastern Palearctic to the north; in these cases the Eastern Palearctic species, or populations of species, tend to be migratory while those to the south are sedentary. This pattern is not repeated in the Central and Western Palearctic, where tropical forests to the south (in India and Africa) are disconnected from the north by high mountains, seas and deserts.

Cettia

There are three Palearctic *Cettia* warblers out of 15 species in a largely Indo-Malayan and Pacific genus (Olsson *et al.*, 2006). These species are bioclimatically specialised, but two species are bioclimatic moderates (the most tolerant species in the genus) and are species of the Palearctic. Cetti's Warbler *C. cetti* is the sole species in the Western Palearctic. Its range is centred on 40°N (Latitude Category D) and it is a species that has managed survive in the wetlands of the Western and Central Palearctic, presumably from a common ancestor of wider distribution as this species is not found in the east. Its southern and western populations are sedentary but those of the continental areas of the Central Palearctic are migrants. The Manchurian (*C. canturians* – bioclimatically moderate migrant) and Japanese (*C. diphone* – bioclimatically semispecialist partial migrant) Bush Warblers are restricted to the Eastern Palearctic and follow the pattern established for the stubtails. In this regard Cetti's Warbler is an exceptional species, having made it to the Western and Central Palearctic along the wetlands of the mid-latitude belt.

Bradypterus

This genus is closely related to *Locustella* and the two require revision, as one cluster of the latter (with Savi's, Eurasian River, Lanceolated and Grasshopper Warblers) includes two Asian *Bradypterus* species (Drovetski *et al.*, 2004). This bush warbler genus differs from *Cettia* in that, in addition to an Indo-Malayan group, there is a large number of African species. Three of the 20 species reach the Palearctic, none in the Western Palearctic. These warblers are bioclimatic specialists of warm and humid forests and mixed, bushy, habitats. The three species include the most bioclimatically tolerant in the genus – the Chinese Bush Warbler

B. tacsanowskius (Latitude Category C) and the Spotted Bush Warbler *B. thoracicus* (Latitude Category B) – which are migrants to the Eastern Palearctic. A third species – the Large-billed (or Long-billed) Bush Warbler *B. major* – is unusual in that it breeds in mountain forests of the Central Palearctic. Together with the previous genus, the bush warblers are therefore specialist species of the tropics with a very small number having reached the Palearctic, most of them in the east; their success seems to have been heavily dependent on a combination of bioclimatic tolerance and migratory behaviour.

Locustella

This genus is not closely related to *Acrocephalus,* as had been previously thought on ecological, plumage and morphological grounds (Helbig and Seibold, 1999). It forms, instead, a cluster with the south-east Asian and Australasian grassbirds (*Megalurus*) and the *Bradypterus* bush warblers of the Palearctic, south-east Asian and African regions (Alström *et al.,* 2006). The genus *Locustella* is a uniquely Palearctic genus with all 10 species breeding in the region. There is a species gradient from east to west so that the Western Palearctic has the fewest (three species, River, Grasshopper and Savi's), followed by the Central (5) and Eastern Palearctic (7). Bioclimatically, these warblers are more tolerant than the bush warblers, with one species that is a semi-generalist (Lanceolated), four species at Category C (moderate), four at Category B and a single specialist (A). It seems that the degree of bioclimatic tolerance has been insufficient to produce a pan-palearctic species, however, most species occupying two regions (West-Central – River, Grasshopper, Savi's; East-Central – Lanceolated, Pallas's Grasshopper). Four species – Gray's Grasshopper, Middendorff's, Styan's and Sakhalin – are exclusive to the Eastern Palearctic. The ten species are insectivores and omnivores of wetlands and associated mixed habitats; most (seven of the ten) are distributed around 50°N (Latitude Category C) with two others centred on 60°N (Latitude Category B). Only Styan's occupies 40°N to the south. These are warblers that have succeeded by exploiting the wetlands, damp meadows and related habitats of the temperate and boreal Palearctic. We can infer that the ancestors of these birds would have initially been sedentary (probably somewhere in the Eastern Palearctic), becoming short-distance migrants as seasonality increased, and eventually long-distance migrants where desert and other barriers intervened. Being highly migratory, they would have been able to continue exploiting the Palearctic habitats, which would have been extensive at times, during the Pleistocene glaciations. The limitations of their bioclimatic tolerance (more than the bush warblers, which enabled a more extensive occupation of the Palearctic, but never as generalised as other groups, e.g. the hirundines) would have exposed them to repeated range fragmentation and speciation during the Plio-Pleistocene.

Acrocephalus

This genus forms a cluster with the closely related genera *Hippolais* of the Palearctic and the yellow warblers (*Chloroptera*) of Africa (Alström *et al.,* 2006). It is a genus that appears to have gone through a rapid radiation, and three main groups are identifiable within it (Helbig and Seibold, 1999). These groups correspond with body size and plumage characteristics: (a) a small-bodied, streaked-plumage cluster that includes the Palearctic Aquatic *A. paludicola*, Sedge *A. schoenobaenus*, Black-browed Reed *A. bistrigiceps* and Moustached *A. melanopogon* Warblers; (b) a small-bodied unstreaked group, including the Palearctic Reed *A. scirpaceus*, Marsh *A. palustris*, Blyth's Reed *A. dumetorum* and Paddyfield *A. agricola* Warblers, with Blyth's Reed appearing as a separate lineage within this group; and (c) a large-bodied, unstreaked group that divides into a tropical African and a Eurasian-Australasian group; the latter includes the Great Reed Warbler *A. arundinaceus*. The Thick-billed Warbler *A. aedon* occupies a basal position to all *Acrocephalus* (Helbig and Seibold, 1999).

Fourteen of the 36 (38.89%) species in this genus breed in the Palearctic. Many species are warm, wet climate specialists, which form part of a radiation of species on Pacific islands. These warblers also seem to have done well in warm and temperate, dry climates and the Palearctic species are part of this latter group. These species are the most bioclimatically tolerant in a genus otherwise dominated by specialists. The only semi-generalist is the Palearctic Sedge Warbler; 6 of 7 (85.71%) moderate and 5 of 7 (71.43%) semi-specialists are Palearctic species. In contrast only two of 21 specialists (9.52%) are found in the Palearctic. The success of Palearctic *Acrocephalus* has been heavily dependent on migratory strategy; of 19 sedentary species only one (5.26%), the Cape Verde Swamp Warbler *A. brevipennis* from the extreme south-west of the Palearctic is sedentary, while 12 of the 14 (85.71%) fully migratory *Acrocephalus* are Palearctic species.

As was the case in *Locustella,* tolerance levels seem to have been insufficient to promote wide, pan-palearctic ranges; only the Paddyfield Warbler, surprisingly a bioclimatic semi-specialist of the mid-latitude belt (Latitude Category D) is pan-palearctic, though its range is fragmented throughout. Unlike in *Locustella,* however, the bias in this genus favours the Western and Central Palearctic (9 species in each against only 6 in the Eastern Palearctic). Seven species share the Western and Central Palearctic: Sedge, Eurasian Reed, Marsh, Aquatic, Moustached, Blyth's Reed and Great Reed; by contrast, none share the Eastern and Central Palearctic. If to this we add the observation that four species are unique to the Eastern Palearctic (against none in the Central and only the Cape Verde Swamp Warbler in the west), it would seem that the Western and Central Palearctic have been linked more closely with each other, and the Eastern isolated, for *Acrocephalus* warblers. Part of the answer may lie in the broadly more southerly distribution of *Acrocephalus* ranges compared to *Locustella*: 60°N, 2 *Locustella* and 0 *Acrocephalus*; 50°N 6 and 6; 40°N 1 and 6; and 30°N 0 and 2. We have already seen how fragmentation, especially along 40°N, has tended to isolate the Eastern Palearctic, and this seems a valid explanation for the difference between *Locustella* and *Acrocephalus.*

Hippolais

This genus is close to *Acrocephalus* and is divided genetically into two groups of closely related species: (a) Icterine *H. icterina*, Melodious *H. polyglotta*, Upcher's *H. languida* and Olive-tree *H. olivetorum* Warblers; and (b) Eastern Olivaceous *H. pallida*, Western Olivaceous *H. opaca*, Booted *H. caligata* and Sykes's *H. rama* Warblers (Helbig and Seibold, 1999). Like *Locustella,* this genus is truly Palearctic; all eight species in the genus are found in the region.

These are bioclimatic specialist (A, 3 species) and semi-specialist (B, 4) species, with only Icterine Warbler reaching moderate (C) status. They are omnivorous birds of mixed, largely savanna-type habitats, and all are highly migratory, the exceptions being the south-western populations of Eastern Olivaceous Warbler living within the Saharan region (Ottasson *et al.*, 2005; Antonov *et al.*, 2007). The majority are species of the mid-latitude belt (Latitude Category D, 40°N), with only Icterine (50°N, temperate – Latitude Category C) and Booted (multi-latitude, Latitude Category F) Warblers breaking the mould. One notable observation is that there are no Eastern Palearctic species in this genus. It seems that the opening up of the mid-latitude forests with aridification, which largely affected the Western and Central Palearctic at these latitudes, benefited the *Hippolais* strategy. Four species breed in the Western and Central Palearctic – Icterine, Booted, Upcher's and Eastern Olivaceous; Western Olivaceous, Melodious and Olive-tree are exclusively Western Palearctic in range, while Sykes's is exclusively Central Palearctic. Ranges tend to be fragmented in parts. The *Hippolais* warblers are representatives of the arid savannas and oases of the mid-latitude Western and Central Palearctic, habitats that came to dominate these lands during the Plio-Pleistocene.

Phylloscopus

Twenty-three (39.66%) of the world's *Phylloscopus* warblers breed in the Palearctic. Even though there are more species than in other genera, the genus is less exclusive to the region than *Hippolais* and *Locustella* and is comparable to *Acrocephalus* in this respect. The reason why there are so many species in this genus seems to be that they are bioclimatically intolerant. There are no generalist or semi-generalist species at all; they are less tolerant than *Locustella, Acrocephalus* or *Sylvia* and on a par with *Hippolais.* Even though there are some *Phylloscopus* in Africa, the majority are Indo-Malayan and, especially, Himalayan. It seems that from a south-east Asian origin, these warblers have speciated most conspicuously around the Himalayan mountain chain (Irwin *et al.*, 2001 a, b; 2005). Not unexpectedly, then, there is a slight bias of species in favour of the Central (13) and Eastern (12) over the Western (10) Palearctic.

The species in the genus have complex inter-relationships and histories, which include polytypic species, cryptic species, hybrids and others that are still being discovered in the forests of south-east Asia (Alström *et al.*, 1992; 2010; Alström and Olsson, 1995; Irwin *et al.*, 2001a; Bensch *et al.*, 2002 a, b; 2009; Irwin, 2002; Olsson *et al.*, 2005; Reeves *et al.*, 2008). The main Western Palearctic species are closely related. Eastern *P. orientalis* and Western Bonelli's *P. bonelli* and Wood *P. sibilatrix* Warblers are sister species that separate from a main cluster that includes Willow Warbler *P. trochilus*, the various chiffchaffs and the Iberian Chiffchaff

P. ibericus (Salomon *et al.*, 2007), and Dusky Warbler *P. fuscatus*. These have sister relationships with each other (Bensch *et al.*, 2006). This group of Western Palearctic species is closest to a second cluster of species that include Green *P. nitidus*, Greenish *P. trochiloides* and Arctic *P. borealis* Warblers (Helbig *et al.*, 1995).

As in *Acrocephalus*, which also has a significant proportion of non-Palearctic species, the Palearctic *Phylloscopus* are the most bioclimatically tolerant in the genus; they include 10 of the 11 (90.91%) Category C species (moderates) but only 9 of the 39 (23.08%) Category A (specialist) species. The *Phylloscopus* warblers are birds of forested habitats, and Palearctic species live in either forest or mixed habitats that include trees; they are insectivores and omnivores. As with *Acrocephalus*, migratory strategy has been a necessity that permitted the occupation of the Palearctic by species that are not particularly bioclimatically tolerant; 21 of 24 (87.5%) migratory *Phylloscopus* are Palearctic, but only one of 24 (4.17%) is sedentary; the sedentary species, like the Cape Verde Swamp Warbler in *Acrocephalus*, is in the extreme south-west of the Palearctic – Canary Islands Chiffchaff *P. canariensis*.

The low bioclimatic tolerance, together with largely mid-latitude belt (40°N, Latitude Category D) ranges (in 13, 56.52%, of the 23 species), has generated few pan-palearctic species. The four have either multi-latitude distributions (chiffchaff) or northerly boreal and Arctic ranges (Willow, Greenish, Arctic). Only the Willow Warbler has a continuous range across the region. Unusually among warblers, there is a number of montane species in the Central (4) and Eastern (1) Palearctic, occupying habitats on the Himalayas and adjacent ranges.

Sylvia

The *Sylvia* warblers have a complex history of differentiation, starting in the Miocene and intensifying in the Pliocene (Blondel *et al.*, 1996). According to Blondel *et al.*'s study, the two atypical large warblers with wide latitudinal ranges (Garden Warbler *S. borin* and Blackcap *S. atricapilla*) separated from the main *Sylvia* line in the Late Miocene, around 6.3–6.8 mya. The ancestor of the Greater Whitethroat *S. communis* split from the lineage that would produce, in time, a group of circum-Mediterranean species. The split is estimated to have occurred between 5.3 and 4.9 mya, which would seem to be the product of the aridification that followed the flooding of the Mediterranean Sea.

The ancestor of a cluster that includes Eastern *S. crassirostris* and Western Orphean *S. hortensis* Warblers, Arabian Warbler *S. leucomelaena*, and Lesser and Small Whitethroats *S. curruca* and *S. minula* diverged between 3.2 and 3 mya and two isolated species – Asian Desert Warbler *S. nana* (and presumably African Desert *S. deserti*) and Barred Warbler *S. nisoria* – also diverged around this time. This coincides, intriguingly, with a major period of Asian aridification. This period of aridification also led to the formation of the cluster of central Mediterranean species – Ménétries's *S. mystacea*, Spectacled *S. conspicillata*, Sardinian *S. melanocephala*, Cyprus *S. melanothorax*, Rüppell's *S. rueppelli* and Subalpine *S. cantillans* Warblers – around 3.4 to 3.1 mya. The internal divisions within this group formed between 2.5 and 2.4 mya, corresponding with the onset of cool and arid conditions across North Africa (and presumably the Mediterranean) and the spread of an arid-adapted flora as a result of the intensifying influence of the growing ice sheets (de Menocal, 1995). Within-lineage differentiation has proceeded subsequently leading to, in the case of the Subalpine Warbler, significant geographical structuring of populations and the emergence of cryptic species (Brambilla *et al.*, 2008).

The split of species within the western Mediterranean group (Marmora's *S. sarda*, Balaearic *S. balearica*, Dartford *S. undata* and Tristram's *S. deserticola*) seems to have started around 2.0–1.8 mya, the time of another arid period across North Africa (de Menocal, 1995), with Tristram's and Dartford apparently separating as recently as 400,000 years ago, the period when the glacial 100,000-year cycles intensified (see p. 24). The radiation and speciation of the *Sylvia* warblers of the mid-latitude belt (of which the Mediterranean is its western part) seems to have followed Plio-Pleistocene aridification on multiple occasions. It is of no surprise, then, that these species are at the arid end of the humidity gradient.

In terms of Palearctic representation, *Sylvia* falls between the fully Palearctic *Hippolais* and *Locustella* on the one hand and the partially represented *Acrocephalus* and *Phylloscopus* on the other. 22 (78.57%) of the world's *Sylvia* warblers breed in the Palearctic. In distributional terms they resemble the *Hippolais* warblers (in which 75% of the species are Category D, with ranges centred on 40°N), with 16 (72.73%) species having ranges centred on 40°N. It is not surprising then, based on what we have already observed about range fragmentation around 40°N, that only one species (Lesser Whitethroat) has a pan-palearctic, albeit discontinuous, distribution. As expected, this species' range is centred on 60°N. Like the *Hippolais* warblers, the genus is practically absent from the Eastern Palearctic (only Lesser Whitethroat breeds there). Blackcap, Garden Warbler, Greater Whitethroat, Barred, Eastern Orphean and Ménétries's share the Western and Central Palearctic; Subalpine, Sardinian, Dartford, Spectacled, Western Orphean, Rüppell's, Cyprus, Tristram's, Marmora's, Balearic and African Desert Warblers are exclusive to the Western Palearctic; and Small, Hume's *S. althaea* and Margelanic *S. margelanica* Whitethroats are found exclusively in the Central Palearctic. This richness of Western Palearctic species with ranges centred on 40°N has generated the erroneous impression that these warblers are "Mediterranean" in character. This interpretation is only relevant in that it describes the bioclimatic characteristics of the tolerances of these species.

A more logical explanation is to consider these species to be part of a radiation of species in the Western and Central Palearctic, similar to that of the *Hippolais* warblers, which has exploited the mixed habitats (in this case shrublands more than the savannas preferred by *Hippolais*) that ensued from the breakdown of the forests during the Plio-Pleistocene aridification. The character of the mid-latitude belt promoted speciation, as in other birds at this time. The absence of south-east Asian *Sylvia,* and their presence in Arabia and Africa, indicates that this genus originated west of the Himalayas and adapted to the drying world of the centre and west; while many species vanished from this region (e.g. Azure-winged Magpies), these warblers thrived. From this range some species became bioclimatically more tolerant than the majority (19 of 28 species are specialists, Category A) and were able to have expanded latitudinal ranges: Blackcap, a semi-generalist, and Greater Whitethroat (moderate), with multi-latitude (F) ranges; Lesser Whitethroat, moderate, centred on 60°N; Garden Warbler, moderate, centred on 50°N. It is worth adding that the most bioclimatically tolerant warblers are all Palearctic. All 13 migratory and five partially migratory *Sylvia* species are Palearctic, while only 4 of 10 sedentary species are represented there.

The Blackcap provides the range of migratory behaviours of the genus within a single species, with birds from the extreme south-west (including the Atlantic Islands) being resident and those from the northernmost, most continental, populations being highly migratory. These birds have been shown to have the ability to change migratory-resident behaviour in a matter of generations in captivity (Berthold and Helbig, 1992), revealing a potential to adapt rapidly to climate change (Pulido and Berthold, 2010). Migratory direction (in Blackcaps, but probably in all *Sylvia*) also seems to be alterable fairly rapidly (Berthold *et al.,* 1992; Helbig, 1996) and has potential feedback effects in terms of reproductive isolation of populations with different migration directions (Rolshausen *et al.,* 2009). In other words, birds from the same breeding population but wintering in different areas may arrive at different times back at the breeding grounds. This means that within the same breeding area, birds from different wintering areas will mate with those from the same wintering grounds (but not from others). Over time this can lead to a genetic split within the population, despite breeding occurring in the same place. The flexibility that allows for rapid changes in migratory behaviour was probably put to good use repeatedly by many Palearctic birds throughout the Plio-Pleistocene.

AEGITHALIDAE

Aegithalos

The long-tailed tits are a south-east Asian group with elements in the Himalayas, the product of *in situ* differentiation and also immigration from south-east Asia, and a single pan-palearctic species. This genus is another example of the launch-pad model (see p. 41). These tits seem to have undergone differentiation during the Late Pleistocene, at species level in the Himalayas and sub-specific level in the Palearctic-Himalayan region (Päckert *et al.*, 2010). Such differentiation, which in this case was followed by geographical expansion and mixing, is typical of many birds at this taxonomic scale and reflects isolation in glacial refugia.

One of six species of this genus of mainly southern Asian and Himalayan species breeds in the Palearctic. They provide us with a further example to the pattern of bioclimatic tolerance within genera. All are bioclimatic specialists and semi-specialists with limited ranges, except for the Long-tailed Tit *A. caudatus*, which is a semi-generalist (multi-latitude, Latitude Category F) with a continuous pan-palearctic distribution.

ALAUDIDAE

The phylogeny of the larks is little studied, but there is evidence of ancient centres of origin in north-east and southern Africa, with subsequent Plio-Pleistocene dispersals (Barnes *et al.*, 2006). Given what we have been finding so far regarding the incidence of aridity in the Old World, starting in the Miocene and intensifying in the Pliocene, it would seem that the conditions favouring the open ground, arid-adapted larks would have optimised their success in the Palearctic.

Climate

Globally, larks are bioclimatic specialists, with a small number of species having succeeded in becoming tolerant of a range of climates. These tolerant species are all found in the Palearctic (Figure 5.7). Larks differ from most groups in that they are predominantly species of warm and dry (instead of wet) bioclimates (Figure 5.8). The Palearctic larks show a tendency towards moving away from the warmest climates, which is not unexpected, and also towards the centre of the humidity gradient, avoiding the driest of environments.

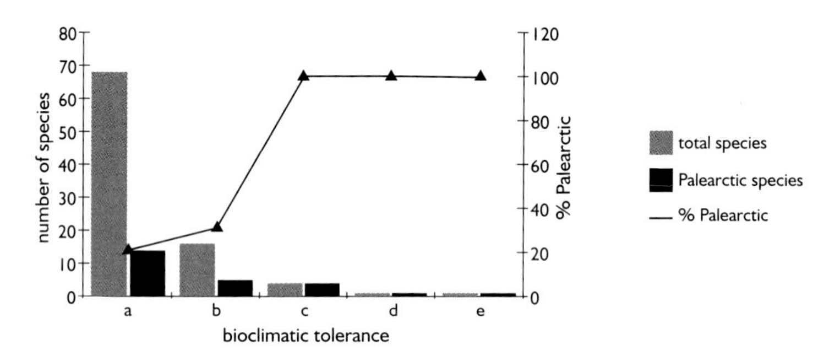

Figure 5.7 *Bioclimatic tolerance of larks. Definitions as on Figure 4.1 (see p. 38).*

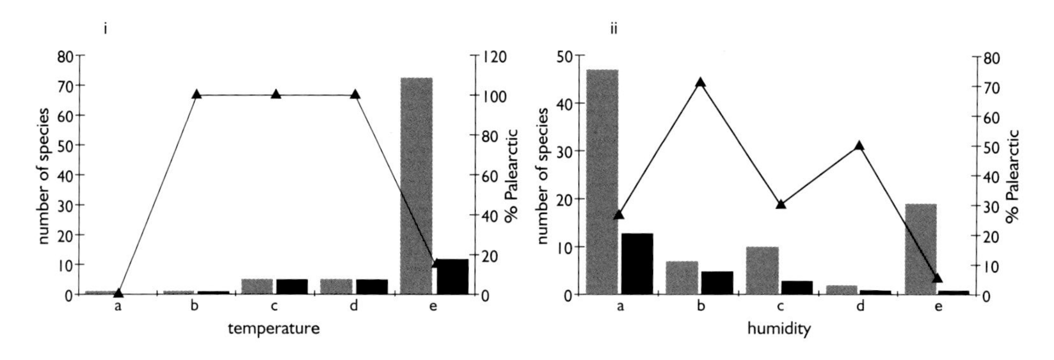

Figure 5.8 *Temperature (i) and humidity (ii) tolerance of larks. Definitions as in Figure 4.2 (see p. 38). Key as in Figure 5.7.*

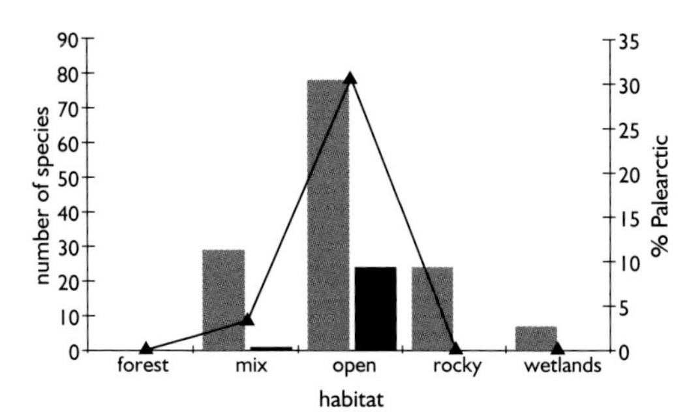

Figure 5.9 *Habitat occupation by larks. Definitions as on Figure 4.3 (see p. 39). Key as in Figure 5.7.*

Habitat

Larks avoid forest altogether, which makes them atypical among most bird families, and are at home preferentially in open habitats with mixed and rocky habitats also occupied (Figure 5.9). They could be regarded as an extreme development of the *Sylvia* adaptation towards arid, open habitats and they have developed extreme adaptations, e.g. in foot and hind claw morphology, for a terrestrial way of life (Green *et al.*, 2009).

Migratory behaviour

This is poorly developed among larks. Among the few cases of migratory species are 10 of the 25 Palearctic larks (Table 5.1). A combination of bioclimatic tolerance and migratory behaviour characterises the Palearctic larks with broadest distribution, many others being confined to southern, desert, situations.

Fossil larks

Some larks are frequent in the fossil record, in numbers comparable to the hirundines and exceeding by far the warblers (Appendix 2). Some species are recorded from the Late Pliocene (Woodlark *Lullula arborea*) and early Pleistocene (Greater Short-toed Lark *Calandrella brachydactyla*, Crested Lark *Galerida cristata*, Eurasian Skylark *Alauda arvensis* and Shore Lark *Eremophila alpestris*) indicating that present-day species survived the entire Pleistocene. The predominant species in the fossil record are Eurasian Skylark, Woodlark, Crested Lark and Shore Lark. The presence of White-winged Lark *Melanocorypha leucoptera* in Germany (Tyrberg, 1998, 2008), if confirmed, would represent a case of westward extension of range, presumably with aridity. The Shore Lark's appearance in a number of lowland western and central European countries might be indicative of a more widespread distribution during cold periods of the Pleistocene. Overall, the larks indicate an abundance of open and mixed habitats across Europe throughout much of the Pleistocene.

Genera with Palearctic species

The larks include 3 (13.64%) pan-palearctic species – Shore Lark, Eurasian Skylark and Crested Lark – but only the Eurasian Skylark has a continuous range. The remaining groups, with few species, are localised in distribution.

Mirafra

Only the Kordofan Bush Lark *M. cordofanica*, out of 27 mostly African but also southern Asian species, reaches the Palearctic at its southernmost fringe in the west. The conditions of the Palearctic therefore appear unsuitable for a group of birds that are specialists in warm, mainly dry, climates, or the available space has been taken up by other larks. The Kordofan Bush Lark occupies open habitats in dry regions and is sedentary, like all its congeners.

Alaemon
Two hoopoe-larks make up this genus, and the Greater Hoopoe-lark *A. alaudipes* is widespread across vast areas of the southern Western and Central Palearctic. The other species breeds to the south and is extralimital. The Greater Hoopoe-lark is a bioclimatic semi-specialist and its congener, the Lesser Hoopoe-lark *A. hemertoni,* is a specialist. Once again the more bioclimatically-tolerant species made it into the Palearctic.

Chersophilus
Dupont's Lark *C. duponti* is placed in a genus on its own. It is a Western Palearctic species, a bioclimatic specialist with its range centred on 40°N (Latitude Category D). We do not know whether its range was more extensive during arid periods of the Plio-Pleistocene, but it does appear to be endemic to the south-west of the Palearctic in a latitude band prone to range fragmentation, as we have repeatedly seen in other groups. It is sedentary.

Eremopterix
One of seven African species, the Black-crowned Sparrow-lark *E. nigriceps* breeds across the southern border of the Western and Central Palearctic. Its distribution recalls that of the Kordofan Lark, though it is better represented in the southern Palearctic, which seems to lie at the limit of this genus of bioclimatic specialists; this species is one of three semi-specialists, which represents the limit of tolerance in the genus. All are sedentary omnivores of open habitats.

Ammomanes
Two of the four Afro-Indian larks of this genus are widely distributed across the southern areas of the Western and Central Palearctic. They are the Bar-tailed Lark *A. cincturus* and the Desert Lark *A. deserti.* All four species in the genus are sedentary, omnivorous, bioclimatic specialists of warm, typically arid, environments. They form part of the cluster of larks that thrive along the southern parts of the Western and Central Palearctic (Latitude Category E).

Ramphocoris
The Thick-billed Lark *R. clotbey* is another species that is alone in its genus. It is distributed across a fringe south of 40°N, from north-west Africa to Arabia, and is therefore a species of the Western Palearctic. It is another bioclimatic specialist, a resident omnivore of open, arid, habitats.

Melanocorypha
We can place this lark genus alongside the *Locustella* and *Hippolais* warblers in being exclusively Palearctic. All six species are found in the region: Calandra *M. calandra* (specialist, Latitude Category D), Bimaculated *M. bimaculata* (semi-specialist, Category D), Tibetan *M. maxima* (specialist, Category D), Mongolian *M. mongolica* (specialist, Category D), White-winged (semi-specialist, Category D) and Black *M. yeltoniensis* (specialist, Category C). So, with the exception of the Black Lark, all are species of 40°N, some on low ground and others on highlands. Unlike the southern larks, which are invariably sedentary, these mid-latitude larks include three species, Bimaculated, White-winged and Black, that have migratory populations in continental areas of their ranges. So lark genera such as *Melanocorypha* do have the potential to become migratory, and this has permitted the exploitation of the seasonal grasslands and steppes of the Central Palearctic.

Calandrella
Four of ten *Calandrella* larks breed in the Palearctic: Greater Short-toed (moderate, Latitude Category D), Lesser Short-toed *C. rufescens* (moderate, Latitude Category D), Hume's Short-toed *C. acutirostris* (specialist, Category D montane) and Asian Short-toed *C. cheleensis* (specialist, Category D) Larks. The Palearctic short-toed larks are all distributed around 40°N. Greater and Lesser Short-toed are additionally the most bioclimatically generalised of the genus, and the most widely distributed. The migratory tendency observed among some *Melanocorypha* larks is seen among the short-toed larks too; it is particularly noticeable among continental populations. Greater Short-toed and Hume's are almost wholly migratory, but the less continental populations of Lesser Short-toed, in the south and west, are sedentary as is the eastern population of Asian Short-toed Larks.

Eremophila

The Shore Lark, a Holarctic species, is the only bioclimatic generalist among the larks. This species has a split latitudinal distribution, a continuous pan-palearctic one centred on 70°N (Latitude Category A) and a discontinuous, pan-palearctic, range centred on 40°N (Category D). The northern populations are migratory, the southern ones sedentary. This species must have had a wider distribution during the colder glacials of the Plio-Pleistocene, with populations retreating north and up mountains with global warming, a situation that is not infrequent among Palearctic birds. The other species in the genus, Temminck's Lark *E. bilopha,* replaces Shore Lark to the south in the Western Palearctic, and is sedentary. It is a bioclimatic specialist with a range just south of 40°N (Category D).

Eremalauda

Dunn's Lark *E. dunni* typifies a group of ground-dwelling omnivores that have done well from the aridification of large zones of the Western and Central Palearctic. The lark family, particularly several genera including *Eremalauda*, can be equated to *Hippolais* and *Sylvia* as representatives of birds that have thrived in the Western and Central Palearctic in regions with seasonal climates containing an arid component and habitats that range from open desert to mixed shrublands and savannas. Dunn's Lark is a bioclimatic specialist that has a discontinuous distribution in the Western Palearctic centred on 30°N (Latitude Category E). A sedentary omnivore of open habitats, its range appears limited to the southernmost parts of the sub-region.

Galerida

Two of six species in this genus breed in the Palearctic. The Crested Lark is among the most bioclimatically tolerant of larks (moderate, Latitude Category F) and has a pan-palearctic discontinuous distribution. It is largely sedentary across much of its range but more northerly and continental populations are migratory. The populations north and south of the Sahara show significant genetic distinctiveness (Guillamet *et al.*, 2006) which would seem to indicate that the Sahara has acted as a barrier to this species, and it must have once lived right across what is now desert. In the Palearctic, the Thekla Lark *G. theklae* occurs exclusively in the West, but populations of this species also occur in north-east Africa; these populations are currently disconnected from each other. The Thekla Lark is a bioclimatic semi-specialist with its range centred on 40°N (Category D). These two species together occur across large areas of the Palearctic and Africa; other species in the genus breed in south Asia. The two Palearctic species are the most bioclimatically tolerant (along with the semi-specialist Sun Lark *G. modesta* from Africa) in the genus, characteristics that have enabled success in the Palearctic with minimal need of migration.

Alauda

Two of four species in this genus breed in the Palearctic. The Eurasian Skylark is the second most bioclimatically tolerant lark (semi-generalist, Category D), after Shore Lark, with a continuous pan-palearctic range and a multi-latitude distribution (Latitude Category F). Other, less tolerant, species in the genus include the Oriental Skylark *A. gulgula* – a bioclimatic semi-specialist with a narrow distribution south of the Eurasian Skylark in the Eastern Palearctic; the Japanese Skylark *A. japonica* – a bioclimatic specialist – and the other Palearctic species, the Razo Lark *A. razae,* a sedentary specialist from the Cape Verde Islands. Many populations of Eurasian Skylark, from northern and continental areas, are migratory, but those from the milder south-west are sedentary or short-distance migrants. The success of the most cosmopolitan species in the genus has been through a combination of omnivory (common to all larks), bioclimatic tolerance and migratory behaviour from areas that would be uninhabitable in the winter. As in many other genera, a single species has come to dominate geographically over localised species.

Lullula

The Woodlark is unusual among larks because of its association with trees. It is a bioclimatic moderate, the only species in its genus, with a multi-latitude (Latitude Category F) distribution. It is restricted to the Western Palearctic, where it must have been successful in the transitional (mixed) habitats between forest and open plain that took over large areas of the region during glacial-interglacial cycles, at times when more continental areas, dominated by arid, open, habitats, would have prevented the Woodlark way of life. Its sensitivity to continental climates is shown by the migratory behaviour of the northernmost and easternmost populations, while those to the south-west are sedentary.

CONCLUSIONS

(a) **Bioclimatic generalisation.** Proportionately fewer species (9, 7.63%) in the Sylvioidea are bioclimatically tolerant (Categories E and D) compared to the Corvoidea (8 species, 22.22%). There are many more species overall (118 against 36) and the difference lies in the greater number of bioclimatically restricted species. Since tolerant species have bigger ranges, often restricting less tolerant ones, the larger number of species may be, at least in part, due to the presence of so many specialised species. The bioclimatically tolerant species (generalists and semi-generalists) are mainly multi-latitude species (Latitude Category F: Barn Swallow, Sand Martin, House Martin, Blackcap and Eurasian Skylark) but there is only one with a 50°N distribution (Category C: Sedge Warbler); there is, additionally, a 60th parallel species (Category B: Lanceolated Warbler). As with the Red-billed Chough in the Corvoidea, the Red-rumped Swallow, associated with rocky habitats, is a Category D bioclimatically-tolerant species; the last species is the Shore Lark with a split Category A and D distribution. As among the Corvoidea, bioclimatic generalists (and here also semi-generalists) appear only once in each genus.

(b) **High mid-latitude belt representation.** As among the Corvoidea, there is a high representation of species (64, 51.2%; 66 if we add two split-category species) with ranges centred on 40°N. These birds are mainly specialists (32 species, 50% of Category D), a much higher proportion than among the Corvoidea (7, 33.33%); in contrast there are fewer semi-specialists (18, 28.13%) than among the Corvoidea (9, 42.86%). The number of moderates (13, 20.31%) and generalists (1, 1.56%) resembles the Corvoidea. As in that group, there are no semi-generalists. This reinforces the pattern of bioclimatic specialisation and fragmentation that we have observed throughout this book. The degree of fragmentation can also be measured by the number of species that are exclusive to a single sub-region of the Palearctic. Among the Corvoidea it is 17 species (47.22% of Category D) but in Sylvioidea it is much higher (49 species, 76.56%). This must reflect the higher number of bioclimatic specialists. Many more species are unique to the Western Palearctic (23, 46.94%) than among Corvoidea (6, 35.29%). The comparable figures for the other sub-regions are: Central (12, 24.49% versus 3, 17.65%) and Eastern (14, 28.57% versus 8, 47.06%).

As among the Corvoidea, birds of open (19 species, 29.69%) and mixed (24, 37.5%) habitats dominate among the mid-latitude species; there are also some wetland species (7, 10.94%) and rocky dwellers (3, 4.69%). But there are also more forest birds than in the Corvoidea, which only has one. These are all warblers (Sylviidae) of three genera – *Phylloscopus* (8 species), *Bradypterus* (2) and *Urosphena* (1). Only *Phylloscopus* has Western Palearctic forest-dwellers with ranges centred on 40°N: Western and Eastern Bonelli's and Iberian Chiffchaff. These are species that would have survived in localised Iberian and Balkan glacial refugia. Four other species (Long-billed Bush Warbler *Bradypterus major*, Hume's *Phylloscopus humei*, Plain Leaf *P. neglectus* and Brooks's Leaf *P. subviridis* Warblers) are exclusively Central Palearctic, and they would have survived in montane forest refugia, and four others occur in the Eastern Palearctic and would have retained connections with the tropics to the south. This shows that forest habitats became heavily fragmented in the Western and Central Palearctic during glacials; the Central Palearctic, which was especially arid, only retained forests in mountain regions. The Eastern Palearctic forest-dwellers are, on the other hand, birds of lowland forests.

(c) **Hirundinidae, Sylviidae and Alaudidae.** When we compare the three main Palearctic families in the Sylvioidea we find important differences. The Hirundinidae have the fewest species (8) but they are spread over 5 different genera giving a ratio of 1.6 species per genus. The Alaudidae has 25 species spread across 13 genera (1.923 species/genus) but the warblers (Sylviidae) have 83 species in 8 genera (10.375 species/genus). This suggests that the hirundines and larks have survived in the Palearctic by keeping to distinct body plans (reflected at the genus level) and a high degree of exclusivity (few species per genus). The warblers, on the other hand, seem to have been able to provide multiple options (species) for each body plan (genus). This may reflect the specialised nature of hirundines (aerial insectivores) and larks (open ground omnivores) against the less restricted character of the warblers (insectivores, omnivores across a range of habitats, restricted ranges with similar species in different parts of the range). There are certainly many more warblers with single sub-region ranges (51 species, 61.45%) than larks (13, 52%) or hirundines (2, 25%).

These findings suggest that survival in the Palearctic may have involved an early (Miocene/Pliocene) adaptation of basic body plans (hirundines in rocky habitats and feeding in the air, larks on the arid, open habitats and warblers mainly in mixed habitats with trees but also wetlands). Subsequent climate-driven habitat fragmentation would have affected the warblers much more than the other families, which simply altered ranges as their habitats moved without losing connectivity between populations to the degree that the warblers of woodland, savanna, shrubland and wetland would have done.

(d) **Uniquely Palearctic genera.** Seven genera contain a single species and are exclusive to the Palearctic: *Hypocolius, Rhopophilus, Scotocerca, Chersophilus, Ramphocoris, Eremelauda and Lullula.* The last four, all lark genera, are unique to the Western Palearctic. Three other (multi-species) genera are unique to the Palearctic and are well-represented in the Western Palearctic. They are the *Locustella* and *Hippolais* warblers and the *Melanocorpyha* larks.

(e) **Representation in the Western Palearctic.** 83 species (43.23%) of Sylvioidea are found in the Western Palearctic, proportionately fewer than the 23 (62.16%) Corvoidea. Curiously, only three genera (all warblers) are missing from the Western Palearctic – *Urosphena, Bradypterus* and *Rhopophilus.*

(f) **Multi-species and single-species genera.** There are six genera that are rich in species. They are the *Melanocorypha* larks (6) and five warbler genera – *Locustella* (9), *Acrocephalus* (14), *Hippolais* (8), *Phylloscopus* (23) and *Sylvia* (22). Between them they occupy the range of habitats available in the Palearctic, from open, treeless expanses to forest. They contrast with the Corvoidea in that the single-species genera that have dominated large areas of the Palearctic by becoming bioclimatically tolerant are aerial insectivores (*Hirundo, Cecropis*) and not birds of forest. To these we may add two other aerial insectivore genera (*Delichon* and *Riparia*) and two open ground genera (*Galerida* and *Alauda*) where the genus is dominated by a single species and the remainder have limited ranges (as in *Garrulus* and *Pyrrhocorax* in the Corvoidea). This supports the earlier conclusion that bioclimatic generalisation in the Palearctic is largely the domain of single species within each genus; when achieved, all available ecological room is taken up, and congeners must adopt more specialised strategies.

CHAPTER 6

Paroidea – tits

Group 1, subgroup a (iii)

The penduline tits are basal to the other tits (Gill *et al.*, 2006). The genus *Remiz* is Palearctic and is sister to the African *Anthoscopus*. The tits of the Paridae evolved in the Old World during the Tertiary, almost certainly within the south-east Asian-Himalayan area, which has the greatest diversity today. This family, and some of the genera within, seem to have followed a 'launch-pad model' of initial dispersion (see p. 41), spreading across the Palearctic to enter Africa and also North America. The latter was penetrated on two occasions in the Pliocene, at 4.0 and 3.5 mya (Gill *et al.*, 2005). Gill *et al.* note that the blue (*Cyanistes*) and great (*Parus*) tits, which are among the oldest lineages, are widespread and show little regional diversity. We will see below that they are also bioclimatically generalised. Put together, this evidence may indicate that the single-generalist model may be reflecting ancient patterns of dispersion and, possibly, early diversification that was then lost or restricted geographically as one widespread generalist became dominant.

Climate

The global picture reveals that tits are bioclimatic specialists, and that Palearctic species are a small subset of bioclimatically tolerant birds that have managed to succeed outside the tropics (Figure 6.1). Globally, tits are birds of warm and wet bioclimates; Palearctic species have shifted their position towards the cool part of the gradient and have also managed to move away from the wettest climates, though preferentially occupying humid ones (Figure 6.2).

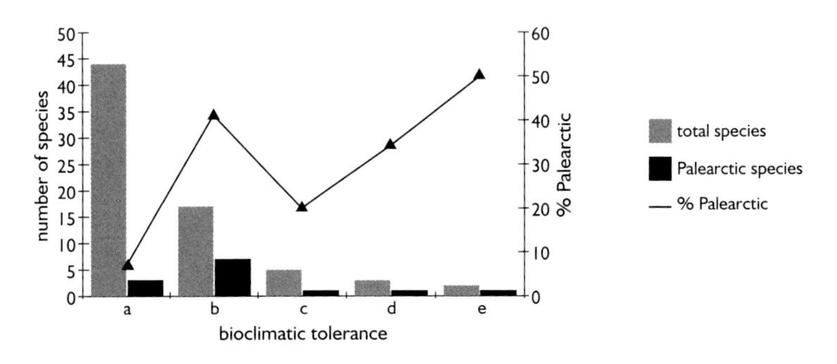

Figure 6.1 *Bioclimatic tolerance of tits. Definitions as on Figure 4.1 (see p. 38).*

Habitat

Tits are birds of forest and associated mixed habitats. Only the ancestral and highly divergent Hume's Groundpecker *Pseudopodoces humilis* of the Tibetan Plateau has departed significantly to live on open ground. The other exception is the penduline tits that live in reed beds, usually close to water. The Palearctic tits do not vary therefore from the global forest-dwelling pattern. This is an exceptional case at family level, since most Palearctic families show departures from global habitat patterns. The tits have been able to retain

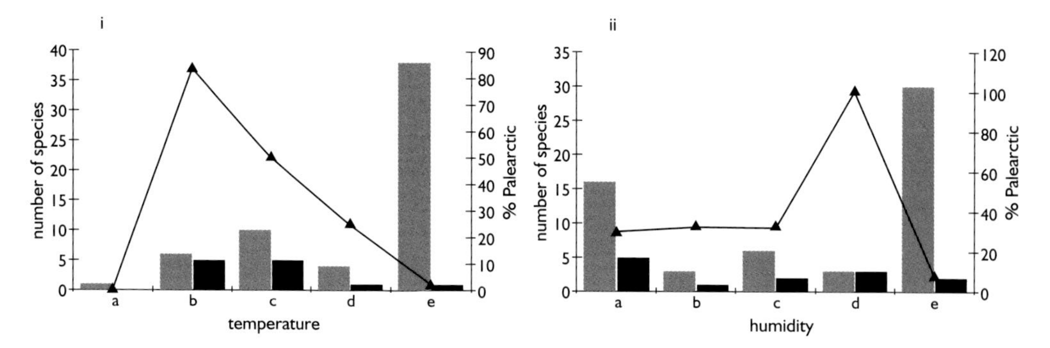

Figure 6.2 *Temperature (i) and humidity (ii) tolerance of tits. Definitions as in Figure 4.2 (see p. 38). Key as in Figure 6.1.*

their habit of living in forests, and persist in the Palearctic by having adapted to a range of bioclimates and corresponding forest types. This may explain their morphological conservatism noted, among others, by Gill *et al.* (2006).

Migratory behaviour

Tits are characterised by sedentary behaviour; the Palearctic tits, unlike many other groups, have not developed significant migration abilities (some species perform seasonal movements within certain populations). Coupled with habitat conservatism, we can conclude that the tits have managed to survive the conditions of the Tertiary and Quaternary in the Palearctic by their ability to tolerate a wide range of bioclimates. They are among the few groups that have achieved this; there are similar examples among a minority of forest birds, most notably the woodpeckers.

Fossil tits

Fossil tits are present in Palearctic sites from the early Pleistocene. They are poorly represented, probably because of their small size (leading to poor preservation) and general absence from cave sites.

Taxa in the Palearctic

Two families have Palearctic representatives. They are the Remizidae (3 species, 100% of the world's species) and Paridae (15 species, 29.71%). The only exclusively Palearctic family in the Paroidea is the Remizidae. The Paroidea resemble the Corvoidea, and not the Sylvoidea, in having an even distribution of species across the Palearctic: 12 species in the Western, 13 in the Central Palearctic, and 11 in the Eastern. There is a high representation of bioclimatically-tolerant species among the Paroidea, comparable to the corvids (Appendix 1). It therefore comes as no surprise to find that seven species have pan-palearctic ranges, three of them continuous. The species with continuous ranges are Coal Tit *Periparus ater* and Great Tit *Parus major* (generalists, Latitude Category F); those with discontinuous pan-palearctic ranges are Penduline Tit *Remiz pendulinus*, Willow Tit *Poecile montana* (semi-generalist, Latitude Category F), Azure Tit *Cyanistes cyanus* (semi-specialist, Latitude Category C) and Siberian Tit *Poecile cinctus* (semi-specialist, Latitude Category B). They are all either bioclimatically tolerant, or specialists with ranges centred on 50°N and 60°N. By contrast, five species with 40°N distributions (White-crowned Penduline Tit *R. coronatus*, Varied Tit *Parus varius*, Sombre Tit *Poecile lugubris*, Rufous-naped Tit *Parus rufonuchalis* and Turkestan Tit *Parus bokharensis*) have fragmented ranges. Eight species occur in a single sub-region – they are White-crowned Penduline Tit, Rufous-naped Tit, Turkestan Tit and Yellow-breasted Tit *Cyanistes flavipectus* (Central), and Chinese Penduline Tit *R. consobrinus,* Yellow-bellied Tit *Periparus venustulus* and Varied Tit *Parus varius* (Eastern). The African Blue Tit *Cyanistes teneriffae* is the only species unique to the Western Palearctic.

REMIZIDAE

Remiz

The bioclimatically moderate Penduline Tit has a discontinuous pan-palearctic range (Temperate, Latitude Category C). The other penduline tits are localised in the Central (White-crowned Penduline Tit, semi-specialist) and the Eastern (Chinese Penduline Tit, specialist) Palearctic. The pattern follows the general trend of a single bioclimatically tolerant species dominating the geographical range of each genus. The penduline tits are sedentary omnivores that occupy wetland habitats. They seem to have succeeded in colonising the temperate and mid-latitude belts of wetlands that are spread across the Palearctic, resembling in this respect the *Locustella* and *Acrocephalus* warblers. A more omnivorous diet than that of the warblers, in which insects predominate, may have been the key to permitting residency in these tits.

PARIDAE

Here I follow Gill *et al.'s* (2005) classification with modifications from Clements (2007).

Poecile

Four species, all sedentary forest omnivores, of this Holarctic genus of 15 species breed in the Palearctic. Two are semi-generalists – Marsh Tit *P. palustris* (Latitude Category C, 50°N) and Willow Tit (Latitude Category B, 60°N); one, Siberian Tit, is a semi-specialist (Latitude Category B); and the Sombre Tit is a specialist (Latitude Category D, 40°N). Willow and Siberian Tits, both Latitude Category B, have discontinuous pan-palearctic ranges; Marsh Tit is found in the west and the east but, having a more southerly range than the previous species, is missing from the dry Central Palearctic. The Sombre Tit, with an even more southerly distribution than Willow or Siberian, is missing from the Eastern Palearctic. The Marsh Tit-Willow Tit pair of semi-generalists within the same genus is unusual; there are differences in the range of the two, but it is also true that they overlap widely.

Periparus

Three of seven *Periparus* tits, a Palearctic/south-east Asian genus, breed in the Palearctic. They are all sedentary forest omnivores. The Coal Tit is one of two bioclimatic generalist tits, and has a continuous pan-palearctic (multi-latitude, Category F) range. The two other species are the Yellow-bellied Tit, a semi-specialist with a discontinuous Eastern Palearctic range centred on 30°N (Latitude Category E), and the Rufous-naped Tit of the mountains of Central Asia. This genus follows the pattern observed in other genera (e.g. *Garrulus* jays, *Alauda* larks and *Delichon* martins) of a widespread generalist along with restricted-range specialists. Because of the Coal Tit's wide latitude range, this species has been able to maintain a west-east trans-palearctic connection, while species further south have become isolated.

Lophophanes

This is a genus of two species. The Crested Tit *L. cristatus*, a bioclimatic moderate with a multi-latitude range, breeds in the Western and Central Palearctic (discontinuous range), while the Grey-crested Tit *L. dichrous,* which is extralimital, is a montane specialist of the southern slopes of the Himalayas. Both are sedentary forest omnivores. The disjunct distribution of these two congeners recalls the more extreme situation of the azure-winged magpies, both cases resulting from the loss of forest and woodland to aridity in the Central Palearctic.

Parus

Three of 22 species in this genus breed in the Palearctic, with only one in the west. They are predominantly African and southern Asian, and are resident omnivores of forest and woodland. They are all bioclimatic specialists and semi-specialists with one exception: the Great Tit has a continuous pan-palearctic range and is a generalist. It provides a demonstrative example of the impact of bioclimatic tolerance on success in the Palearctic, and it also shows how such generalisation is rare within each genus. The other two Palearctic species are the Varied Tit, a semi-specialist (Latitude Category D) in the extreme east of the Palearctic, and the Turkestan Tit, a montane specialist (also Latitude Category D) with a localised distribution in the Central Palearctic. This situation recalls that of the coal and crested tits.

Cyanistes

This is an exclusively Palearctic genus of four sedentary forest omnivores. The Blue Tit *C. caeruleus* is a bioclimatic semi-generalist (multi-latitude, Category F) with a range that is continuous in the Western and discontinuous in the Eastern Palearctic. The Azure Tit is a semi-specialist (Temperate, Category C) that has a pan-palearctic range, discontinuous and marginal in the west. The ranges of the Blue and Azure Tits overlap minimally. The Yellow-breasted Tit is a semi-specialist with a restricted range along 40°N in the mountains of Central Asia. Finally, the African Blue Tit is an island specialist living in the extreme west of the Palearctic. Again, this tit genus illustrates well the patterns I have been describing; here the Blue Tit is the most generalised species, followed by Azure, Yellow-breasted and African Blue, with increasingly restricted ranges.

CONCLUSIONS

(a) **Bioclimatic generalisation**. The proportion of bioclimatically tolerant (Categories E and D) members of the Paroidea (6 species, 31.58%) is the highest of the passerine groups that we have seen so far (8 species, 22.22%, in Corvoidea and 9 species, 7.63%, in Sylvoidea). Close to a third of the tits are bioclimatically tolerant, a figure that rises above 42% if we add bioclimatic moderates. This result suggests that tits have been particularly good at achieving high tolerance levels, a feat that has allowed them to spread widely and colonise the New World on two occasions.

The generalists and semi-generalists are mainly multi-latitude species (latitude Category F): Coal, Great and Blue Tits, with one 50°N (latitude Category C – Marsh Tit) and one 60°N (latitude Category B – Willow Tit) species. It is significant that there are no mid-latitude belt species among them. With the exception of the Marsh/Willow Tit pair of semi-generalists, bioclimatically tolerant tits are restricted to one species per genus. The only exceptions are the two semi-generalist congeners, Willow and Marsh Tits.

(b) **Low mid-latitude belt representation**. The tits also provide a clear contrast with previous groups in that the proportion of mid-latitude species, with ranges centred on 40°N, is low, with only six species (31.58%); they are all bioclimatic semi-specialists and specialists. This figure is much lower than the 58.33% of the Corvoidea or the 51.2% of the Sylvoidea. All six species have, as we would expect, fragmented ranges: Sombre Tit (Western and Central Palearctic), White-cheeked Penduline, Yellow-breasted, Rufous-naped and Turkestan Tits (Central) and Varied Tit (Eastern).

(c) **A recurring pattern**. The pattern in tit genera is for a single, widely distributed bioclimatic generalist and a range of more specialised species with restricted ranges. These restricted ranges are either exclusively Eastern Palearctic or lie in the mountains of the Central Palearctic. *Remiz* also shows restricted distribution in the wetlands of the Central Palearctic. Only *Poecile* departs from this trend, having widely distributed boreal-temperate species. In terms of habitat all species, bar the wetland-habitat *Remiz*, are sedentary or mixed-habitat omnivores. With the exception of the wetland *Remiz,* the remaining genera repeat a common theme – they are all sedentary forest or mixed-habitat omnivores. This strategy has repeatedly allowed tits from different genera to succeed as residents over large areas of the Palearctic, which is unusual among birds.

(d) **Uniquely Palearctic genera**. *Remiz* (three species) and *Cyanistes* (four species) are uniquely Palearctic genera. A single species in the former and two (Blue and Azure Tits) in the latter have dominated the temperate regions of the Palearctic, in wetlands and forest respectively. The remaining species are localised, including an island form (African Blue Tit). The remaining genera have species in Africa, south-east Asia or North America but *Remiz* and *Cyanistes* may be considered endemic Palearctic genera. No species is unique to the Western Palearctic, although this sub-region is the domain of the Blue Tit.

CHAPTER 7

Muscicapoidea – chats, thrushes, flycatchers and allies

Group 1, subgroup a (iv)

The Muscicapoidea is a complex superfamily and here I adopt a wide definition based on recent analyses, especially those of Voelker and Spellman (2004) and Treplin *et al.* (2008). I include the two "core" families Turdidae (thrushes) and Muscicapidae (chats, flycatchers and allies); the latter may be separated into two tribes – Saxicolini and Muscicapini (Voelker and Spellman 2004). Closest to these are the starlings (Sturnidae). We then have a series of families whose phylogenetic position is far from clear but which appear to have some relationship with the Muscicapoidea. It may turn out that these belong elsewhere but, in the absence of further information, I will include them with this superfamily. These families are the wrens (Troglodytidae), creepers (Certhiidae and Tichodromidae), nuthatches (Sittidae), dippers (Cinclidae), waxwings (Bombycillidae) and kinglets (Regulidae). This is a total of ten families.

Climate

The muscicapoids are bioclimatic specialists of warm and wet bioclimates. The Palearctic subset is strikingly different from the remainder and includes the most bioclimatically tolerant species (which includes generalists and semi-generalists; Figures 7.1–7.2). Palearctic species are strongly biased towards cool bioclimates, a position which they presumably achieve by being migratory, while the majority of species are concentrated at the warmest end of the temperature gradient (Figure 7.2i). They also show a clear shift away from the most humid bioclimates and ranging even into the driest (Figure 7.2ii).

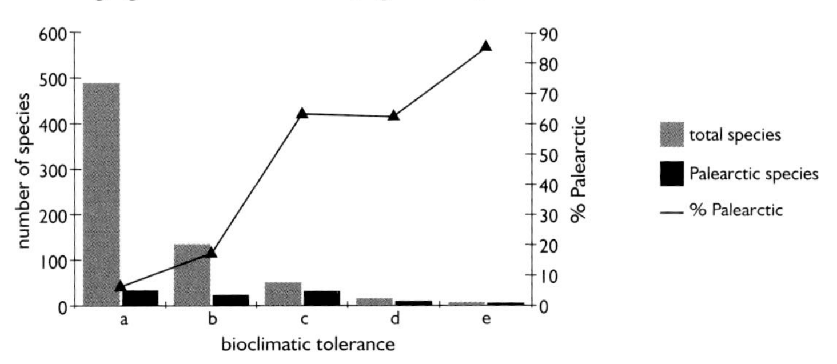

Figure 7.1 *Bioclimatic tolerance of muscicapoids. Definitions as on Figure 4.1 (see p. 38).*

Habitat

Muscicapoids occupy a range of habitats but are preferentially forest and mixed-habitat birds (Figure 7.3). The Palearctic species are predominantly birds of open habitats, a feature that is most conspicuous in the Muscicapidae (Figure 7.4i), but also retain a forest component, especially the Turdidae (Figure 7.4ii). Part of the success of this group in the Palearctic has to do with their ability to occupy habitats other than forest. In this respect the colonisation of rocky habitats (Figure 7.3) is noteworthy.

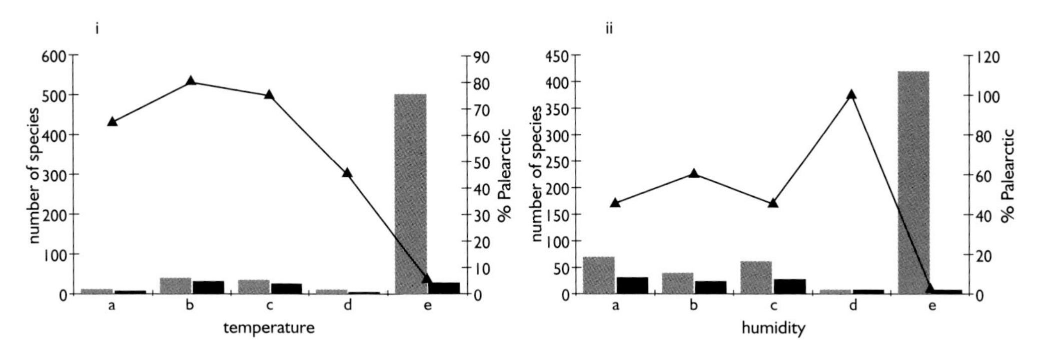

Figure 7.2 *Temperature (i) and humidity (ii) tolerance of muscicapoids. Definitions as in Figure 4.2 (see p. 38). Key as in Figure 7.1.*

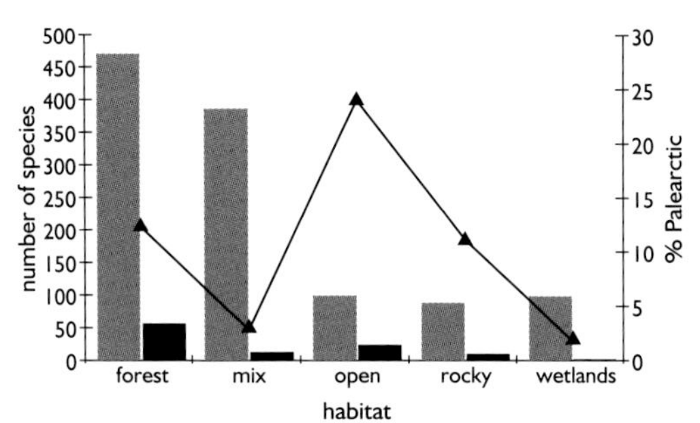

Figure 7.3 *Habitat occupation by muscicapoids. Definitions as in Figure 4.3 (see p. 39). Key as in Figure 7.1.*

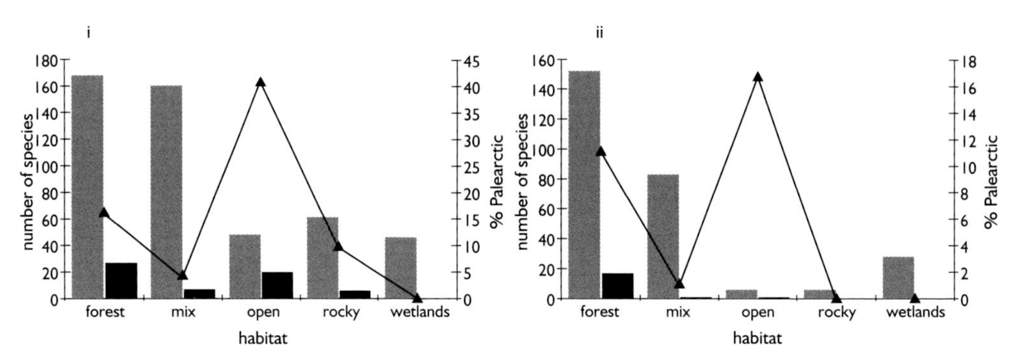

Figure 7.4 *Habitat occupation by muscicapids (i) and turdids (ii). Definitions as in Figure 4.3 (see p. 39). Key as in Figure 7.1.*

Migratory behaviour

A strong feature of the Palearctic muscicapoids is their ability to migrate (Table 7.1). This is in stark contrast with the superfamily as a whole, which is dominated by sedentary species. Migratory behaviour, and an ability to move away from forest and live in a range of bioclimates including cool and dry ones, are the hallmark of the group's success in the Palearctic.

Fossil species

A range of species have been found in Pleistocene deposits in Europe, but they tend to be scarce in the fossil record and provide little by way of geographical range shift information. The Late Pleistocene record of

	Global	Palearctic	Global %	Palearctic %	Palearctic of global %
Migratory	135	76	19.31	71.7	56.3
Sedentary	631	48	90.27	45.28	7.61
Total Species	699	106			

Table 7.1 *Migratory behaviour in the Muscicapoidea.*

Turdus is, on the other hand, important, especially for Eurasian Blackbird *Turdus merula*, Mistle Thrush *T. viscivorus* and Fieldfare *T. pilaris*; there are also good numbers of Redwing *T. iliacus*, Song Thrush *T. philomelos* and Ring Ouzel *T. torquatus* (Appendix 2). These results suggest that the conditions of Late Pleistocene Europe were particularly favourable for thrushes.

Taxa in the Palearctic

The families have the following numbers of Palearctic species: Muscicapidae (65, 22.95% of the world's species), Turdidae (19, 11.18%), Sturnidae (8, 6.96%%), Cinclidae (2, 40%), Sittidae (7, 29.17%), Tichodromidae (1, 100%), Certhiidae (3, 33.33%), Troglodytidae (1, 1.3%), Regulidae (3, 50%) and Bombycillidae (2, 66.67%). Tichodromidae, a single-species family, is the only exclusively Palearctic family in the Muscicapoidea. Bearing in mind the small number of species involved, Cinclidae, Sittidae, Certhiidae, Regulidae and Bombycillidae are fairly well represented. Of the two large families, Muscicapidae is well represented, better than the Sylviidae warblers but less so than the larks or tits. The Turdidae are poorly represented among the large Palearctic families. The Muscicapoidea are distributed evenly across the Palearctic but there is a tendency towards fewer species in the Eastern Palearctic, due to a general scarcity of chats, especially *Oenanthe* wheatears, in that sub-region.

There are very few (five) bioclimatically generalist species among the Muscicapoidea. The number is higher (12) in the case of semi-generalists (Appendix 1). There are no bioclimatically tolerant species in the Sturnidae or the Bombycillidae. The highest number is in the Muscicapidae (nine species, generalists plus semi-generalists), which represents 15% of the species; the Turdidae, on the other hand, only have two, which is 10.53% of the species in that family. Of the smaller families, the two dippers (Cinclidae) are semi-generalists, one of three kinglets (Regulidae) is a generalist and one of three treecreepers (Certhiidae) is a semi-generalist. One of seven nuthatches (Sittidae) is a generalist, which represents 14.29% of the species, a figure comparable to the Muscicapidae although the overall number of species is much lower. Overall, the proportion of bioclimatically tolerant species (16.19%) is lower than in the tits (36.84%) and corvids (22.22%) but higher than in the sylviids (7.2%).

All the families comprising the Muscicapoidea have at least one species with a pan-palearctic range. Overall, 18 species (17.14%) have pan-palearctic ranges and six have continuous ranges: Blue Rock Thrush *Monticola solitarius*, Rufous-tailed Rock Thrush *M. saxatilis*, Northern Wheatear *Oenanthe oenanthe*, Eurasian Nuthatch *Sitta europaea*, Eurasian Treecreeper *Certhia familiaris* and Bohemian Waxwing *Bombycilla garrulus*. These highlight three ecological groups that have succeeded in maintaining wide geographical distributions: (a) chats of open and rocky terrain, akin to the choughs and crag martins, of which the rock thrushes and the wheatear are examples and to which we can add Desert *Oenanthe deserti* and Isabelline *O. isabellina* Wheatears and the Wallcreeper *Tichodroma muraria*; (b) sedentary birds of temperate and boreal forests, akin to the tits, jays and nutcrackers – to the nuthatch and treecreeper we can add the Goldcrest *Regulus regulus*; and (c) birds of boreal forest and edge habitats (including species with mountain outposts to the south) – to the waxwing we can add Bluethroat *Luscinia svecica*, Red-flanked Bluetail *Tarsiger cyanurus*, Fieldfare, Redwing and Eurasian Starling *Sturnus vulgaris*.

MUSCICAPIDAE

Cercotrichas

The Rufous Bush Robin *C. galactotes* is the sole representative of the genus in the Palearctic. The Black Bush Robin *C. podobe* penetrates into the Eastern Sahara and Arabian Peninsula and could potentially be considered a second species, but is marginal. The eleven species of the genus are sedentary, warm-climate specialists; they are all African except the Rufous Bush Robin, which has sedentary populations in the Sahara and migratory ones to the north, from the Iberian Peninsula east to Pakistan. This distribution recalls that of the Eastern Olivaceous Warbler. From this we can infer a warm-climate, Palearctic-African distribution for the genus that shrunk with Plio-Pleistocene climate change, leaving a single, uniquely migratory species in the genus, on the southern fringes of the Palearctic.

Erithacus

This is a single-species Palearctic genus, the Japanese and Ryukyu Robins often placed here belonging with *Luscinia* instead (*L. akahige* and *L. komadori*; Seki, 2006). The European Robin *E. rubecula*, a bioclimatic semi-generalist, has a wide distribution (latitude Category F) across the Western and Central Palearctic; the more northerly and continental populations are migratory but the south-western ones are sedentary. *Erithacus* is probably a Miocene-Pliocene solution to life in the humid forests of the Western Palearctic, representing an early divergence from *Luscinia*.

Luscinia

This is an interesting genus of 13 Palearctic-Himalayan species. One, the Bluethroat, has penetrated the Nearctic and occurs in western Alaska. Eight of the 13 *Luscinia* are Palearctic; the remaining species are Himalayan montane (4) and wet tropical forest (1, Ryukyu Robin). Four species are bioclimatic moderates, the most tolerant in an otherwise specialist genus, and all four are Palearctic as we would expect: Common Nightingale *L. megarhynchos*, Siberian Rubythroat *L. calliope*, Bluethroat and Siberian Blue Robin *L. cyane*. The Bluethroat, a bioclimatic moderate (Category C), is the only pan-palearctic species. Its range is split between the Boreal (60°N, Latitude Category B) and the mid-latitude (40°N, Latitude Category D) belts, a recurring pattern among boreal species that had a wider distribution before the glaciations and whose ranges were severed by global warming, leaving high-altitude populations along the mountains of the mid-latitude belt. In the case of the Bluethroat, it has been shown that this population split is reflected genetically (Zink *et al.*, 2003). The Bluethroat's range is fragmented within the Western Palearctic but continuous elsewhere.

There are no exclusively Western Palearctic *Luscinia* species, but two have ranges that partly overlap in the Western and Central Palearctic: Common Nightingale is a bioclimatic moderate (range centred on 50°N, Latitude Category C) with a continuous range in the Western Palearctic but fragmented in the Central; and Thrush Nightingale *L. luscinia,* a bioclimatic specialist (range also centred on 50°N, Latitude Category C) with a continuous range in the Central but discontinuous in the Western Palearctic. One species, White-tailed Rubythroat *L. pectoralis* (bioclimatic specialist, Latitude Category D) occurs in the Central Palearctic; another, Siberian Rubythroat (bioclimatic moderate, Latitude Category B/D), occurs in the Central and Eastern Palearctic with a Bluethroat-lke split range; and three occur solely in the Eastern Palearctic – Siberian Blue Robin (bioclimatic moderate, Latitude Category C); Rufous-tailed Robin *L. sibilans* (bioclimatic semi-specialist, Latitude Category B) and Japanese Robin (bioclimatic moderate, Latitude Category D). Migratory behaviour typifies all species except the White-tailed Rubythroat, which only undertakes altitudinal short-range movements, and the partially migratory Japanese Robin.

In the absence of a complete phylogeographic study of this genus, I suggest tentatively that *Luscinia* emerged on the slopes of the Himalayas as a forest omnivore of cool and humid climates. Its range spread naturally into the Palearctic during cool periods and the genus established itself across the region. Subsequent climatic swings rearranged, fragmented and fused species ranges.

Irania

Like *Erithacus,* this is a single-species Palearctic genus. The White-throated Robin *I. gutturalis* is a migratory mixed-habitat omnivore with a localised distribution along 40°N in eastern parts of the Western Palearctic into the Central Palearctic.

Tarsiger

This genus is close to *Luscinia,* with their precise relationships in need of further study (Seki, 2006). One of five species, the Red-flanked Bluetail breeds in the Palearctic. This is a genus of southern Asian-Himalayan bioclimatic specialists; as in *Luscinia* the highest level of bioclimatc tolerance is moderate (C) and is only achieved by the sole Palearctic species, which is highly migratory across most of its range except in the oceanic south-east. Red-flanked Bluetail has a pan-palearctic range that is only continuous in the Eastern Palearctic, and is marginal in the Western Palearctic. Like the Bluethroat, it has a Boreal (Latitude Category B) and Warm (mid-latitude, Category D) distribution, although this is not as disjointed as that of the Bluethroat.

Ficedula

This genus is closest to *Tarsiger* and contains 30 species, 10 (32.26%) of which breed in the Palearctic (Outlaw *et al.,* 2006). The genus is divisible into two major lineages, with Palearctic birds represented in both. One group has species that are exclusively Eastern Palearctic: Mugimaki Flycatcher *F. mugimaki,* Narcissus Flycatcher *F. narcissina* and Yellow-rumped Flycatcher *F. zanthopygia.* The second group includes Western Palearctic species, along with the Taiga Flycatcher *F. albicilla* of the Central and Eastern Palearctic. Of the Western Palearctic species, the Atlas Flycatcher *F. speculigera* is exclusive to the subregion; the remaining species are all of the Western and Central Palearctic. These are Red-breasted Flycatcher *F. parva,* Collared *F. albicollis,* Pied *F. hypoleuca* and Semi-collared *F. semitorquata* Flycatchers. These flycatchers are conservative in habits, being highly migratory forest insectivores (though some may consume fruit in small amounts at specific times of the year). Five species – Red-breasted (multi-latitude, Category F); Collared, Mugimaki (Temperate, Category C); Yellow-rumped, Narcissus (Warm, Category D) and Pied (split Boreal/Warm, Category B/D) – are bioclimatic moderates. The remaining species are either specialists or semi-specialists. There is no *Ficedula* flycatcher with a pan-palearctic range. The deep division, forming two major groups within the genus, suggests an ancient split between east and west. The common ancestor of the Taiga and Red-breasted Flycatchers would have presumably had a pan-palearctic range that split, making the Taiga Flycatcher the only species in its group with an eastern distribution.

Muscicapa

This genus forms a second major grouping of Old World flycatchers along with *Niltava* (Lei *et al.,* 2007). This genus contrasts with *Ficedula* in having an important African contribution of species (15 species, 62.5%) alongside a southern Asian-Himalayan component. Five species (20.83%) breed in the Palearctic. The Spotted Flycatcher *M. striata,* a bioclimatic moderate, is the only pan-palearctic flycatcher (multi-latitude, Category F), albeit with a fragmented range in the east; but it is not the most bioclimatically-tolerant species. The Brown Flycatcher *M. dauurica* is a generalist and the Dark-sided Flycatcher *M. sibirica* a semi-generalist – the two are Eastern Palearctic species and are uniquely tolerant among the flycatchers. The other species are Grey-streaked Flycatcher *M. griseisticta,* an Eastern Palearctic semi-specialist, and Rusty-tailed Flycatcher *M. ruficauda,* a montane specialist of the Central Palearctic. The latter is the only mid-latitude belt species, the others having Temperate (Category C) or multi-latitude (Category F) ranges. Like *Ficedula,* this genus is made up of migratory forest insectivores (with a minor omnivorous component). It continues the trend of single-species generalists within a genus.

Niltava

This is a southern Asian-Himalayan genus of eight species, with one, the Blue-and-white Flycatcher *N. cyanomelana,* breeding in the Eastern Palearctic. It is a highly migratory (Warm, Category D) bioclimatically moderate forest insectivore. The other species in the genus are bioclimatic specialists and semi-specialists, so the trend of increased bioclimatic tolerance among Palearctic species is supported.

Terpsiphone

The Japanese Paradise-flycatcher *T. atrocaudata* is the only paradise-flycatcher to reach the Palearctic. It is a specialist migratory forest insectivore. Lei *et al's* (2007) analysis suggests that this genus should be placed in the Monarchidae (rather than within the Muscicapidae), which would make this the sole Palearctic representative of this family.

Phoenicurus

Seven of eleven species breed in the Palearctic. The redstarts, like *Luscinia,* are Palearctic-Himalayan and may have a similar history. They provide an excellent example in support of the arguments put forward in this book regarding the importance of Central Asian aridification, prior to the Pleistocene, in the isolation and formation of lineages. The various redstart lineages, including the sister genera *Chaimarrornis* and *Rhyacornis,* have been isolated at different times since the Late Miocene in eastern, south-eastern and western refugia. Looking at the Palearctic species, the Blue-capped Redstart *P. coeruleocephalus* separated from the Blue-fronted *P. frontalis* around 4.3 mya, and Güldenstädt's *P. erythrogaster* from its sister group around 3.3 mya.

Daurian Redstart *P. auroreus* separated from its sister group (which includes the closely-related Black *P. ochruros* and Common *P. phoenicurus* Redstarts) around 3.0 mya (Voelker, 2010). These three species are the most bioclimatically tolerant redstarts (semi-generalists, Category D); all others are either specialists or semi-specialists. Between them, these three redstarts occupy much of the Palearctic, but no single species is pan-palearctic. All three are multi-latitude (Latitude Category F). The Common Redstart has a continuous range across the Western and Central Palearctic; the Black Redstart is continuous in the west and discontinuous in the east; Daurian is continuous in the Eastern Palearctic. Black and Common Redstarts are largely separated by habitat, the former typically on rocky ground and the latter in forest, and by migratory behaviour; northerly and continental Black Redstarts are migratory and all Common Redstarts are migratory (as is Daurian).

The remaining species include Moussier's Redstart *P. moussieri*, a specialist (Latitude Category D) resident of the Western Palearctic, and three specialists and semi-specialists of Central Asian Mountains (Latitude Category D), Güldenstädt's, Eversmann's and Blue-capped.

Cercomela

One of nine (11.11%) species of this genus of tropical Afro-Asian chats of dry habitats – Blackstart *C. melanura* – just reaches the southern edge of the Western Palearctic along the sub-tropical (Latitude Category E) belt. It is a sedentary bioclimatic specialist.

Saxicola

These bush chats are a southern Asian genus with a small component – five species (38.46%) – in the Palearctic. The most bioclimatically tolerant species is the Siberian Stonechat *S. maura*. It is a migratory bioclimatic generalist with a pan-palearctic, multi-latitude (Category F) range that is only marginal in the west but continuous elsewhere. Being the only generalist in the genus, it follows the trend that we have noted for other genera, of a single bioclimatically-tolerant (and widespread) species occurring in each. Its counterpart in the west is the European Stonechat *S. torquata*, a bioclimatically moderate multi-latitude range species. Its northern and continental populations are migratory, but those in the south-west are short-range migrants or sedentary; in this respect it resembles species like the Blackcap. The separation of these two stonechats recalls that of the redstarts, in this case with no overlap in the central areas of the Palearctic. This suggests a similar history, with the break-up of a pan-palearctic common ancestor of the two species.

The Whinchat *S. rubetra* follows the split-range pattern we have observed in a number of species (including Bluethroat, Red-flanked Bluetail and Siberian Rubythroat). It is a highly migratory bioclimatic moderate that occupies a boreal range (Latitude Category B) but has populations to the south that occupy high ground (Latitude Category D). The Whinchat is essentially a Western Palearctic species with part of its range extending into the Central Palearctic. Similar in ecology is Hodgson's Bush Chat *S. insignis* of alpine meadows of the Central Palearctic; it is a migratory bioclimatic semi-specialist with a distribution centred on 50°N. The fifth species in the genus, the Canary Islands Chat *S. dacotiae,* is a sedentary bioclimatic specialist, an island endemic of the south-western extreme of the Palearctic.

Oenanthe

Fifteen (68.18%) of the 22 wheatear species breed in the Palearctic; the rest breed to the south in adjacent areas of Africa (e.g. Somali Wheatear *O. phillipsi* in the Ethiopian mountains), or in other arid regions of that continent (e.g. Mountain *O. monticola* and Capped *O. pileata* Wheatears in southern and eastern Africa). So this is truly a genus borne from the aridification of the mid-latitude Old World during the Tertiary. No fewer than 11 species (73.33%) have ranges centred on the mid-latitude belt (40°N); three others are to the south and the remaining species, Northern Wheatear, has a multi-latitude (Category F) distribution that includes the mountains and high ground of the mid-latitude belt. Three main lineages have been recognised by genetic analysis, which has resolved earlier incorrect arrangements based on plumage and morphology. These lineages are (a) Northern and Isabelline Wheatears; (b) Desert *O. deserti,* (Western) Pied *O. pleschanka* and Black-eared *O. hispanica* Wheatears; and (c) other Palearctic wheatears.

These analyses are extremely interesting in the light of my bioclimatic analysis. The first lineage, that of Northern and Isabelline Wheatears, includes the only bioclimatically tolerant wheatear in a genus of specialists and semi-specialists; this is the Northern Wheatear (semi-generalist, multi-latitude, Category F), which has a continuous pan-palearctic range and, like the Bluethroat, reaches north-western America. Being the only semi-generalist in the genus, the Northern Wheatear follows the trend of a single bioclimatically tolerant (and widespread) species in each genus. The Isabelline Wheatear is a semi-specialist of the mid-latitude belt (Latitude Category D), also with a pan-palearctic range, continuous only in the Central Palearctic. The ability to hold pan-palearctic ranges in species with ranges centred on 40°N is a feature of birds of rocky terrain (e.g. choughs, Crag Martin, Wallcreeper). This particular lineage has therefore produced two of the three pan-palearctic wheatears.

The second lineage has the third pan-palearctic wheatear – Desert – a semi-specialist of the mid-latitude belt (Latitude Category D) with similar range characteristics to the Isabelline. It includes the only other semi-specialist wheatear – Pied – which has a Western (fragmented) and Central Palearctic range centred on 40°N (Latitude Category D). The third species is a specialist – Black-eared – with a continuous distribution in the Western and a discontinuous one in the Central Palearctic (also Latitude Category D).

The third lineage includes the remaining Palearctic wheatears; they are all bioclimatic specialists with ranges that occupy the Western and Central Palearctic (Mourning *O. lugens,* Red-tailed *O. xanthopryma,* Hume's *O. alboniger* and Hooded *O. monacha*), Western Palearctic (Red-rumped *O. moesta,* Black *O. leucura,* White-crowned Black *O. leucopyga* and Cyprus *O. cypriaca*), and Central Palearctic (Eastern Pied *O. picata).*

All species or populations that breed around, or north of, the mid-latitude belt are migratory: Isabelline, Northern, Black-eared, Desert (populations), Red-tailed (populations), Finsch's *O. finschii* (populations), (Western) Pied, Cyprus, Eastern Pied, Mourning (populations). Species or populations that largely breed on the southern flanks or to the south of the mid-latitude belt are sedentary: Desert (populations), Red-rumped, Red-tailed (populations), Finsch's (populations), Mourning (populations), Hume's, Black, Hooded and White-crowned Black.

Monticola

A recent analysis (Zuccon and Ericson, 2010) subsumes the genera *Thamnolaea* and *Pseudocossyphus* within *Monticola.* This means that three (17.64%) of the 17 species in this genus breed in the Palearctic. The genus arose in the Pliocene around 5.5 mya (Outlaw *et al.,* 2007) and has its origins somewhere in the huge area of the Palearctic and the mountain belts that stretch south across East Africa to the Cape (Zuccon and Ericson, 2010). This is admittedly a huge area and part of the discussion regarding *Monticola* origins has centred on whether the group is ancestrally African or Palearctic. Following the themes developed in this book, the question becomes irrelevant. *Monticola* have their origins somewhere in the mid-latitude belt (40°N) or the southern extension of mountains down the Rift Valley to South Africa. These chains formed a highway along which species at home in the dry, rocky terrain that became extensive along them during the Pliocene spread, became disconnected, speciated or rejoined as climatic conditions fluctuated.

Monticola is dominated by bioclimatic specialists (nine species) and semi-specialists (seven species). One species, the Palearctic Blue Rock Thrush, is a semi-generalist. Once again, a single bioclimatically tolerant and widespread species stands alone in a multi-species genus. The three Palearctic rock thrushes are typically birds of the mid-latitude belt (Latitude Category D). The Blue and the Rufous-tailed Rock Thrushes are further examples of rocky-habitat species of this belt that have continuous pan-palearctic distributions. In this case the more bioclimatically tolerant species (Blue) is sedentary in the western part of its range, though populations in the Central and Eastern Palearctic are migratory, while the more specialised Rufous-tailed Rock Thrush is highly migratory. The third species, the White-throated Rock Thrush *M. gularis,* is a migratory species of the mountains of the Eastern Palearctic; its range overlaps with the Blue but not with Rufous-tailed. The general pattern (which can be seen between congeners in other families, e.g. Sardinian and Subalpine Warblers) is for the less-migratory species (in this case Blue) to occupy lower ground, while the migratory one, in this case Rufous-tailed or White-throated depending on the region, takes the higher ground.

TURDIDAE

Zoothera

Only two (5.41%) species of this large genus breed in the Palearctic. Most species are southern Asian–Himalayan but others are tropical African. They are primarily resident forest omnivores, and are bioclimatic specialists and semi-specialists of warm and wet climates. This genus is not represented in the Western Palearctic at all. The two Palearctic species are the most bioclimatically tolerant of the genus, a recurring theme as we have seen; Scaly Thrush *Z. dauma* is a bioclimatic moderate (Temperate Latitude Category C) of the Eastern Palearctic, with a small outpost in the Central Palearctic; Siberian Thrush *Z. sibirica* is similar but occurs exclusively in the Eastern Palearctic. Both are highly migratory, the only species to do so in an otherwise mostly sedentary genus. The *Zoothera* thrushes are typical of birds with a mainly southern Asian distribution that have managed to succeed in the Eastern Palearctic (which has connections with the tropical south) but have been unsuccessful in the west, where deserts and mountains in the south have prevented connectivity with the tropics.

Turdus

This is one of the more species-rich bird genera, with 15 of them (22.73%) breeding in the Palearctic. The ancestral species in the genus are Palearctic, indicating an early radiation leading to the Mistle Thrush *T. viscivorus* and then the Song Thrush *T. philomelos.* Subsequently from the common ancestor of these species four main groups of thrushes emerged – an Eurasian group, an African group, a Central American–Caribbean group and a largely South American group (Voelker *et al.,* 2007). The Eurasian group is sorted into five lineages (other than the ancestral Mistle and Song Thrush lineage): (a) a lineage exclusive to the Eurasian Blackbird *T. merula;* (b) an Eastern Palearctic group – Tickell's *T. unicolor,* Black-breasted *T. dissimilis,* Grey-backed *T. hortulorum* and Japanese *T. cardis* Thrushes and Grey-winged Blackbird *T. boulboul*; (c) a boreal group composed of Naumann's *T. naumanni* and Red-throated *T. ruficollis* Thrushes and Ring Ouzel *T. torquatus*; (d) a pair consisting of the Fieldfare *T. pilaris* and Kessler's Thrush *T. kessleri*; and (e) another eastern group, related to the Pacific Island Thrush *T. poliocephalus*, including Brown-headed *T. chrysolaus,* Eyebrowed *T. obscurus,* Grey-sided *T. feae* and Pale *T. pallidus* Thrushes. Missing from this classification is the Redwing *T. iliacus* which, oddly, appears to be best placed within the Central American group (Voelker *et al.,* 2007).

The Eurasian thrush group (a) contains just the Eurasian Blackbird. This is the only bioclimatic generalist in the genus, with a Western and Central Palearctic (multi-latitude Category F, discontinuous) distribution. It is a partially migratory forest omnivore. Group (b) is a largely south-east Asian–Himalayan group of bioclimatic specialists. The Grey-backed Thrush breeds in the Eastern Palearctic and is a migratory bioclimatic moderate with a Temperate (Latitude Category C) distribution. The Japanese Thrush is a migratory bioclimatic specialist (Latitude Category D). Group (c) is interesting because it is exclusively Palearctic, and the three species have largely non-overlapping ranges in the Eastern (Naumann's – bioclimatic moderate, Boreal, Latitude Category B), Central (Red-throated – semi-specialist, multi-latitude Category F), and Western (Ring Ouzel – moderate, split Boreal/Warm Category B/D) Palearctic. Naumann's and

Red-throated Thrushes are forest species, while Ring Ouzel occupies open montane habitats. All three are highly migratory.

Group (d) includes one of only two pan-palearctic thrushes, the Fieldfare (bioclimatic moderate – Boreal, Latitude Category B) and a localised extralimital montane species – Kessler's Thrush in the Himalayas. Group (e) contains Eastern Palearctic species: Brown-headed (semi-specialist, Warm, Latitude Category D), Eyebrowed (moderate, Boreal, Latitude Category B), Grey-sided (specialist, Warm montane, Latitude Category D) and Pale Thrush (moderate, Temperate, Latitude Category C). These species occur along a latitude-altitude gradient. All are migratory.

There are three other Palearctic species. The Mistle Thrush is ancestral to the other thrushes. It is a bioclimatic semi-generalist with a multi-latitude (Latitude Category F) range. It is a Western and Central Palearctic species with migratory populations in continental and northern areas, and sedentary ones in the south-west. Next is the Song Thrush, a bioclimatic moderate with a split Boreal/Warm (Latitude Category B/D) distribution in the Western and Central Palearctic. Finally, the apparently anomalous Redwing is a second species with a pan-palearctic range; it is a bioclimatic moderate with a Boreal (Latitude Category B) range.

Overall, the thrushes conform, within the Eurasian group and subgroups, with the pattern of limited bioclimatically tolerant species. In this case the Eurasian Blackbird is the most tolerant, followed by Mistle Thrush. The Palearctic thrushes are mainly forest dwellers. All are omnivorous and all except the Chinese Thrush *T. mupinensis* are fully or partially migratory.

Voelker (2010) provides a range of estimates for the divergence of a number of thrush lineages, and they strongly indicate early radiations in the Pliocene: Fieldfare from its sister group 4.2 mya; Kessler's Thrush from Fieldfare 4.0 mya; Grey-winged Blackbird from its sister group 3.9 mya; Island Thrush from its sister group 2.8 mya; Ring Ouzel from its sister group 2.6 mya; Tickell's Thrush from its sister group 2.0 mya; and Red-throated from Naumann's Thrush 1.1 mya.

Myophonus

The Blue Whistling Thrush *M. caeruleus* is the sole representative of this genus of nine south-east Asian bioclimatic specialists. It penetrates the southern margins of the Central and Eastern Palearctic, along the Himalayan fringes, and is the only migratory species of its genus.

Catharus

The Grey-cheeked Thrush *C. minimus* is one of 12 New World thrushes. It is a semi-specialist that penetrates the north-eastern Palearctic (Latitude Class A) from Alaska. It is migratory.

STURNIDAE

Sturnus

Six (37.5%) of the species of this largely southern Asian genus breed in the Palearctic. They are omnivores of open and mixed habitats. The Eurasian migratory starlings form a homogeneous group that is closely related to two African groups. The Eurasian Starling *S. vulgaris* and its close relative the Spotless Starling *S. unicolor* are ancestral to and separated from the other species, which radiated more recently (Lovette and Rubenstein, 2007). *Sturnus* are mainly bioclimatic specialists but three of the Palearctic species are moderates. The Eurasian Starling is the only pan-palearctic (multi-latitude, Category F – all other Palearctic starlings are Latitude Category D) species, though its range is marginal in the east; the Daurian *S. sturninus* and White-cheeked *S. cineraceus* Starlings are also bioclimatic moderates, but they are restricted to the Eastern Palearctic and do not overlap with the Eurasian Starling. Of the other three Palearctic species the Rose-coloured Starling *S. roseus* is a bioclimatic semi-specialist of the Central Palearctic, with marginal populations in the Western Palearctic; the Chestnut-cheeked Starling *S. philippensis* is a semi-specialist in the Eastern Palearctic; and the Spotless Starling occurs in the south-western Palearctic. Unusually for a starling of our region, Spotless Starling is sedentary; the others are migratory (Daurian, Rose-coloured, Chestnut-cheeked) or have northern and continental migratory populations with sedentary ones in the oceanic south-west (Eurasian) or south-east (White-cheeked).

Acridotheres

This southern Asian genus is close to *Sturnus*; one species (12.5%), the Common Myna *A. tristis,* reaches southern areas of the Central Palearctic. It is a semi-specialist sedentary omnivore of mixed habitats. All other species are specialists.

Onychognathus

One species (10%) of this African genus, Tristram's Starling *O. tristramii,* reaches the southern edge of the Western Palearctic. It is a specialist of warm, arid bioclimates and is a sedentary omnivore of open habitats.

CINCLIDAE

Cinclus

Two of the five species of dipper are Palearctic. The other three occur in the New World. The genus appears to have originated in Eurasia around 4 mya, with a rapid dispersal into America shortly after (Voelker, 2002a). The Eurasian species – White-throated Dipper *C. cinclus* and Brown Dipper *C. pallasii* – appear to have split from a common ancestor around 2.5 mya. The White-throated Dipper is a bioclimatic semi-generalist with a pan-palearctic range (split Arctic-Warm A/D distribution). This range seems to have been markedly affected subsequently, in the Western Palearctic at least, by Pleistocene isolation and redistribution of birds in glacial refugia, accentuated by ecological and biogeographic barriers during interglacials. Five distinct regional and genetic groups have been identified in the Western Palearctic (Hourlay *et al.*, 2008), and their history is difficult to unravel. These groups include (a) a large western European lineage that may have survived in an Italian glacial refugium; (b) an eastern European lineage that may have survived in a Balkan glacial refugium; (c) a north-east European lineage; (d) a lineage that includes the isolated Moroccan population; and (e) a mixed group that includes Asian, Corsican and Irish dippers. It thus seems that the dippers from the mid-latitude belt (Category D in the split category range of this species) are those most affected by isolation during the Pleistocene, as we would have expected.

The Brown Dipper is a strictly Eastern Palearctic species. It is also a semi-generalist and has a multi-latitude (Latitude Category F) range. The two dippers overlap in range in the Eastern Palearctic, presumably due to a post-glacial range spread of the White-throated eastwards into Brown Dipper territory. It is interesting that both species have maintained the same bioclimatic tolerance status but that range overlap is limited to a specific region. The two species are sedentary, which is not surprsing given the specialised morphology of dippers, which would not seem suited to migration; that they survived the Pleistocene glaciations has to be due to their broad bioclimatic tolerance, though the Western Palearctic population structure does show that it was a survival within limited refugial areas, from which other regions were re-colonised subsequently.

SITTIDAE

Sitta

Seven (29.17%) of the world's nuthatches breed in the Palearctic. This is a largely southern Asian genus with many species clustered around the Himalayas (Menon *et al.,* 2008). This genus follows the trend that we have observed in many genera so far, with a single bioclimatic generalist with a large range (in this case Eurasian Nuthatch *S. europaea,* Latitude Category F, continuous pan-palearctic range) and a range of specialists with smaller, localised, ranges. Three main genetic lineages have been recognised in this species, in the Caucasus, Western Europe and north-east Europe and northern Asia (Zink *et al.,* 2006). They probably represent population isolation during moments of forest breakdown at the expense of steppe-tundra during Pleistocene glacials; the Caucasus seems to have provided a refuge for a number of other Palearctic species, such as Wren *Troglodytes troglodytes*, White Wagtail *Motacilla alba* and Common Rosefinch *Carpodacus erythrinus* (Zink *et al.*, 2006).

The Eurasian Nuthatch is related to the southern Asian and Himalayan White-tailed Nuthatch *S. himalayensis* and the North American White-breasted Nuthatch *S. carolinensis* (Pasquet, 1998), indicating an ancient Holarctic lineage that produced two species of bioclimatic generalists (Eurasian and White-breasted) that occupied large areas of the Palearctic and Nearctic respectively, and a semi-specialist, White-tailed Nuthatch, that remained localised in southern Asia. This latter species occupies a forest belt between 1,800 and 3,500 metres and its geographical range is limited by this band. Menon *et al.* (2008) noted that mountain species are restricted in range. This example illustrates the 'launchpad' model (see p. 41), a situation that may have occurred repeatedly in Palearctic (even Holarctic) birds, whereby a tropical southern Asian group diversifies altitudinally on the Himalayan slopes; a species with a climatic tolerance that matches a corresponding latitude band to the north spreads into that area. Unrestricted by the montane habitat, it spreads across the latitude belt through the Palearctic and, if its bioclimatic tolerance allows, into the Nearctic. Global warming (or cooling) may then sever the Himalayan and Palearctic populations. Examples where this phenomenon may have occurred include Eurasian and Oriental Skylarks *Alauda arvensis* and *A. gulgula*, Spotted and Clark's Nutcrackers *Nucifraga caryocatactes* and *N. columbiana*, and species of leaf warblers, tits and woodpeckers, among others.

The sister group to Eurasian, White-tailed and White-breasted Nuthatches provides a beautiful natural experiment. This group consists of six species with localised ranges in the Palearctic and southern Asia (unlike the Eurasian Nuthatch), and a widespread North American species (Red-breasted Nuthatch *S. canadensis,* a bioclimatic moderate). Four breed in the Palearctic: Algerian *S. ledanti* (Bioclimatic specialist, Latitude Category D), Corsican *S. whiteheadi* (Bioclimatic specialist, Latitude Category D), Krüper's *S. krueperi* (Bioclimatic specialist, Latitude Category D) and Chinese *S. villosa* (Bioclimatic semi-specialist, Latitude Category D) Nuthatches. The remaining species in the group, Yunnan Nuthatch *S. yunnanensis,* breeds on the slopes of the Himalayas and is also a bioclimatic specialist. This species occupies high elevations, but the others tend to occupy lower altitudes (in some cases down to sea level). These birds, though still montane, would seem to be derived from a warmer altitude belt than the White-tailed Nuthatch. It seems that there was a similar geographical spread to that seen in the previous nuthatch group, which reached North-west Africa in the west and North America in the east. Being a group of warmer climates, the range expansion appears to have followed the forests of the mid-latitude belt (all species are currently Latitude Category D), which would have had a more northerly extension in the Late Miocene than today. This would be borne out by the colonisation of North America by this group at a time of warm conditions. The fragmentation of these widespread populations appears to have started with the Messinian Salinity Crisis (the time when the Mediterranean dried up; see p. 21) and consequent aridification around 5 mya (Pasquet, 1998), and later glacials may have restricted populations even further. Occupation of the mid-latitude belt led to fragmented ranges among these forest birds (but not in the nuthatches of higher latitudes). So birds of this second group are all confined to reduced areas within a single Palearctic sub-region. The Red-breasted Nuthatch had no such impediment in North America, where the mountain ranges run north–south, and became widespread on that continent.

The corollary to this story is provided by two species that modified their behaviour and became species of rocky habitats. Having made the habitat-jump, the Western Rock *S. neumayer* and the Eastern Rock *S. tephronota* Nuthatches (or their common ancestor) occupied a wide range across the mid-latitude belt of the Western and Central Palearctic, in a similar fashion to other unrelated species with similar habits (e.g. the choughs, crag martins or wheatears). The absence of a habitat corridor into North America confined these species to the Palearctic.

TICHODROMIDAE

Tichodroma

The Wallcreeper *T. muraria* is a true one-off. This species is sometimes placed with the nuthatches, sometimes with the treecreepers or, as here, in a family of its own. The Wallcreeper resembles the rock nuthatches as a species of the rocky habitats of the mid-latitude belt. Like other such species it has a widespread distribution (it is a bioclimatic moderate), in this case pan-palearctic.

CERTHIIDAE

Certhia

Three species (33.33%) breed in the Palearctic. The remaining species in the genus are southern Asian and Himalayan. The Eurasian Treecreeper *C. familiaris* is a bioclimatic semi-generalist (Latitude Category C). It is one of two bioclimatically tolerant species in the genus; the other is the American Treecreeper *C. americana.* This observation, along with the example of the nuthatches, indicates that the 'single bioclimatically tolerant species per genus' rule need not apply to Holarctic species, where it becomes possible to have non-overlapping tolerant species in the Palearctic and Nearctic. The Eurasian Treecreeper is closely related to Hodgson's Treecreeper *C. hodgsoni* of the Himalayas (Tietze *et al.*, 2006). So, as in the nuthatches, we could hypothesise a geographical expansion across a latitudinal belt, equivalent in climatic characteristics to a Himalayan altitudinal belt. This trans-Palearctic expansion did not reach the Nearctic, though. The American Treecreeper is instead most closely related to the Short-toed Treecreeper *C. brachydactyla* of the Western Palearctic. This is a multi-latitude (Category F) species whose ancestor, perhaps as in the nuthatches of the Red-breasted group, must have had a pan-palearctic range and spread into North America in a warm period. This would be borne out by the subsequent range restriction to the mild oceanic south-west of the Palearctic. The American species, like the Red-breasted Nuthatch, became widespread on that continent. The third Palearctic species, Bar-tailed Treecreeper *C. himalayana,* is a semi-specialist of the mid-latitude belt that penetrates the southern fringe of the Central Palearctic.

TROGLODYTIDAE

Troglodytes

The Wren *T. troglodytes* is a unique species in the Palearctic. The Troglodytidae is a New World family (Barker, 2004) and the Wren, a bioclimatic multi-latitude (Category F) generalist, has spread to have a Holarctic range. It is discontinuous, however, missing from large areas of the Central Palearctic but with a continuous range in the Eastern and Western Palearctic and the mountain ranges in the south of the Central Palearctic. It is largely sedentary in the oceanic south-west and south-east but the northernmost and most continental populations are migratory. This sensitivity to cold and dry climates may explain its absence from the interior of the Palearctic.

REGULIDAE

Regulus

Three (50%) of the species in this Holarctic genus breed in the Palearctic. The Goldcrest *R. regulus* has a pan-palearctic range, which is discontinuous in the east, and is a bioclimatic generalist (Temperate, Latitude Category C). This is the most widespread species of the genus. It is largely sedentary although northernmost populations are migratory. The Firecrest *R. ignicapillus* is a bioclimatic moderate with a multi-latitude (Category F) distribution. The situation recalls that of the Eurasian Treecreeper/Short-toed Treecreeper (see above). The Canary Islands Goldcrest *R. teneriffae* is an island endemic of the extreme south-west of the Palearctic and is derived from the mainland Goldcrest (Sturmbauer *et al.*, 1998). The Canary Islands Goldcrest and the extralimital Flamecrest *R. goodfellowi* of south-east Asia are bioclimatic specialists of warm and wet climates. The remaining species in the genus are Nearctic: Ruby-crowned *R. calendula* (a semi-generalist) and Golden Crowned *R. satrapa* (a moderate). It illustrates another tendency that is observable in other groups with Palearctic and Holarctic birds (such as nuthatches); within a genus, Palearctic birds tend to be more bioclimatically tolerant than their Nearctic counterparts. This may be the result of the connectivity of Nearctic areas with tropical ones to the south (as in the Eastern Palearctic), which would have allowed greater possibilities of range shifts to the south during glacials than in the Palearctic (where this would have required greater climatic tolerance).

BOMBYCILLIDAE

Bombycilla

The affinities of the waxwings remain unclear. Two bioclimatically moderate (Category C) species take up a large Boreal range across the Holarctic, Bohemian *B. garrulus* in the Palearctic and Cedar *B. cedrorum* in the Nearctic. A third species, the Japanese Waxwing *B. japonica* is a Temperate (Latitude Category C) semi-specialist. Waxwings are migratory omnivores, with Japanese being only partially migratory. They would seem to fit into the group of birds with large Boreal ranges that would have survived glacial periods by shifting ranges southwards, though never far enough south to experience the range fragmentation typical of species in the mid-latitude belt.

CONCLUSIONS

(a) **Bioclimatic generalisation.** Eighteen species (16.22%) are bioclimatically-tolerant (Categories E and D); this is higher than in the Sylviodea but lower than in the Corvoidea. These are mainly multi-latitude species (Latitude Category F): Siberian Stonechat, Common Redstart, Black Redstart, Daurian Redstart, European Robin, Northern Wheatear, Eurasian Blackbird, Mistle Thrush, Eurasian Nuthatch, Firecrest, Wren and Brown Dipper. Four species have Temperate (Latitude Category C) ranges – Brown Flycatcher, Dark-sided Flycatcher, Goldcrest and Eurasian Treecreeper. There is a mid-latitude (Latitude Category D) species – Blue Rock Thrush – and a split category (A/D) one, White-throated Dipper. The tendency remains for single generalists in each genus, but the situation is relaxed among semi-generalists, although in most cases species have largely separate ranges: Common–Daurian Redstart, White-throated–Brown Dipper. Several genera are characterised by having more than one bioclimatically tolerant species, usually at different levels of tolerance: Goldcrest (E)–Firecrest (D); Eurasian Blackbird (E)–Mistle Thrush (D); Brown Flycatcher (E)–Dark-sided Flycatcher (D)–Spotted Flycatcher (C); Eurasian Treecreeper (D)–Short-toed Treecreeper (C). Among genera with Palearctic and Nearctic ranges, Palearctic species tend to be more bioclimatically generalised than Nearctic ones.

(b) **Mid-latitude belt representation.** The representation of mid-latitude species falls between that of the Corvoidea and Sylvoidea, which is high, and that of the Paroidea, which is low; 47 (42.34%), to which we may add eight split-category species. There are many specialist (24, 51.06%) and semi-specialist (14, 29.79%) species; by contrast there is a single semi-generalist and no generalist. So these birds are specialised and have fragmented ranges, excepting species of rocky habitats (e.g. Wallcreeper). 18 (38.3%) of the species are only found in a single sub-region: nine in the Eastern, two in the Central and seven in the Western Palearctic. This pattern follows the trend observed in other superfamilies.

There is a high representation of forest birds (20 species, 42.55%), while open- and mixed-habitat groups dominate in the other superfamilies. Open-ground birds are well represented (11, 23.4%), though these are largely wheatears. The rocky species are also well represented (8, 17.02%) but wetland species are practically missing.

(c) **Families of Muscicapoidea.** A striking feature of this superfamily is the high number of families represented. Seven of the 10 families have only one genus and the number of species ranges from one in Tichodromidae and Troglodytidae to 65 in Muscicapidae. The Muscicapidae stands out as the family with most genera (13) and an average of 4.9 species per genus which, though high, is much lower than in the Sylviidae. This family has several large genera: *Oenanthe* (15), *Luscinia* (13), *Ficedula* (10) and *Phoenicurus* (7). Turdidae has fewer genera (4) but a similarly high species/genus ratio (4.75), though this is in large measure due to the large number of *Turdus* species (15).

(c) **Geographical nature of the genera.** Several genera from the Muscicapidae are southern Asian-Himalayan with a significant representation in the Palearctic – *Luscinia, Tarsiger, Ficedula, Phoenicurus* and *Saxicola*. The route of geographical expansion in these and other genera in this superfamily (and it is suggested in other superfamilies too) seems to have been via occupation of Himalayan altitude belts by tropical Asian species; once established in these belts species would have had opportunities to launch into similar climates in the Palearctic (latitude substituting for altitude) and expand ranges significantly. Three genera – *Oenanthe, Monticola, Tichodroma* – are made up of rocky dwellers of the mid-latitude belt and, in the case of the first two, the southern extension of this belt down the Rift Valley to southern Africa. This is not unique to this genus; other groups (e.g. *Gypaetus*) have similar distributional patterns. Four genera have become Holarctic from Palearctic origins – *Cinclus, Sitta, Certhia* and *Regulus. Troglodytes* entered the Palearctic from the Nearctic and the origin of the Holarctic *Bombycilla* is uncertain. Only *Erithacus* and *Irania* (one species in each) are uniquely Palearctic.

(d) **Representation in the Western Palearctic**. 63 species (56.76%) of the Muscicapoidea are found in the Western Palearctic, proportionately fewer than in the Corvoidea but more than in the Sylvoidea. Seven Palearctic genera are missing from the Western Palearctic – *Niltava, Terpsiphone, Zoothera, Myophonus, Catharus, Acridotheres* and *Onychognathus*.

CHAPTER 8

Passeroidea – sparrows, finches, pipits and buntings

Group 1, subgroup a (v)

Five families make up this superfamily: Prunellidae (accentors), Motacillidae (pipits and wagtails), Passeridae (sparrows), Fringillidae (finches) and Emberizidae (buntings). Following Treplin *et al.'s* (2008) analysis of phylogenetic relationships, the five families that include Palearctic species cluster together. The accentors are basal to this cluster. The sparrows separate next followed by the pipits and wagtails, leaving the finches and buntings as sister families.

Climate

The passeroids are bioclimatic specialists, with Palearctic species part of a minority of species that are bioclimatically tolerant (Figure 8.1). Part of the success of these Palearctic species seems to have been the ability to occupy a range of bioclimates, especially the colder end of the gradient, while the global picture is one of species of warm situations (Figure 8.2i). They have also spread across the humidity range and have succeeded in dry bioclimates (Figure 8.2ii), which became increasingly widespread during the Miocene and Pliocene.

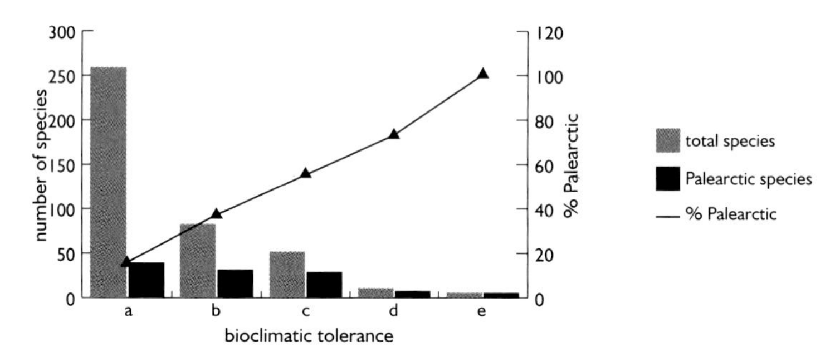

Figure 8.1 *Bioclimatic tolerance of passeroids. Definitions as on Figure 4.1 (see p. 38).*

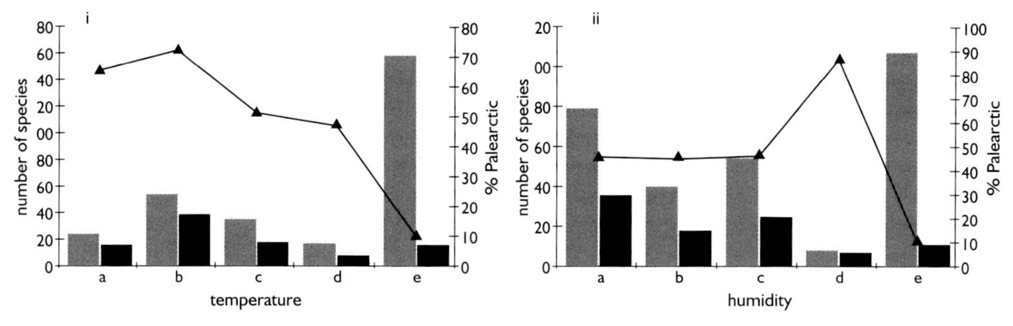

Figure 8.2 *Temperature (i) and humidity (ii) tolerance of passeroids. Definitions as in Figure 4.2 (see p. 38). Key as in Figure 8.1.*

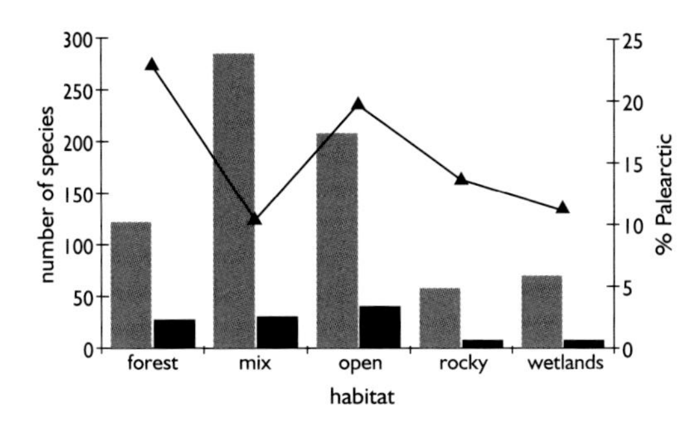

Figure 8.3 *Habitat occupation by passeroids. Definitions as in Figure 4.3 (see p. 39). Key as in Figure 8.1.*

Habitat

Passeroids have occupied a wide range of habitats and are particularly successful in forest and in open habitats (Figure 8.3). Open habitat is taken up largely by sparrows and buntings, many of which live in arid bioclimates. In this respect they resemble the larks. The ability to occupy non-forested habitats, among them rocky ones, has been a key to the success of this superfamily.

Migratory behaviour

Migratory behaviour is well-developed among Palearctic species in this group (Table 8.1), and contrasts with the generally sedentary behaviour of the superfamily as a whole. A combination of bioclimatic tolerance, an ability to occupy cold and dry bioclimates, the ability to live in open and rocky habitats and a strong migratory predisposition have made the passeroids highly successful in the Palearctic.

Fossil species

Passeroids are not conspicuous in the fossil record (Appendix 2) but all groups are present by the Late Pliocene-Early Pleistocene. There is a hint from presence in unusual sites that some montane species – Alpine Accentor, White winged Snowfinch – descended to lower ground during glacials. There is also an indication of the presence of Arctic species – Lapland Longspur, Snow Bunting – in locations where we would not expect them today.

Taxa in the Palearctic

The families in the Passeroidea have the following number of Palearctic species: Prunellidae (10 species, 76.92% of the world's species), Passeridae (13, 36.11%), Fringillidae (42, 30.44%), Emberizidae (32, 18.93%) and Motacillidae (20, 30.77%). There are more passeroids in the Western Palearctic (67) than in the Central (60) or Eastern Palearctic (53). There are six bioclimatically generalist and eight semi-generalist species (see Appendix 1). Prunellidae has no bioclimatically tolerant species, Emberizidae has only one (Reed Bunting *Emberiza schoeniclus*) and Passeridae has two (Eurasian Tree *Passer montanus* and House

	Global	Palearctic	Global %	Palearctic %	Palearctic of global %
Migratory	176	82	42.82	70.69	46.59
Sedentary	356	71	86.62	61.21	19.94
Total Species	411	116			

Table 8.1 *Migratory behaviour in Passeroidea.*

P. domesticus Sparrows. The Motacillidae with five (23.81% – White *Motacilla alba*, Grey *M. cinerea* and Yellow *M. flava* Wagtails along with Olive-backed *Anthus hodgsoni* and Richard's *A. richardi* Pipits) and Fringillidae with six (14.63% – Eurasian Linnet *Linaria cannabina*, European Goldfinch *C. carduelis*, Common Rosefinch *Carpodacus erythrinus*, Common Crossbill *Loxia curvirostra*, Hawfinch *Coccothraustes coccothraustes* and Eurasian Bullfinch *Pyrrhula pyrrhula*) are the families with most tolerant species. The overall proportion of tolerant species (12.2%) is low in comparison with other passerine superfamilies, except for the sylviids (7.2%). All the families in the Passeroidea have species with pan-palearctic ranges; overall, 23 species (20%) have pan-palearctic ranges but only four have continuous ranges: White Wagtail, Eurasian Tree Sparrow, Pine Grosbeak *Pinicola enucleator* and Eurasian Bullfinch.

PRUNELLIDAE

Prunella

The accentors are the basal (closest to the ancestral) family of the core group of the Passeroidea (Johansson *et al.*, 2008). All are in a single genus. Ten of the 13 species are Palearctic, the other three having adjacent ranges to the south of the region, in the Himalayas and the Arabian Peninsula. To all intents and purposes the accentors are a Palearctic family, with no tropical representatives. They are sedentary or partially migratory (including altitudinal displacement) omnivores, the only exception being the migratory Siberian Accentor *P. montanella.* Accentors are bioclimatic specialists and semi-specialists with the exception of Dunnock *P. modularis* (Latitude Split Category B/D) and Alpine Accentor *P. collaris* (Latitude Category D), which are moderates; these are the main accentors of the Western Palearctic. The montane Alpine Accentor is the only species with a pan-palearctic range, discontinuous in the Central Palearctic, a feature that, as we have seen, is typical of rock-dwelling birds of the mid-latitude belt (see pp. 46, 53, 78, 82). The Dunnock's split-category distribution recalls that of other species (e.g. Bluethroat *Luscinia svecica*) that we have previously described (see p. 75). The difference in this case is that Alpine Accentor is strictly Western Palearctic in its distribution.

With the exception of the Siberian Accentor, a species with a split Arctic-Temperate montane distribution (Latitude Category A/C), all other Palearctic accentors have mid-latitude belt (Latitude Category D) ranges. Together with the remaining non-palearctic species they form a coherent genus of species that have adapted and succeeded along the mountain chains of this belt. As we would expect, especially as all species are of open or mixed habitat, the ranges of the remaining species are fragmented but no species is exclusive to the Western Palearctic (other than Dunnock, as described above): Radde's *P. ocularis* (Western and Central); Black-throated *P. atrogularis* and Altai *P. himalayana* (Central) Accentors; Brown *P. fulvescens,* Siberian and Robin *P. rubeculoides* Accentors (Central and Eastern); Japanese *P. rubida* and Koslov's *P. koslowi* Accentors (Eastern).

PASSERIDAE

Passer

Passer behaves like many of the multi-species genera that we have already discussed in that it has a single generalist species – the Eurasian Tree Sparrow – which has a continuous pan-palearctic (multi-latitude, Category F) range. There is also a single semi-generalist – House Sparrow – also with a multi-latitude pan-palearctic range, though discontinuous in the east. These are the only bioclimatically tolerant sparrows in a genus of specialists and semi-specialists. Eight species (34.78%) are Palearctic, and most of the remaining species are part of the dry Saharo-Arabian belt, with a few to the south, in Africa and southern Asia. Six species breed in the Western Palearctic. The Saharo-Arabian belt is the most likely centre of origin of the genus, probably in the Miocene, with the Saxaul Sparrow *P. ammodendri* belonging to a subgroup that seems to include the oldest species (Allende *et al.,* 2001). The remaining Palearctic species are indeed semi-specialists (Saxaul Sparrow, Russet Sparrow *P. rutilans*) or specialists (Spanish *P. hispaniolensis,* Dead Sea *P. moabiticus,* Cape Verde *P. iagoensis* and Desert *P. simplex* Sparrows). They are mostly omnivores of warm, dry climates and occupy open or mixed habitats. Migratory tendency is not pronounced, with most species being sedentary, partially migratory or short-distance migrants. Most sparrows have fragmented ranges and only the Cape Verde and Dead Sea Sparrows are unique to the Western Palearctic. Spanish and Desert Sparrows occur in the Western and Central Palearctic; Saxaul Sparrow in the Central and Eastern Palearctic; and Russet Sparrow is exclusively eastern.

Petronia

The rock sparrows are bioclimatically similar to the sparrows and occupy a similar dry region across the Sahara and Arabia, as well as areas of Africa and India to the south. The two Palearctic species are semi-specialists, the most bioclimatically tolerant of the genus. The Rock Sparrow *P. petronia* has the widest distribution. It is typical of other rocky-habitat species of the mid-latitude belt in having a pan-palearctic (albeit discontinuous) range. The Chestnut-sided Petronia *P. xanthocollis* has a very limited range within the region, on the southern margins of the Central Palearctic. These birds are largely sedentary or partially migratory.

Carpospiza

The Pale Rockfinch *C. brachydactyla* is the only species in its genus. It breeds in the southern areas of the Central Palearctic, barely penetrating into the Western Palearctic. An omnivore of dry, open, habitats, this bird is a short-distance migrant.

Montifringilla

The snowfinches make up a Himalayan-Tibetan Plateau genus of sedentary montane omnivores, which seems to have split from *Petronia* around 2.5–3.0 mya (Qu *et al.*, 2006). Subsequent splitting of the snowfinch lineages seems linked to periods of uplift of the Tibetan Plateau. The White-winged Snowfinch *M. nivalis*, like other rocky-habitat species, has managed to spread right across the mid-latitude belt to the Pyrenees in the west. It is the only semi-specialist in a genus of specialists so, once more, the most bioclimatically tolerant species is the one that is the most widespread in the Palearctic (discontinuous pan-palearctic range, Latitude Category D). Another species, Père David's Snowfinch *M. davidiana,* occupies montane regions of the southern Central Palearctic. The snowfinches have managed to survive in the Palearctic and adjacent montane regions of similar climate as climatic specialists; their successful strategy has enabled them to remain sedentary or short-distance altitude migrants. This strategy is unusual among Palearctic birds.

FRINGILLIDAE

Fringilla

These finches are forest and mixed-habitat omnivores, and form are a uniquely Palearctic genus. They represent the basal (nearest the common ancestor) genus of the family (Liang, 2008). The Chaffinch *F. coelebs* is a bioclimatic moderate (Latitude Category F) with a Western (continuous) and Central Palearctic (discontinuous) range. The Brambling *F. montifringilla,* a semi-specialist, has a Boreal distribution (Latitude Category B) and, like other species of this category, is pan-palearctic with a continuous range. The northernmost and continental populations of the two species are migratory, while those to the south and west are sedentary.

The north-African *F. coelebs africana* and Canary Island *F. c. canariensis* are subspecies of Chaffinch at an intermediate stage of speciation, while the Blue Chaffinch *F. teydea* of Tenerife and Gran Canaria has reached species status. The process of island colonisation seems to have taken place within the last one million years (Marshall *et al.*, 1999) and the population structure of Chaffinches on the Canary Islands suggests the possibility of hidden taxa within this group (Suarez *et al.*, 2009).

Coccothraustes

The Hawfinch *C. coccothraustes*, together with the grosbeaks of the genera *Eophona* and *Mycerobas,* constitute a group of related species that separate out from the Fringillidae after the chaffinches (Arnaiz-Villena *et al.*, 2007). The Hawfinch is the only species in *Coccothraustes*, a Palearctic genus. It is a bioclimatic generalist (Temperate, Latitude Category C) with a continuous pan-palearctic range. The northernmost and continental populations are migratory but those from the south and west are largely sedentary. This genus seems to represent an extreme version of the 'single generalist' model.

Eophona

This is a uniquely Eastern Palearctic genus of forest omnivores with two highly migratory, bioclimatically moderate, species – Yellow-billed *E. migratoria* and Japanese *E. personata* Grosbeaks.

Mycerobas

This is a genus of four Himalayan resident forest omnivores. One species, the White-winged Grosbeak *M. carnipes,* penetrates the southern fringes of the Central and Western Palearctic.

Pinicola

Two species form this genus. The Pine Grosbeak *P. enucleator* is a bioclimatic moderate of coniferous forests (Boreal, Latitude Category B) with a continuous pan-palearctic range. It is also found in North America. In contrast, the Crimson-browed Finch *P. subhimachalus* is restricted to high ground in the Himalayas. It is a bioclimatic specialist and its range falls outside the Palearctic. The ranges of the two species are discrete but the genus range must have once been more widespread, perhaps during colder periods, when Palearctic and Himalayan *Pinicola* must have been connected. *Pinicola* is most closely related to the bullfinches, *Pyrrhula.*

Pyrrhula

The Eurasian Bullfinch *P. pyrrhula* exemplifies many of the patterns that we have observed in other genera. It is a semi-generalist (Temperate, Latitude Category C) with a continuous pan-palearctic range. It is the only bioclimatically tolerant species in a genus of specialists and semi-specialists. The remaining six species breed in south-east Asia or the Himalayan forests. The Eurasian Bullfinch is therefore an example of the longitudinal spread of a species along a latitude band, presumably from a Himalayan species at an equivalent altitude ban. A recent molecular study suggests that the Azores Bullfinch should be considered a species in its own right, as *P. murina*. The situation is less clear for the Eastern Palearctic forms *cineracea* and *griseiventris*, which may well be distinct species, or subspecies of *P. pyrrhula* (Töpfer *et al.*, 2011). Since the split between these closely related forms is fairly recent, the argument put forward here regarding the spread of the lineage into the Palearctic from south-east Asia (confirmed by Töpfer *et al.*, 2011) remains valid.

Rhodopechys

Pending a full taxonomic revision I have ascribed to this genus the species described as 'arid zone cardueline finches' by Arnaiz-Villena *et al.* (2008). This genus now includes three species formerly in the genus *Leucosticte* (Plain Mountain Finch *R. nemoricola,* Black-headed (or Brandt's) Mountain Finch *R. brandti* and Asian Rosy Finch *R. arctoa*) along with the Dark-breasted Rosefinch *R. nipalensis* (formerly in *Carpodacus*). Five (71.43%) species are Palearctic, the other two being close, in the Himalayan-Tibetan region. Two breed in the Western Palearctic. This group of species of arid environments seems to have arisen around 13.5 mya in Asia, a time when vast areas of Africa and Asia became arid. Some species then expanded their range during warm periods, even reaching North America. Subsequent climatic fluctuations have rearranged the ranges of the component species of this genus.

The Asian Rosy Finch is a bioclimatic moderate, the most tolerant of the genus, with a Boreal (Latitude Category B) Eastern Palearctic range. Three 'classic' *Rhodopechys* (sometimes also placed in *Bucanates*) are bioclimatic semi-specialists: Mongolian Finch *R. mongolica* (Latitude Category D) and Trumpeter Finch *R. githagineus* (Latitude Category E) are east-west geographical counterparts, the former species occupying higher ground than the latter, but both are species of arid, stony habitats; Crimson-winged Finch *R. sanguinea* (Latitude Category D) of areas of the Western and Central Palearctic is a montane counterpart of Trumpeter Finch. The remaining species, Black-headed Mountain Finch (specialist, Latitude Category D), is a high-altitude species of the Central Palearctic. Between them, the species of this genus are aligned on a south-west (Trumpeter) to north-east (Asian Rosy) axis of arid environments, with most species being in the mountains of the mid-latitude belt. All are omnivorous and three (Mongolian, Trumpeter and Crimson-winged Finches) are sedentary; Black-headed Mountain Finch is partially migratory, moving to lower ground in winter, while the northernmost species, Asian Rosy, is fully migratory.

Carpodacus

Eight (38.1%) of the *Carpodacus* rosefinches have Palearctic ranges. In this classification I follow Arnaiz-Villena *et al.* (2008) and Nguembock *et al.* (2009) and add *Uragus* and *Haematospiza* to this genus. Carpodacus is largely Himalayan, where it seems to have originated, with a few species reaching the Palearctic and two in North America (though these may require taxonomic revision). Three of the eight Palearctic species have relatively large ranges. The Common Rosefinch *C. erythrinus* is a multilatitude (Category F) bioclimatic semi-generalist with a pan-palearctic range that is fragmented in the west; it is a migratory omnivore of mixed habitat. The Long-tailed Rosefinch *C. sibiricus* is a semi-specialist of Temperate (Latitude Category C) latitudes. It is also a migratory omnivore of mixed habitats. Pallas's Rosefinch *C. roseus* is a semi-specialist of Boreal (Latitude Category B) latitudes. This rosefinch is a largely migratory omnivore of forests.

The remaining species in the genus are essentially Himalayan-Tibetan montane bioclimatic specialists that reach the margins of the Palearctic: Great Rosefinch *C. rubicilla* (Central Palearctic, Latitude Category D, open habitat and sedentary); Red-breasted (or Red-fronted) Rosefinch *C. puniceus* (Western and Central Palearctic, Latitude Category D, rocky habitat, resident); Red-mantled Rosefinch *C. rhodochlamys* (Central Palearctic, Latitude Category D, partially migratory, forest); and Beautiful Rosefinch *C. pulcherrimus* (Eastern Palearctic, Latitude Category D, mixed habitat, resident). There is one further species, Sinai Rosefinch *C. synoicus,* a specialist of arid habitats in the Western Palearctic (Latitude Category E, rocky habitats, resident) – this bird and the Common and Great Rosefinches are the only Western Palearctic *Carpodacus* species.

Loxia

The crossbills are a Palearctic genus of sedentary birds of coniferous forests, well-known for their irregular irruptions. All breed within the Western Palearctic. Two of the four species have Holarctic ranges. They are the most bioclimatically tolerant and geographically widespread species: the Common Crossbill *L. curvirostra* is a generalist (multi-latitude, Category F) with a continuous pan-palearctic range that stretches to the slopes of the Himalayas and beyond, while the Two-barred Crossbill *L. leucoptera* is bioclimatically moderate (Boreal, Latitude Category B) with a pan-palearctic range that is discontinuous in the west. The pattern of having only one bioclimatic generalist species is repeated in this genus. The Parrot Crossbill *L. pytyopsittacus* is a semi-specialist of the boreal forests (Latitude Category B) of the Western and Central Palearctic. Finally, the Scottish Crossbill *L. scotica,* a bioclimatic specialist, is a Western Palearctic endemic with a restricted range. This genus is a possible candidate for the 'launch-pad' model (see p. 41), with origins in the Himalayas.

Acanthis

The taxonomy of the *Carduelis* and *Serinus* genera follows Nguembock *et al.*'s (2009) revision; *Carduelis* is split into seven genera – *Chloris, Spinus, Sporagra, Pseudomitris, Carduelis, Acanthis* and *Linaria. Serinus* is split into five genera – *Dendrospiza, Serinus, Chloroptila, Ochrospiza* and *Crithagra.* Not all genera have Palearctic representatives.

Two species, related to the crossbills, form the genus *Acanthis.* Both species have pan-palearctic ranges that are fragmented in the west. The Common Redpoll *A. flammea* is a bioclimatic semi-specialist omnivore of open habitats with a Boreal (Latitude Category B) range. Northernmost populations are migratory but more southerly ones are resident. The Arctic Redpoll *A. hornemanni* is a bioclimatic specialist with an Arctic (Latitude Category A) range that extends into the Nearctic. It is a resident omnivore of open habitats. These species have therefore evolved as specialists of cold and dry climates. This specialisation has restricted them to the northern Palearctic.

Spinus

This Holarctic genus is most closely related to *Loxia* and *Acanthis.* It contains two species – the Palearctic Eurasian Siskin *S. spinus* and the Nearctic Pine Siskin *S. pinus.* They are partially migratory omnivores of forests (generally coniferous) and are subject, like the crossbills, to irregular irruptions. The Eurasian Siskin is a split category (Boreal/Warm, B/D) forest species, pan-palearctic, but fragmented in the east, of the type (e.g. Pied Flycatcher *Ficdula hypoleuca*, Bluethroat *Luscinia svecica*) that we have described previously; these are species with once-continuous ranges during glacials, with a subsequent expansion north, and up mountain ranges in the south, leaving discontinuous ranges today.

Serinus

Most closely allied to the *Loxia-Acanthis-Spinus* cluster is a group of genera formerly placed under *Serinus* (*Serinus* itself) and *Carduelis* (*Carduelis, Linaria, Sporagra* and *Pseudomitris*). Not all genera have Palearctic representatives but an absence from the Eastern Palearctic is conspicuous. The genus *Serinus* seems to have emerged within the Palearctic (presumably somewhere near the junction of Europe with Asia) with a radiation starting around 10 mya and proceeding rapidly in the Pliocene (Arnaiz-Villena *et al.*, 1999). This pattern and timing of radiation is repeated in other cardueline finches (Arnaiz-Villena, 1998).

Serinus has species scattered across the Palearctic, in Arabian and north-east African mountain ranges, in tropical Africa and Asia and in the Himalayas. They are absent from the Eastern Palearctic. Four (33.33%) species are Palearctic, all being bioclimatic specialists except the Serin *S. serinus,* which is bioclimatically moderate. This genus keeps to the 'single bioclimatically tolerant species per genus' pattern (see p. 46). It is an exclusively Western Palearctic mixed-habitat omnivore (multi-latitude, Category F). The northernmost and continental populations are highly migratory, but the southern and western ones are sedentary. Its range is limited to the mild Western Palearctic, those furthest north and east within the sub-region having to migrate. The remaining serins are species of warm climates, a number occupying montane habitats.

The Red-fronted *S. pusillus* (Western and Central Palearctic) and Syrian *S. syriacus* (Western Palearctic) Serins are resident bioclimatic specialists of the mountains of the mid-latitude belt (Latitude Category D). The Canary *S. canarius* of the islands of the south-western Palearctic is a forest bird that (typically for island species) occupies all altitude belts.

Chloroptila

The taxonomic position of the Citril Finch *C. citrinella* is unclear but it seems that it is sufficiently distinct to warrant separation into a genus of its own (Nguembock *et al.*, 2009). The closely-related Corsican Finch *C. corsicana* has recently been split from Citril Finch and is now considered a distinct species (Forschler *et al.*, 2009). Both are bioclimatic specialists with restricted ranges.

Carduelis

This genus is reserved for the European Goldfinch *C. carduelis.* It is a bioclimatic semi-generalist (multi-latitude, Category F) with a continuous range across the Western and Central Palearctic. It is an omnivore of open habitats and, like the Serin, it has highly migratory northern and continental populations but sedentary south-western ones. Like *Coccothraustes,* this genus seems to represent an extreme version of the 'single generalist' model (see p. 46).

Linaria

Two (50%) linnets are Palearctic, the other two being montane specialists of the Arabian Peninsula and north-east Africa. The Eurasian Linnet *L. cannabina* is a bioclimatic semi-generalist (multi-latitude, Category F) with a Western (continuous) and Central (discontinuous) Palearctic range. It follows the 'single tolerant species' pattern as seen in many other gebera described in this book (e.g. *Oenanthe*, *Alauda*, *Periparus*). Like the Goldfinch, it is an open habitat omnivore with similar migratory behaviour. The Twite *L. flavirostris* is a bioclimatic moderate with a split category (B/D) range, resembling other species of open habitats (e.g. Ring Ouzel *Turdus torquatus*). Its distribution along the mid-latitude belt is montane. Like the Linnet, the Twite's range is limited to the Western and Central Palearctic. It is a classic example of the effect of the heterogeneous nature of the mid-latitude belt (where the range is fragmented) and the more homogeneous nature of the Boreal belt (where the range is continuous). Like *Carduelis* and *Serinus*, this genus is absent from the Eastern Palearctic.

Chloris

Formerly in *Carduelis,* the genus *Chloris* incorporates the greenfinches. This genus of six species includes the Desert Finch *C. obsoleta* (formerly in *Rhodopechys*), which is the ancestral species in this group (Zamora *et al.*, 2006). Of the six species, one is south-east Asian and two are Himalayan and extralimital. The other three have complementary ranges: Oriental Greenfinch *C. sinica* is a multi-latitude (Category F) bioclimatic moderate of the Eastern Palearctic; European Greenfinch *C. chloris,* also a multi-latitude moderate, is its equivalent in the Western (continuous) and Central (discontinuous) Palearctic; and Desert Finch occurs to the south of the other two.

The ancestor of these forest omnivores must have had a continuous pan-palearctic range, from an eastern origin, which was severed by loss of forests with the aridification of the Central Palearctic. Curiously, the Desert Finch seems to be a 'greenfinch adaptation' to this aridification in the Central Palearctic (bioclimatic specialist, Latitude Category D). The Oriental Greenfinch is largely sedentary except for migratory northern populations; the situation is similar in the European Greenfinch, which behaves similarly to the European Goldfinch, Serin and Eurasian Linnet. The more continental populations of Desert Finch are also migratory.

EMBERIZIDAE

Emberiza

This large genus is divisible into four groups, one African and three Eurasian ones (Alström *et al.*, 2008). Thirty (76.92%) species breed in the Palearctic. Only one Palearctic species belongs to the African group; this is the House Bunting *E. striolata,* a specialist of warm, dry bioclimates, which has a discontinuous distribution across the southern Western and Central Palearctic (Latitude Category E). It is a sedentary omnivore of open, rocky country.

The first of the Palearctic lineages has just two species, Black-headed *E. melanocephala* and Red-headed *E. bruniceps* Buntings. They are semi-specialists of the mid-latitude belt (Latitude Category D), are migratory omnivores of mixed and open habitats, and occupy the Western (discontinuous) and Central (continuous) Palearctic. Both species winter in south Asia. The two species are remarkably similar, which may explain why they have non-overlapping ranges. An origin of this lineage in the Central Palearctic seems likely.

The second lineage is dominated by mid-latitude belt species (9, 69.23%). These tend away from bioclimatic tolerance and range from moderates (Rock *E. cia,* Chestnut-eared *E. fucata* and Meadow *E. cioides*), through semi-specialists (Cirl *E. cirlus,* Grey-necked *E. buchanani* and Jankowski's *E. jankowskii*) to specialists (White-capped *E. stewarti,* Cretzschmar's *E. caesia* and Godlewski's *E. godlewskii*). We may add the Ortolan Bunting *E. hortulana*, a moderate with a split Boreal/Warm range (Latitude Category B/D), to this cluster of species. They are omnivores of open, rocky or mixed habitats and exhibit a range of behaviours from mainly sedentary (Cirl) through altitude migrants (Godlewski's), partial migrants (White-capped, Rock, Meadow and Chestnut-eared), short-distance (Jankowski's) and long-distance (Ortolan, Cretzschmar's and Grey-necked) migrants.

Also in this second lineage are three other species with more northerly ranges. The first is the Yellowhammer *E. citrinella,* a part-migratory bioclimatic moderate of open habitats with a Temperate (Latitude Category C) range; the second is the Pine Bunting *E. leucocephalus,* a migratory bioclimatic moderate of Boreal forests (Latitude Category B); and the third is the Corn Bunting *E. calandra,* a part-migratory semi-specialist of open habitats with a multi-latitude (F) range. Overall, there are no species with pan-palearctic ranges in this group and there is a dominance of species occupying the Central (9) over Western (6) or Eastern (5) Palearctic. Two species (White-capped and Grey-necked) are exclusive to the Central Palearctic, two others (Jankowski's and Chestnut-eared) to the Eastern Palearctic, and Cretzschmar's Bunting is exclusive to the Western Palearctic. Four species are shared by Western and Central Palearctic – Yellowhammer, Ortolan, Rock and Corn Buntings– and three are shared between Central and Eastern Palearctic – Pine, Godlewski's and Meadow Buntings.

The third lineage is predominantly one of the Eastern Palearctic. All 13 species breed in this sub-region against only four each in the Central and Western Palearctic. These four species are pan-palearctic: two (Yellow-breasted Bunting *E. aureola,* bioclimatic moderate, and Rustic Bunting *E. rustica,* bioclimatic semi-specialist) have Boreal (Latitude Category B) ranges; one (Little Bunting *E. pusilla,* bioclimatic specialist) has an Arctic (Latitude Category A) range; and the fourth (Reed Bunting *E. schoeniclus,* bioclimatic generalist) has a multi-latitude range (Latitude Category F). These species illustrate that pan-palearctic ranges are heavily dependent on latitude. If the ranges are at high latitudes, then bioclimatically intolerant species that are matched with a particular latitude band are able to spread widely across it. Species from lower latitudes, or with multi-latitude ranges, need to be bioclimatically tolerant in order to have pan-palearctic ranges.

The remaining species in this third Eurasian bunting group are exclusive to the Eastern Palearctic. They are largely bioclimatically intolerant: moderates (Black-faced Bunting *E. spodocephala,* multi-latitude, Category F and Yellow-throated Bunting *E. elegans,* Warm, Category D); semi-specialists (Tristram's Bunting *E. tristrami,* Temperate, Category C, Grey Bunting *E. variabilis,* Warm, Category D, Yellow-breasted Bunting *E. aureola,* Boreal, Category B, Chestnut Bunting *E. rutila,* Boreal, Category B and Pallas's Reed Bunting *E. pallasi,* Boreal, Category B); and specialists (Yellow-browed Bunting *E. chrysophrys,* Boreal, Category B, Japanese Yellow Bunting *E. sulphurata,* Warm, Category D, and Japanese Reed Bunting *E. yessoensis,* Temperate, Category C). The buntings of this third group are birds of forest and mixed habitats, in contrast with those of the second group that do not include forest species and include open and rocky habitats. They provide a beautiful natural experiment; in the Eastern Palearctic, where forests were never severed from the tropics, a group of forest-dwelling buntings persisted but did not expand westwards, where forests became discontinuous because of the presence of the Himalayas and other high mountain chains.

There is a bunting that has not been allocated to these major clusters. Cinerous Bunting *E. cineracea* is a mid-latitude belt (Latitude Category D) specialist of the Western Palearctic. It probably belongs with the second group.

Calcarius

This is a genus of four North American buntings that are typical of cold, arid, open and tundra habitats. One species, Lapland Longspur *C. lapponicus,* crossed into the Palearctic and has a pan-palearctic range. It is a bioclimatic semi-specialist with an Arctic (Latitude Category A) range. It is highly migratory.

Plectrophenax

The snow buntings resemble the longspurs. There are two North American species, one of which, Snow Bunting *P. nivalis,* colonised the Palearctic and has an Arctic (Latitude Category A) pan-palearctic range. It is a highly migratory bioclimatic specialist.

MOTACILLIDAE

Motacilla

Eight (61.54%) of the species in this genus are Palearctic. The wagtails first appeared in the Eastern Palearctic around 4.5 mya and rapidly dispersed across the Palearctic, reaching Africa, which became a second source of wagtail diversity from around 2.9 mya. More recent dispersals into North America and north Africa from Eurasia characterise wagtail distribution (Voelker, 2002b). These findings have led to significant recent taxonomic revisions; 'yellow wagtail' is split into three species, Western Yellow Wagtail *M. flava,* Eastern Yellow Wagtail *M. tschutensis* and Asian Yellow Wagtail *M. taivana*; Citrine Wagtail is split into Western Citrine *M. werae* and Eastern Citrine *M. citreola* (Voelker, 2002b; Pavlova *et al.*, 2003), with Black-backed Wagtail retained as subspecies *lugens* of White Wagtail *M. alba* (Pavlova *et al.*, 2005). In terms of the Palearctic species of the genus, the Western Yellow Wagtail is closest to the ancestral species. The African Pied Wagtail *M. aguimp* is part of this Palearctic cluster and separates out next, placing it in an intermediate position between the Western Yellow and the other Palearctic wagtails. The remaining species sort themselves into two groups: (a) the remaining yellow wagtails, the citrines and the Grey Wagtail *M. cinerea*; and (b) White Wagtail and the south White-browed Wagtail *M. madaraspatensis.*

The wagtails are unusual in that two congeners, Grey and White, are both generalists. They belong to different lineages within the genus (see above), and this may indicate rather deep-rooted differences between them. Nevertheless, this as an exceptional situation. The two species have multi-latitude (Category F) continuous pan-palearctic ranges, bar the discontinuous distribution of Grey Wagtail in the Central Palearctic.

The group that includes the yellow wagtails has the Grey as its sole generalist. The Western Yellow is a semi-generalist (multi-latitude Category F, Western and Central Palearctic); the Western Citrine is a moderate (multi-latitude Category F, Western and Central Palearctic); Eastern Yellow is a semi-specialist (Boreal Category B, Central and Eastern Palearctic); Eastern Citrine (Warm Category D, Central and Eastern Palearctic) and Asian Yellow (Warm Category D, Eastern Palearctic) are both specialists. So the yellow wagtail group does conform to the bioclimatic tolerance structure that is typical of most genera.

One other species, the Japanese Wagtail *M. grandis,* is a semi-specialist with a mid-latitude range (Category D) and an Eastern Palearctic distribution. It is the only sedentary species in the genus, the yellows and citrines being highly migratory and the Grey and White Wagtails being partially migratory with large numbers of migratory northern and continental birds.

Dendronanthus

This is a single-species genus. The Forest Wagtail *D. indicus* is an Eastern Palearctic migratory wagtail; a bioclimatic moderate, its range is centred on the mid-latitude belt (Latitude Category D).

Anthus

Twelve (27.27%) of the species in this genus of omnivores and insectivores are Palearctic. The pipits are a more ancient group than the wagtails, having apparently arisen in the Eastern Palearctic around 7.0 mya. By around 5.0 mya they had dispersed across the Palearctic and into Africa and the Americas (Voelker, 1999a). There seems to have been significant diversification and speciation in the Pliocene but little during the Pleistocene. The net result is four major lineages within the genus – (a) a group of small-bodied African pipits; (b) a largely Palearctic group of migratory species; (c) a mainly South American group; and (d) a group with African, Eurasian and Australasian species (Voelker, 1999b).

Eight species cluster together in the Palearctic migratory group. Olive-backed *A. hodgsoni* and Tree *A. trivialis* Pipits are closely related and part of a subgroup, along with Pechora Pipit *A. gustavi.* Olive-backed and Tree are eastern and western counterparts with some range overlap in the Central and Eastern Palearctic. Olive-backed is a semi-generalist and Tree a moderate with pan-palearctic ranges but, while Olive-backed has managed a multi-latitude (Category F) range (marginal in the west), Tree has a split (B/D) range (discontinuous in the east); the latter has probably been imposed by the nature of the topography in the west with southern, mid-latitude populations remaining in appropriate climatic belts on mountains. The Pechora Pipit is a bioclimatic specialist, also with a pan-palearctic range, which is marginal in the west and is centred in the Arctic (Latitude Category A) belt. Olive-backed and Tree Pipits are associated with trees, unusual among pipits, while Pechora occupies open habitats. All three are highly migratory.

The basal species of the second subgroup is the Red-throated Pipit *A. cervinus.* It resembles Pechora Pipit, being a bioclimatic specialist with a pan-palearctic Arctic (Latitude Category A) range that is marginal in the west. It is also a migratory species of open habitats. The next species in the subgroup is the extralimital Rosy Pipit *A. roseatus* of the Himalayas. The remaining species are Water *A. spinoletta* and Rock *A. petrosus* Pipits, Meadow Pipit *A. pratensis* and Buff-bellied Pipit *A. rubescens,* all of which are bioclimatic moderates. Like the Tree Pipit, the Water Pipit has a split B/D pan-palearctic range which, in this species, is fragmented in the Central Palearctic. The Rock Pipit, on the other hand, is exclusive to the Western Palearctic and has a Boreal (Latitude Category B) range. The Water Pipit is migratory but the Rock Pipit, living in the oceanic west, is only partially migratory. The Meadow Pipit, like the Rock, is also a Western Palearctic species with a Boreal (Latitude Category B) range and is also part-migratory. Buff-bellied Pipit resembles these two species in also having a Boreal (B) range but it is confined to the Eastern Palearctic; its range continues into the Nearctic.

The four remaining pipits are in the African-Eurasian-Australasian group. Two are very closely related. They are Tawny Pipit *A. campestris* and the island endemic Berthelot's Pipit *A. berthelotii.* The Tawny Pipit is a bioclimatic moderate migrant with a Western and Central Palearctic mid-latitude belt (Latitude Category D) range. Berthelot's is a resident specialist of warm, wet climates in the Canary Islands. Blyth's Pipit *A. godlewskii* is another mid-latitude belt (Latitude Category D) specialist migrant, differing from the other species in this group in being restricted to the Eastern Palearctic. Finally, Richard's Pipit *A. richardi* is a Temperate (Latitude Category C) semi-generalist migrant of the Eastern Palearctic.

CONCLUSIONS

(a) **Bioclimatic generalisation.** This group is characterised by having few bioclimatically tolerant species (14, 12.2%). One family, Prunellidae, has none. These are, as in the Muscicapoidea, mainly multi-latitude species (Latitude Category F): Eurasian Tree Sparrow, House Sparrow, Eurasian Linnet, European Goldfinch, Common Rosefinch, Common Crossbill, Reed Bunting, White Wagtail, Grey Wagtail, and Olive-backed Pipit. The remaining three have Temperate (Latitude Category C) ranges: Eurasian Bullfinch, Hawfinch and Richard's Pipit.

(b) **Mid-latitude belt representation.** The representation of mid-latitude species is similar to the Muscapoidea. There are 49 species (42.61%) in this category compared with 47 (42.34%) in Muscicapoidea. The latter superfamily has eight split-category species, while the Passeroidea has seven. There is a high representation in the Prunellidae, Passeridae and Emberizidae. As in Muscapoidea most species are specialists (26, 53.06%) and semi-specialists (14, 28.57%); in contrast there are no semi-generalists or generalists. Only three species, moderates and semi-specialists of open, rocky habitats have pan-palearctic ranges. They are Alpine Accentor, Rock Sparrow and White-winged Snowfinch. But the degree of isolation seems to have been higher than among the Muscicapoidea: 28 (57.14%) of species occupy only one sub-region of the Palearctic, which contrasts with 18 (38.3%) in Muscicapoidea. Another novelty is the dominance of the Central Palearctic (28 species) over the Eastern (23) and Western (20), which is not surprising as many species in this superfamily are birds of the open and arid habitats that came to dominate the Central Palearctic. This conclusion is supported by the few forest species (6, 12.24%) represented, which contrasts with 20 (42.55%) in Muscicapoidea. The dominant habitat categories are, instead, open (20 species, 40.82%) and mixed (15, 30.61%). As in Muscapoidea, wetland species are rare.

(c) **Families of Passeroidea.** The number of genera in the five families ranges from one in Prunellidae to 16 in Fringillidae, which has 41 Palearctic species. The genera of Muscicapoidea typically have few species – 17 genera (62.96%) have three of fewer species. By contrast *Emberiza* has 30 species, making it the most species-rich Palearctic passerine genus. There are several other genera with considerable number of species: *Anthus* (12), *Prunella* (10), *Passer* (8), *Carpodacus* (8) and *Motacilla* (8).

(d) **Geographical nature of the genera.** The major influence of the Himalayan-Tibetan massif is evident in a number of genera that have radiated within and out of this area. The accentors, snowfinches and rosefinches are the best examples within the Passeroidea. The bullfinches and rosefinches additionally provide further evidence of the altitude-latitude 'launch-pad' pattern that we observed in a number of Muscicapoid genera (see pp. 75–79). A distinctive feature of a number of genera is their capacity to survive in arid regions, and which seem to have emerged from within the Sahara/Arabian-Central Asian desert belt; the *Passer* and *Petronia* sparrows and the *Rhodopechys* finches are cases in point.

Two of the three subgroups within the genus *Emberiza* appear to have radiated within the mid-latitude belt, in mixed and open habitats and many occur in the Central and Western Palearctic. Several genera of finches, notably *Serinus*, *Carduelis* and *Linaria*, also seem to have kept to the Western and Central Palearctic. A third bunting group, however, radiated within forests in the east. Two widespread genera, *Motacilla* and *Anthus,* emerged in the Eastern Palearctic and dispersed widely. The pipits are separable into two lineages; one, as in some of the buntings, seems to have been a mid-latitude belt radiation, and the second is dominated by boreal species with large (sometimes Holarctic) ranges. The finch genus *Acanthis* and the bunting genera *Calcarius* and *Plectrophenax* have few species, with these being Arctic or boreal and having pan-palearctic or Holarctic ranges. Two genera, *Loxia* and *Spinus*, became Holarctic specialists of boreal coniferous forests.

(e) **Representation in the Western Palearctic.** 70 species (60.87%) of Passeroidea are found in the Western Palearctic, a figure that is higher than in Muscicapoidea or Sylvoidea and is comparable to the Corvoidea. Only three genera with Palearctic species are missing from the Western Palearctic: *Eophona, Mycerobas* and *Dendronanthus*.

CHAPTER 9
Falcons

Group 1, subgroup c

The centre of origin of the falcons appears to have been the Neotropics (Griffiths, 1999). Genera and species that occupied areas outside this region are all part of the subfamily Falconinae. These are all members of the genus *Falco* (Appendix 1). The initial divergence of *Falco* probably dates back to the Miocene (Groombridge *et al.*, 2002), which would coincide with the earliest fossil *Falco* in Europe, which date to the Early Pliocene and possibly the Late Miocene (Mlikovsky, 2002). Within the genus *Falco*, genetic analyses separate four major groups: the hierofalcons (Gyrfalcon *F. rusticolus*, Saker *F. cherrug*, Lanner *F. biarmicus* and Laggar *F. jugger*) appear as a sister group to the rest of the genus and separated from them at an early stage of the *Falco* adaptive radiation. The Australian Black Falcon *F. subniger* appears to cluster with the hierofalcons and may be a geographic counterpart (Wink *et al.*, 2004a).

The next group that separates out is the kestrels, then the hobbies and finally the peregrines. The Merlin *F. columbarius* is part of a large cluster that includes the hobbies and the peregrines but is distinct from these two groups (Seibold *et al.*, 1993; Helbig *et al.*, 1994).

The hierofalcons, effectively a 'superspecies', are thought to have originated in Africa, with the Lanner close to the common ancestor, and subsequent waves of emigration into Eurasia led to the Saker, Laggar and Gyrfalcon (Nittinger *et al.*, 2005, 2007). It seems that incomplete sorting of the species lineages and hybridisation occurring during alternating periods of isolation and merging of populations during the Pleistocene may have contributed to the low genetic differentiation. The Lanner and the Laggar, for example, may not warrant being treated as separate species. I suggest a different interpretation to the origins of this superspecies. It probably represents an early entry, and separation, of a large, open plains falcon that came into Eurasia from North America. This species would have occupied large areas of Eurasia and Africa, probably the total range covered today by the different species. With increasing climatic deterioration (colder and drier) in the Late Pliocene and during the Pleistocene, the various populations separated. During brief periods of climatic amelioration populations met again. Isolation was long enough to differentiate a temperate steppe species (*F. cherrug*) and a cold tundra species (*F. rusticolus*). The separation was less clear between the African and Indian tropical forms and even less so among the African populations. The result is a hierarchical split within the superspecies: differentiation of Lanner into subspecies within Africa, differentiation of the Lanner and Laggar between Africa and south Asia, and differentiation of the Lanner/Laggar, Saker and Gyrfalcon between Africa, south Asia and the Palearctic. The populations occupying the increasingly seasonal Pleistocene environments of the Palearctic became migratory, the others retaining ancestral sedentary behaviours.

Taxonomic relationships among the kestrels are complicated. Some authors have suggested an African origin for the group, on the basis that most species occur in that continent (Groombridge *et al.*, 2002), but this need not be the only interpretation of such evidence. The small Red-footed Falcon *F. vespertinus* and its closely related eastern counterpart, the Amur Falcon *F. amurensis*, are distinct from the rest of the kestrels, from which they diverged in the Miocene (Groombridge *et al.*, 2002; Wink *et al.*, 2004b). The American Kestrel *F. sparverius* also separates out from the Old World kestrels but it may be closely related to the Red-footed and Amur Falcons (Wink *et al.*, 1998). The Lesser Kestrel *F. naumanni* is the basal, and quite distinct, species of the Old World Kestrels. This species appears to have diverged from the rest of the kestrels in the Late Miocene (Groombridge *et al.*, 2002) and is divided into distinct populations, identifiable genetically as having diverged between 100-200,000 years ago, across a west-east gradient from Europe to Asia (Wink *et al.*, 2004b). The divergence of the Indian Ocean island kestrels seems to have occurred early in the Pleistocene (Groombridge *et al.*, 2002).

The picture of kestrel diversification is, in my view, not dissimilar to that of the hierofalcons; in this case an ancient group dating back to the Miocene with periodic separation of populations related to the increasing cooling and aridification of the northern hemisphere, first separating the New World (*F. sparverius*) from the Old World, then steppe species (*F. vespertinus/amurensis*, then *F. naumanni*) from the rest, as the Palearctic became increasingly differentiated into wooded savanna and steppe. The greater diversification of the kestrels compared to the hierofalcons is illustrated by the Eurasian Kestrel *F. tinnunculus*, which covers the geographical range of the entire hierofalcon superspecies. It subdivided into local species, particularly in Africa where the equivalent hierofalcon, *F. biarmicus*, has only differentiated to subspecies level.

The five small, tree-nesting hobbys are geographical counterparts across the Palearctic, Africa, south and south-east Asia, Australia and New Zealand. The two larger, rocky habitat-nesting species, Eleonora's *F. eleonorae* and Sooty *F. concolor* Falcons, are also geographical counterparts, the former in north Africa and the Mediterranean and the latter down the length of the eastern side of Africa. Both are late-breeding species that specialise on catching migratory birds on their southbound migration and are unusual in that all populations winter in Madagascar. The hobbys have therefore diverged on two occasions: one, probably ancient, which separated the two hobby groups, and a later period of diversification within the two lineages.

The peregrine group is made up of the Peregrine Falcon *F. peregrinus*, which occurs on all continents and is split into a number of subspecies that are regional replacements, and two species that replace it in open (including arid) habitats, the Prairie Falcon *F. mexicanus* in western North America and the Barbary Falcon *F. pelegrinoides*, which has a fragmented distribution across the southern mid-latitude belt.

Climate

The falcons are species of warm (predominantly wet and humid) climates, which represents the climate in which they presumably originated in the Neotropics, but there is a good representation of species in dry bioclimates (Figures 9.1–9.2). Only three genera have broken away from warm climates, the *Falco* falcons and the caracaras of the genera *Phalcoboenus* and *Polyborus*. *Falco* has occupied all bioclimates, doing particularly well in temperate and cool dry ones. It is this genus that provides the Palearctic species, which includes three species that are among the most globally climatically tolerant of all raptors – Eurasian Kestrel, Eurasian Hobby *F. subbuteo* and Peregrine Falcon (Figure 9.1). Many of the remaining Palearctic falcons are noteworthy as being among the species that have managed to penetrate the driest of climates for which they appear to have become specialised: Lesser Kestrel, Amur Falcon, Red-footed Falcon, Sooty Falcon, Lanner Falcon, Saker and Barbary Falcon. Of the remaining species, two are notably specialists of cold climates – Gyrfalcon and Merlin. Eleonora's Falcon is a species of warm climates.

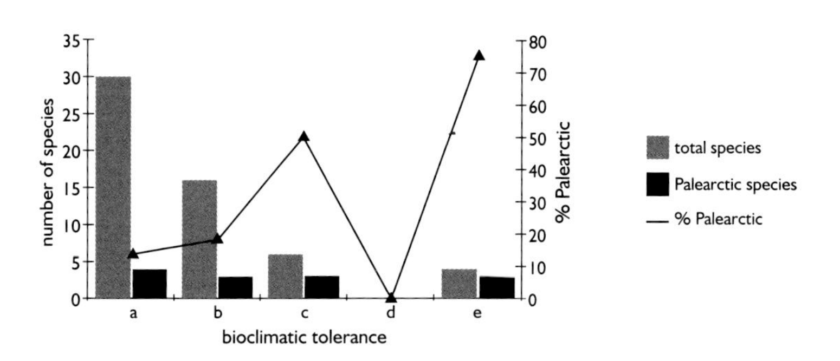

Figure 9.1 *Bioclimatic tolerance of falcons. Definitions as on Figure 4.1 (see p. 38).*

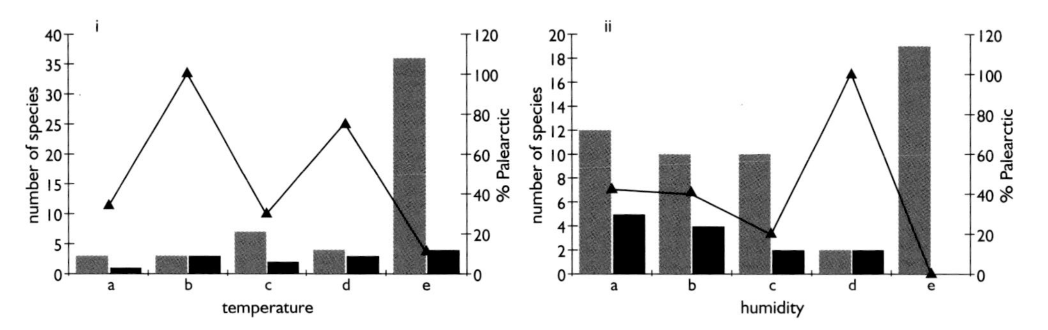

Figure 9.2 *Temperature (i) and humidity (ii) tolerance of falcons. Definitions as in Figure 4.2 (see p. 38).Key as in Figure 9.1.*

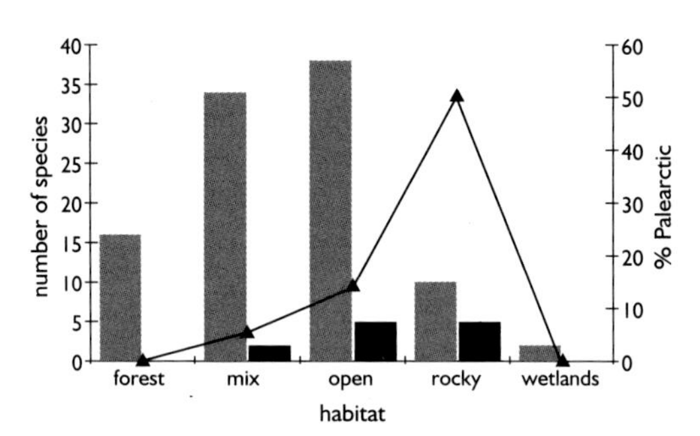

Figure 9.3 *Habitat occupation by falcons. Definitions as in Figure 4.3 (see p. 39). Key as in Figure 9.1.*

Habitat

The falcons are unusual among raptors in that they are dominant not in forests but in savannas and open grassland (Figure 9.3). In this respect they resemble another subfamily with New World origins, the buzzards, and the Old World vultures. All other habitats, except marine and wetlands (where they are rare), have been occupied, with deserts and rocky environments being important habitats. The underlying theme is avoidance of dense vegetation; the long, pointed wings of the falcons are clearly adapted to fast flight in open environments.

Western Palearctic falcons reflect this global trend, with nine species of open grassland, six of savanna, five of desert and four of rocky habitats. In contrast there are no forest *Falco*. This preference for open terrain would have predisposed the falcons for the opening of the Tertiary forests of the Palearctic with the advent of the Plio-Pleistocene glaciations.

Migratory behaviour

Globally, falcons are predominantly sedentary but Palearctic species are characterised by migratory behaviour (Table 9.1). Falcon morphology, adapted for open environments, may have facilitated migratory

	Global	Palearctic	Global %	Palearctic %	Palearctic of global %
Migratory	15	10	23.81	76.92	66.67
Sedentary	57	6	90.48	46.15	10.53
Total Species	63	13			

Table 9.1 *Migratory behaviour in the Falconidae.*

behaviour, especially if species were colonising highly seasonal (e.g. steppe and tundra) habitats. The Western Palearctic falcons exemplify this observation. Having colonised a wide range of highly seasonal environments, many were forced down the migratory route. Ten of the twelve Western Palearctic species have fully migratory populations, six have partially migratory populations and only four have resident ones. As in other raptors, in species with a range of migratory behaviours the least migratory populations are always in the south and west of the geographical range.

Fossil falcons

Falcons are well represented in the fossil record of the Western Palearctic, largely due to the dominance of the Eurasian Kestrel but with important remains of Eurasian Hobby, Peregrine Falcon, Merlin, Lesser Kestrel and Red-footed Falcon (Appendix 1). A number of species are present from the Early Pleistocene, with the rest entering the record during the Middle Pleistocene. As with other raptors, there is a dominance of Late Pleistocene records, reflecting a preservation bias in favour of more recent sites.

Although most fossils are within the present geographical range of the different species, a number of exceptions are worthy of highlight. The presence of Gyrfalcon in countries (France, Spain, Italy, Hungary, Croatia; Tyrberg, 1998, 2008) well south of the bird's present geographical range during the Late Pleistocene illustrates the southward extension of tundra at this time. The Merlin may represent another example, but it is less obvious since it winters today further south than the Gyrfalcon. Similarly, the westward extension of two steppe species is evident during the Late Pleistocene. Red-footed Falcon fossils have been recorded from France, Switzerland, Croatia, Italy and Spain, while Saker has been reported from Poland, Italy and Spain (Tyrberg, 1998, 2008; Sánchez Marco, 2007b).

Taxa in the Palearctic

All the Palearctic Falconidae belong to the genus *Falco*. Twelve species (30.77% of the world's species) breed in the Palearctic. Three (25%) are bioclimatic generalists and the remainder range from moderates to specialists. Within the region this family has most species in the Western (11) followed by the Central (8) and Eastern (6) Palearctic.

Falco

The falcons resemble the wagtails in one way: they seem not to observe the 'single generalist per genus' rule but, like the wagtails, these species are distributed (one in each) between distinct lineages within the genus. In this case one is in the kestrel lineage (Eurasian Kestrel), another in the hobby lineage (Eurasian Hobby) and a third in the peregrine line (Peregrine).

Three of the four hierofcalons breed in the Palearctic. They are bioclimatic specialists (Lanner – semi-tropical Latitude Category E) and semi-specialists (Gyrfalcon – Arctic, Latitude Category A; Saker – Warm, Latitude Category D) of arid climates. The radiation of this group thus seems to have been the product of cooling and aridification of the Palearctic in the Tertiary. One species – Gyrfalcon – adapted to the open habitats of the Arctic and achieved a continuous pan-palearctic range. It also breeds in the Nearctic, from where the common ancestor of the falcons dispersed into the Palearctic. It seems likely that a large falcon then adapted to the range of arid climate situations that emerged in the Palearctic. The Saker became the mid-latitude species of Central Palearctic arid habitats, reaching the eastern parts of the Western Palearctic, but prevented from entering the Eastern Palearctic (as in so many other species) by mountains and forests. Lanner (south-western Palearctic and also Africa) and Laggar (south Asia) completed the radiation in the south. The Gyrfalcon is highly migratory, as are the bulk of the Sakers, those living in the continental interior. In contrast, Sakers in the south and all Lanners and Laggars are residents.

The kestrels are species of the mid-latitude (Latitude Category D – Lesser Kestrel and Amur Falcon) and Temperate (Latitude Category C – Red-footed Falcon) belts. The Lesser Kestrel is a bioclimatic moderate with a pan-palearctic range that is continuous in the Central Palearctic, discontinuous in the west, and marginal in the east. It is a species of open steppe habitats. The closely-related Red-footed and Amur

Falcons have non-overlapping ranges, with Red-footed in the Western (marginal) and Central Palearctic and Amur in the east. They are also species of open habitats and, like the Lesser Kestrel, are highly migratory. The fourth species, the Eurasian Kestrel, is a bioclimatic generalist with a continuous pan-palearctic multi-latitude (Latitude Category F) range. Many Eurasian Kestrel populations are migratory but those of the Western Palearctic and southern Central Palearctic are sedentary or short-distance migrants.

One species of hobby dominates in the Palearctic. The Eurasian Hobby *F. subbuteo* is a second bioclimatic generalist in the genus, and has a pan-palearctic (discontinuous in the west) multi-latitude (Latitude Category F) range. It is highly migratory. The two other Palearctic hobbys are bioclimatically intolerant: Eleonora's Falcon is a semi-specialist of the mid-latitude belt (Latitude Category D) and the Sooty Falcon is a subtropical specialist (Latitude Category E). Both species are highly migratory and exclusively Western Palearctic. The hobbys follow the pattern of having one geographically widespread species with others of restricted ranges occurring at the periphery.

A similar situation arises with the peregrines. The Peregrine Falcon is a bioclimatic generalist with a continuous, multi-latitude (Latitude Category F) pan-palearctic range. Its sister species, the Barbary Falcon, is a bioclimatic specialist of the subtropical belt (Latitude Category E) of the Western and Central Palearctic, and is a bird of arid climates. In this respect it seems to converge with the Lanner. Many populations of northern and continental Peregrines are migratory but many in the Western Palearctic are sedentary or short-distance migrants. Barbary Falcons, living in the southern Palearctic, are sedentary.

The Merlin, rather distantly related to the hobbys and peregrines, is a bioclimatic moderate migrant, with a Boreal (Latitude Category B) pan-palearctic range that continues eastwards into North America.

CONCLUSIONS

(a) **Bioclimatic generalisation**. This group is characterised by having a significant proportion of bioclimatically tolerant species (3, 25%). All three have multi-latitude (Latitude Category F) pan-palearctic ranges.

(b) **Mid-latitude belt representation**. One-third of the falcons have mid-latitude ranges and there are no split-category species. Overall, the falcons are overwhelmingly birds of the mild (Temperate, Warm, Subtropical) latitudes of the Palearctic, with Gyrfalcon (Arctic) and Merlin (Boreal) being exceptional.

(c) **Representation in the Western Palearctic**. There are more species of falcons in the Western (11) than in the Central (8) or Eastern Palearctic (6). Since these are species of open habitats (where they are able to hunt most efficiently), it is no surprise that they should have become more prominent in the Western and Central Palearctic than in the more forested east.

CHAPTER 10

Terrestrial non-passerines

Group 1, subgroup d

In this chapter I discuss a group of related terrestrial non-passerines that are most closely related, within the same grouping, with the accipitrine raptors and owls. The falcons also form a branch of this large cluster of bird orders, but are part of a distinct sub-group that also includes the parrots and passerine birds (Hackett *et al.,* 2008). The group includes seven orders of birds, of which three are represented in the western Palearctic and are of interest here. They are the hoopoes, woodpeckers and coraciiforms (including the bee-eaters, rollers and kingfishers). Only six of the extant 22 families, 11 of 120 genera, and 18 of 669 species are represented in the Western Palearctic.

This impoverished situation was not always so. The Eocene and Oligocene of Europe, with its tropical and paratropical (seasonal forests with a dry winter season) forests (Janis, 1993), was home to a great diversity of birds from these orders and families, some of which were abundant (Mayr, 2005). They included mousebirds (Coliidae, now confined to sub-Saharan Africa), trogons (Trogoniidae, today mainly Neotropical but also from the tropics of Africa and Asia), cuckoo-rollers (Leptosomidae, today confined to Madagascar), motmots and todies (Momotidae and Todidae, today New World tropical species), kingfishers (Alcedinidae), early forms of rollers (Coraciidae), hoopoes (Messelirrisoridae; Mayr, 2000), woodpeckers (Mayr, 2006) and forms ancestral to woodpeckers, jacamars and puffbirds (Galbulidae and Bucconidae, today of tropical America). To these we must add the ground hornbills (Bucerotidae) that have been reported from two Late Miocene Western Palearctic localities, in Morocco (Brunet, 1971) and Bulgaria (Boev and Kovachev, 2007). This means that all seven orders, and many families, once lived in what is today the Western Palearctic. Only a few managed to survive the drastic climate changes of the Late Tertiary. The warm-humid climate forest forms fared worst.

UPUPIDAE

The two extant species of hoopoe, the Eurasian Hoopoe *Upupa epops* and the African Hoopoe *U. africana,* are probably best considered allospecies. Together they span much of the middle and lower latitudes of the Palearctic, sub-Saharan Africa and south and south-east Asia. Hoopoes are absent from much of the Sahara, Arabia, the Central Asian deserts and the Himalayas and Tibet. It would seem that this distribution reflects a widespread former distribution, with subsequent fragmentation of populations that became separated by arid environments. The African Hoopoe, physically separated from the rest by the Sahara, seems to have diverged sufficiently to be considered a separate species.

The breeding populations of Western Palearctic Hoopoes must have been severely restricted to the southern peninsulas of Europe and other refugia, e.g. the Crimea (Finlayson and Carrión, 2007), at the height of the glaciations, at which time we must assume that even the southern populations would have been migratory. The ability to migrate would have permitted survival in areas that would have otherwise been abandoned.

Climate

Hoopoes are widely tolerant of most bioclimates except the coldest and driest (Appendix 1). The Eurasian Hoopoe is more tolerant than the African, having a broad bioclimatic tolerance and occupying a Central position in the temperature gradient with a tendency towards the wet end of the humidity gradient. The intolerance of hoopoes to very dry climates explains the isolation of populations by deserts today.

Habitat

Hoopoes are primarily birds of open country and savanna. This habitat preference would have permitted expansion into many parts of the Palearctic with the reduction of forests, at the expense of savannas and grasslands, during the Plio-Pleistocene. With the subsequent desertification of significant parts of the large range some of these populations would have been lost.

Migratory behaviour

The distribution of migratory behaviour, if seen in the context of the two species, is a very clear one. Most Palearctic populations are highly migratory. Only those in the extreme south-west (Finlayson, 1992; Thévenot *et al.*, 2003), south and south-east Asia, and the African Hoopoe are resident.

Fossil hoopoes

Hoopoes are not abundant in the European fossil record but they do occur, in situations not dissimilar to today, in the Middle and Late Pleistocene.

Taxa in the Palearctic

Upupa

The Eurasian Hoopoe is the only Palearctic hoopoe. It is a pan-palearctic, multi-latitude (Latitude Category F) bioclimatic semi-generalist. It behaves as an extreme version of the 'single generalist rule', in which there are no congeners (as happens in other cases, e.g. Hawfinch *Coccothraustes coccothraustes*). The Hoopoe is a mixed-strategy carnivore of mixed habitats and is highly migratory, although populations in the extreme south-west, in southern Iberia and Morocco, are sedentary.

PICIDAE

The wrynecks are basal to the rest of the woodpeckers, from which they diverged in the Late Oligocene–Early Miocene (around 25 mya) (Benz *et al.*, 2006; Fuchs *et al.*, 2006). Hole-excavating and communication via drumming, characteristic of all other woodpeckers, are behaviours that are absent in the wrynecks. Like the hoopoes, the two wrynecks are probably allospecies; the Eurasian Wryneck *Jynx torquilla* occurs across a wide belt of Eurasian temperate forest from the British Isles to east Asia, and the Red-throated Wryneck *J. ruficollis* occurs across sub-Saharan Africa. The Saharan and Arabian deserts have become an effective barrier between the two species.

The woodpecker subfamily Picinae, with 24 genera and 185 species, is represented in the Western Palearctic by 10 species in four genera. The diversification of modern Picinae seems to have started around 13.7 mya, soon after the mid-Miocene climatic optimum (Fuchs *et al.*, 2006; Zachos *et al.*, 2001) from an ancestral group that evolved in tropical environments within Eurasia. It is with the break up of forests, associated with the climatic changes of the late Miocene (*c.* 7-8 mya), that we see the differentiation of African, Indo-Malayan and New World woodpeckers from each other (Fuchs *et al.*, 2006). In the Palearctic the colonisation of temperate habitats, in areas formerly occupied by tropical forests, took place during a relatively brief period of geological time around the Miocene–Pliocene boundary. The Palearctic woodpeckers are therefore part of an adaptive radiation of tropical forest woodpeckers into a new world of temperate forests. The subsequent history of these woodpeckers has involved survival through habitat tracking of these forests during the Plio-Pleistocene.

The Black Woodpecker *Dryocopus martius* of the Western Palearctic is one of seven species in the genus *Dryocopus*. This is a genus of large woodpeckers, four of which are New World species, two are south-east Asian and the Black Woodpecker occupies the Palearctic. They are largely geographical counterparts. Closely related are the *Picus* woodpeckers (Benz *et al.*, 2006; Fuchs *et al.*, 2006).This is a large genus of 15 species that are largely south and south-east Asian in distribution with four species outside this tropical forested region: the Green *P. viridis*, Levaillant's *P. vaillanti*, Grey-headed *P. canus* and Japanese *P. awokera* Woodpeckers. They are essentially geographical counterparts although the Green and Grey-headed have a broad zone of overlap within the Western Palearctic. The North African Levaillant's Woodpecker is the geographical counterpart of the Green Woodpecker in the Maghreb.

A separate large cluster in the woodpecker phylogenetic tree includes the remaining Palearctic woodpecker genera, *Picoides* and *Dendrocopos* (Fuchs *et al.*, 2006). *Picoides* consists of 11 New World species of which one, the Three-toed Woodpecker *P. tridactylus*, has penetrated the Old World and has a broad Holarctic distribution (Cramp, 1985). It is therefore the Palearctic geographical counterpart of a range of species that have diversified across the New World. The 22 *Dendrocopos* woodpeckers are, like the *Picus* woodpeckers, south and south-east Asian with some species having managed to live in the Palearctic. The Great Spotted Woodpecker *D. major* occupies a huge area of the Palearctic, only absent from the treeless tundra, steppes and deserts. It overlaps over large areas with the White-backed *D. leucotos* and the much smaller Lesser Spotted *D. minor*. The Middle Spotted Woodpecker *D. medius* appears to be a Western Palearctic counterpart of the more eastern White-backed Woodpecker. The remaining species are geographical counterparts, usually with restricted geographical ranges, in the forests of the mountains of the mid-latitude belt, including the fringes of the Himalayas.

Climate

The wrynecks resemble the hoopoes in their wide bioclimatic tolerance that avoids the coldest and driest bioclimates. The similarity also extends to the wider bioclimatic tolerance of the Eurasian Wryneck compared to its African counterpart. The Eurasian Wryneck is nevertheless not as tolerant as the Eurasian Hoopoe and occupies a cooler position along the temperature gradient. The Red-throated Wryneck is, on the other hand, a warm-wet climate specialist. The woodpeckers are overwhelmingly species of warm, wet-humid climates (Figures 10.1–10.2). Some species have moved along the humidity gradient to occupy warm and dry climates, while fewer have reached montane, temperate and even cool climates. Overall, compared to other groups, woodpeckers appear to be bioclimatically specialised.

Palearctic species are, not surprisingly, drawn from the most bioclimatically tolerant genera (Figure 10.1) and occupy the cool end of the temperature gradient (Figure 10.2i) but are well spaced out on the humidity gradient (Figure 10.2ii). Only the Grey-headed Woodpecker reaches the highest category of bioclimatic

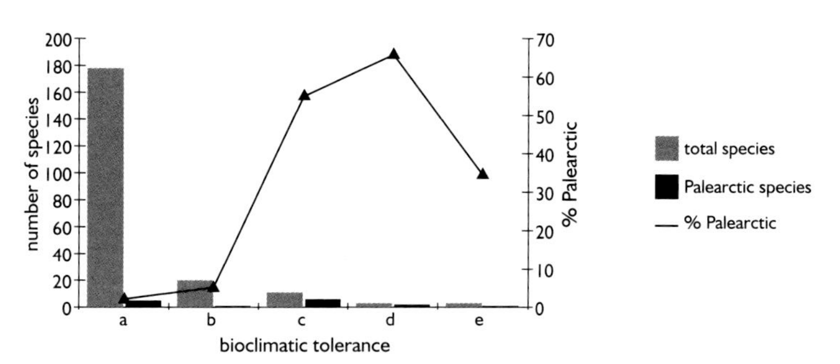

Figure 10.1 *Bioclimatic tolerance of woodpeckers. Definitions as on Figure 4.1 (see p. 38).*

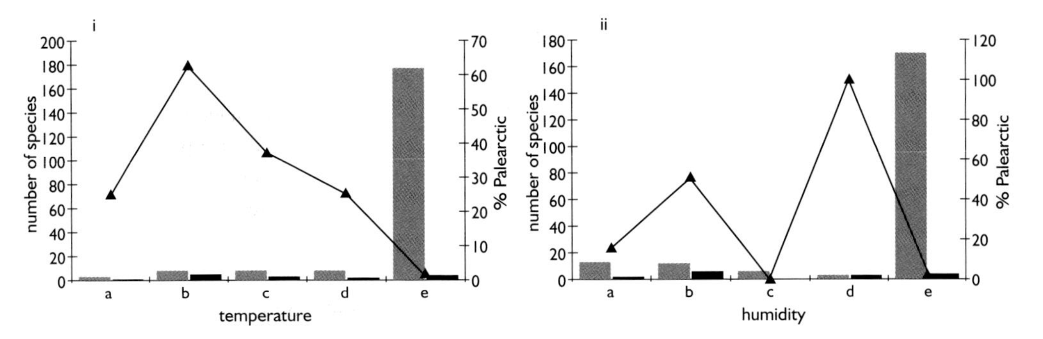

Figure 10.2 *Temperature (i) and humidity (ii) tolerance of woodpeckers. Definitions as in Figure 4.2 (see p. 38). Key as in Figure 10.1.*

tolerance, occupying a position on the cool and wet side close to the centre of the bioclimatic gradients (Appendix 1). The White-backed and Great Spotted Woodpeckers are the next most tolerant species, while the most specialised is the Syrian Woodpecker *Dendrocopos syriacus*, which occupies warm and dry bioclimates.

Habitat

The wrynecks are species of open woodland, savanna and shrubland. This predilection for open woodland would have enabled the wrynecks to survive across large areas of the Old World as dense forests shrank and were overtaken by more open habitats. With further climatic deterioration many areas that became steppe and desert would have been abandoned, giving rise to the current disjunct distribution of the allospecies.

Woodpeckers are birds of forest and woodland, some species also occupying scrub, usually close to woodland. The Palearctic woodpeckers are within the habitat requirements of all woodpeckers and resemble the tits (Paridae) in this respect. The success of the woodpeckers away from the tropical forests has therefore depended on shifts in climatic tolerance and not habitat.

Migratory habitat

Wrynecks include resident, partially migratory and fully migratory populations. Most Eurasian Wrynecks are migratory, wintering in tropical Africa, south and south-east Asia. Those in the extreme south-west, in south-western Iberia and the Maghreb, are largely resident (Finlayson, 1992; Hollom *et al.,* 1988). The Red-throated Wryneck conforms to the pattern that we have seen in other groups, for example the hoopoes, by being sedentary.

Woodpeckers are almost exclusively non-migratory. Palearctic species are all sedentary. Bioclimatic tolerance in forest habitats has been sufficient to permit success away from the tropics, a strategy uncommon in birds, and which we also found among the tits (Paridae).

Fossil woodpeckers

Wrynecks are not abundant in the European fossil record and occur, in situations that are not dissimilar to today, from the Early Pleistocene onwards. As with most forest birds, which show no tendency to accumulate in cave sites, woodpeckers are not especially numerous in the Pleistocene fossil record of the Western Palearctic. All nine species are represented, some from the Early Pleistocene, indicating a presence in the Western Palearctic throughout the Pleistocene (Appendix 2). There is very little that is unusual about the distribution of fossil woodpeckers as they generally occur in locations where we would expect to find them today. Only the presence of Grey-headed Woodpecker in northern Spain and of White-backed Woodpecker in France and Germany might appear unusual and suggestive of westerly range extensions in the past (Tyrberg, 1998, 2008).

Taxa in the Palearctic

Jynx

The Eurasian Wryneck *J. torquilla* is the only Palearctic species. It is a bioclimatic moderate with a multi-latitude (Latitude Category F) pan-palearctic range across the forests of mid-latitude, temperate and boreal Eurasia. Like the Eurasian Hoopoe, the Eurasian Wryneck is highly migratory but retains sedentary populations in the extreme south-west. It is another case of a relatively tolerant and widespread species with no other congeners.

Picus

The four Palearctic *Picus,* representing 26.67% of the world's species, conform to the 'single generalist per genus' rule. The Grey-headed Woodpecker is a bioclimatic generalist with a Temperate (Latitude Category C) pan-palearctic range that is discontinuous in the west. The Green Woodpecker is a multi-latitude (Latitude Category F) moderate of the Western Palearctic, with a marginal range in the Central Palearctic. The other two species are specialists on the subtropical (Latitude Category E) margins of the Palearctic: Levaillant's Woodpecker in North-west Africa and Japanese Woodpecker in Japan. They are all resident forest omnivores.

This genus, along with *Dendrocopus* and *Dryocopus* discussed below, are similar to certain passerine genera (such as nuthatches, treecreepers and bullfinches) in that they exhibit a of a 'launchpad' effect; in each case, one species from a suite of south-east Asian or Himalayan specialists penetrated the Palearctic and became bioclimatically tolerant and widespread. In some cases, such as *Dryocopus*, the expansion was not confined to the Palearctic but reached the Nearctic, too. *Picus*, *Dryocopus* and *Dendrocopos* occupy temperate belts (the latter genus also includes mid-latitude belt and multi-latitude species) in the Palearctic indicating that the range expansion must have occurred during relatively mild climatic conditions.

Dryocopus

One species (14.29% of the world's species) occupies the Palearctic. The Black Woodpecker is a bioclimatic moderate omnivore of temperate (Latitude Category C) forests. It is a resident with a pan-palearctic range that is discontinuous in the Western Palearctic.

Dendrocopus

Eight species (36.36%) have Palearctic ranges. The Great Spotted and White-backed Woodpeckers are semi-generalists that appear to depart from the 'single tolerant species per genus' rule; but we have other examples of genera with a generalist and a semi-generalist so it may be that it is possible for two lower order (semi-generalists) tolerant species to exist in a region like the Palearctic within the same genus. The two species have overlapping ranges, which are pan-palearctic and continuous (except for the White-backed's in the Western Palearctic, which is discontinuous). The White-backed Woodpecker also has a narrower latitudinal range, centred on the Temperate (Latitude Category C) belt, whereas the Great Spotted Woodpecker has a multi-latitude (Latitude Category F) range.

Middle and Lesser Spotted Woodpeckers are bioclimatic moderates of substantially different sizes. Lesser Spotted has a temperate (Latitude Category C) pan-palearctic range that is discontinuous in the west, while the Middle Spotted is a Western Palearctic, multi-latitude (Latitude Category F) species. Next in bioclimatic tolerance is the Syrian Woodpecker, a mid-latitude belt (Latitude Category D) species that has a discontinuous range in the Western and Central Palearctic; like other mid-latitude belt forest species it is absent from the Eastern Palearctic.

The remaining three species are Eastern Palearctic specialists: Rufous-bellied *D. hyperythrus* and Grey-capped *D. canicapillus* Woodpeckers have multi-latitude ranges, and Pygmy Woodpecker *D. kizuki* is a mid-latitude belt species.

Picoides

Picoides differs from the other Palearctic woodpecker genera in that it is a New World genus with a single species (9.1%) in the Palearctic. The Three-toed Woodpecker *P. tridactylus* is a Holarctic bioclimatic moderate with a split B/D range (i.e. it has a geographical range that is Boreal, with isolated relict populations in mountains to the south). It has a pan-palearctic range that is discontinuous in the west. It is the only Palearctic woodpecker with a Boreal-centred range, which may reflect an entry from North America in relatively cooler times than the reverse entry of *Dryocopus.* Its range must have been wider during glacials, with contraction during warmer times leading to the split-category distribution typical of other Palearctic birds.

MEROPIDAE

The bee-eaters are a compact group of Old World birds with a conservative body form that contrasts with a high diversity of ecological and behavioural traits (Marks *et al.,* 2007). Bee-eaters branched off from other families at an early stage, during the Early Eocene over 55 mya (Ericson *et al.,* 2006). Mlíkovský (2002) records an early species of bee-eater, *Merops radoboyensis,* from the Middle Miocene of Croatia, suggesting an early origin for the main genus of bee-eaters. The bee-eater lineages sort themselves out into two big groups, with a few specialised species appearing as sisters to these. One group is composed of resident African bee-eaters that subdivide into forest and savanna groups; the species in the second group are all migratory and share a common ancestor. They include a subgroup of intra-African migrants and another with species that occur outside Africa (Marks *et al.,* 2007).

It is from this latter group that the Western Palearctic species are derived. The widespread Little Green Bee-eater *Merops orientalis* has a continuous distribution across sub-Saharan Africa, Arabia, India and south-east Asia. Its closest relatives are two south-east Asian species, Chestnut-headed *M. leschenaultii* and Blue-throated *M. viridis* Bee-eaters. The Blue-cheeked Bee-eater *M. superciliosus* has a disjunct distribution across North Africa, Arabia, India and Central Asia, and is closely related to the Madagascar Bee-eater *M. superciliosus.* The Blue-tailed Bee-eater *M. philippinus* shares a common ancestry with these bee-eaters and the next two species, the European Bee-eater *M. apiaster,* which has the most northerly distribution of all bee-eaters and occurs across much of the western and Central mid-latitude belt, and the Rainbow Bee-eater *M. ornatus* of Australia (Marks *et al.,* 2007). The bee-eaters in this second group are, to some degree, geographical counterparts, but there is significant overlap among some species in south-east Asia. The three Western Palearctic species are geographical counterparts on a south-north gradient although there is also some overlap; compared to the European Bee-eater, the Blue-cheeked avoids temperate and oceanic regions (Cramp, 1985).

Climate

The bee-eaters are species of warm and wet climates with a significant number also in warm and dry climates: Temperature Category E (24), D (1); Humidity Category E (15), D (1), C (6), A (3). Only the European Bee-eater has made it into the temperate zone, and it occupies a slightly cooler position on the temperature gradient than other bee-eaters. All bee-eaters are characterised by narrow climatic ranges (17 specialists and 9 semi-specialists), the European and Little Green being among a small group with slightly broader climate tolerance (Appendix 1). The specialised diet of the bee-eaters, consisting of large flying insects, has undoubtedly restricted their climatic expansion.

Habitat

In contrast to their narrow climatic tolerance, bee-eaters occupy a broad habitat range, particularly savannas, shrublands (22 of 45 species) and open habitats, including deserts (13 species). The three Western Palearctic bee-eaters are species typical of open habitats including savanna, shrublands, grassland and desert (Appendix 1). These are habitats that would have expanded during the Plio-Pleistocene in the Palearctic, thus opening opportunities for these bee-eaters.

Migratory behaviour

The majority of bee-eaters are sedentary: 21 species have sedentary populations against only seven with migrants. As we have seen, the migratory species belong to a separate and distinct lineage, which suggests that migratory behaviour evolved once in the common ancestor of these bee-eaters. The three Palearctic bee-eaters are included in this group and migratory behaviour would have been a key ingredient in the success of bee-eaters away from tropical areas, giving them the ability to exploit the seasonal emergence of large flying insects and to move away during the cold season. The Little Green Bee-eater conforms to the pattern that we have observed in many species, with the most northerly and continental populations being migratory. In the other two species all populations are migratory. The origins of migratory behaviour in bee-eaters may well go back to intra-tropical migrants that exploited seasonal flushes of insects related to rainfall-drought cycles, as happens within Africa in the Northern Carmine *M. nubicus,* White-throated *M. albicollis* and Rosy *M. malimbicus* Bee-eaters (Curry-Lindahl, 1981a).

Fossil bee-eaters

Fossil bee-eaters are scarce, not surprising given their habits and restricted distribution in the Palearctic. They are present throughout the Pleistocene in areas that are in the breeding range today (Tyrberg, 1998, 2008).

Taxa in the Palearctic

Merops
Three (13.04%) bee-eaters have Western and Central Palearctic ranges and are absent from the east. Two, European and Little Green, are bioclimatic semi-specialists of the mid-latitude (Latitude Category D) and subtropical (Latitude Category E) belts respectively. The Blue-cheeked Bee-eater is a specialist of the mid-latitude belt. Bee-eaters are aerial insect feeders which, like the swifts, nightjars and most hirundines, have to abandon the Palearctic in the winter. They are highly migratory, a trait not shared by many other bee-eaters.

CORACIIDAE

The rollers and their close relatives the ground rollers (Brachypteraciidae) of Madagascar form a branch that falls between the bee-eaters on one side and the todies, motmots and kingfishers on the other (Hackett *et al.,* 2008). Rollers are tropical African, south and south-east Asian birds, with one species (the Dollarbird, *Eurystomus orientalis*) reaching New Guinea, Australia and the Eastern Palearctic and another (the European Roller, *Coracias garrulus*) occurring in large areas of the Western and Central Palearctic. Of the 12 species in the subfamily Coraciinae, it is the only one that does not breed in the tropics.

The European, Abyssinnian *C. abyssinica,* Lilac-breasted *C. caudata* and Indian *C. benghalensis* Rollers would appear, on plumage grounds, to be closely related and are geographical counterparts (with Abyssinian and Lilac-breasted overlapping in some areas). Between them they span the Western and Central Palearctic, Sahel, Arabia, south and south-east Asia and sub-Saharan Africa. Three other rollers, Racquet-tailed *C. spatulata* , Rufous-crowned *C. noevia* and Blue-bellied *C. cyanogaster,* are sub-saharan and are probably separated by habitat. The Purple-winged Roller *C. temmincki* is a south-east Asian island endemic.

Climate

Rollers are birds of warm climates, spread across the entire humidity gradient; Temperature Category E (10), D (1), C (1); Humidity Category E (4), C (6), B (2). Most species specialise within this climatic range (4 specialists, 5 semi-specialists, 1 moderate, 1 semi-generalist). The European Roller stands out as the most bioclimatically tolerant of the *Coracias* rollers, only the Dollarbird being more tolerant among all rollers

(Appendix 1). The European Roller occupies a mid-point along the temperature gradient, thus much cooler than other rollers, and shows a tendency towards the drier end of the climatic spectrum. It is more tolerant than the European Bee-eater, for example, and ranges further north in Europe.

Habitat

Rollers avoid forest and are at home in savanna and treeless habitats (11 species in mixed and 4 in open habitats, out of 12 species). Preference for open habitats would have favoured roller populations during the opening up of the forests of the Palearctic and Africa in the Plio-Pleistocene.

Migratory behaviour

Rollers are largely sedentary (11 of 12 species have resident populations) but tropical species seem to respond to seasonal rainfall by adjusting breeding and performing short-distance migratory movements (Curry-Lindahl, 1981a). The European Roller, living in the most extreme conditions for a roller, has become a full migrant and is one of two migratory species, the other being the Dollarbird in the Eastern Palearctic.

Fossil rollers

Early forms of roller are present in Europe from the Eocene (Mayr, 2005). They are present throughout the Pleistocene but are scarce, though less so than the bee-eaters (Tyrberg, 1998, 2008). The distribution of fossil sites is comparable to the present range of European Roller in the Western Palearctic.

Taxa in the Palearctic

Coracias

One species (12.5%), the European Roller, breeds in the Western and Central Palearctic. It is a bioclimatic moderate with a multi-latitude range and is highly migratory, the only non-resident in the genus. The rollers and bee-eaters are examples of genera with tropical African, and to some degree Asian, ranges that are able to penetrate areas of the southern and western Palearctic (*Cercotrichas* is another example), but depend on migratory behaviour.

Eurystomus

The Dollarbird is one of four species in this genus but is the only one to breed in the Palearctic. It is a migratory bioclimatic semi-generalist with a multi-latitude (Latitude Category F) range in the Eastern Palearctic. Populations to the south, in tropical South-east Asia, are sedentary.

ALCEDINIDAE

One species from each of the the subfamilies Alcedininae and Cerylinae and three from the Daceloninae are represented in the Palearctic. Kingfishers have an Old World centre of origin, with subsequent dispersals into the New World and Australasia, where they radiated into a very diverse assemblage. The Alcedininae is the ancestral kingfisher group. These are largely Afro-Asian with few species in Australia. The Daceloninae are also Afro-Asian but also include most of the Australian and Pacific endemics. Finally, the New World species are all within the Cerylinae, which also includes a number of Afro-Asian species (Moyle, 2006).

The most widespread of the Western Palearctic kingfishers is the Common Kingfisher *Alcedo atthis,* whose closest relative is the Small Blue Kingfisher *A. coerulescens* of South-east Asia. These two species cluster with two African species, Half-collared Kingfisher *A. semitorquata* and Shining Blue Kingfisher *A. quadrybrachys* (Moyle *et al.,* 2007). This suggests that these species may be geographical counterparts. The other Palearctic kingfishers are marginal, in the south. The Pied Kingfisher *Ceryle rudis* is a unique species and the only member of its genus. It is widely distributed in South-east Asia and sub-Saharan Africa. The four dacelonine kingfishers appear to be geographical counterparts, the White-throated Kingfisher *H. smyrnensis* in the south-west and Ruddy *H. coromanda* and Black-capped *H. pileata* in the south-east. To these we could add the Grey-headed Kingfisher *H. leucocephala* of sub-Saharan Africa, which is found in the Cape Verde Islands (Cramp, 1985).

Climate

The kingfishers are almost entirely species of warm and wet bioclimates with a small number also present in warm dry climates; Temperature Category E (87), D (1), C (2), B (1), A (1); Humidity Category E (71), D (3), C (13), B (4), A (1). There are very few species of kingfisher that occur outside these climatic boundaries. The Common Kingfisher is unique among kingfishers in that it exhibits a very wide bioclimatic tolerance, comparable to the most tolerant birds, and the broadest among the families considered in this chapter. This is particularly unusual in a family of highly specialised warmth-loving species (71 specialists, 17 semi-specialists, 2 moderates, 1 semi-generalist, 1 generalist). The Common Kingfisher occupies a Central position along the temperature gradient and shows a slight tendency towards the wet end of the humidity range (Appendix 1). Ruddy Kingfisher is a temperate, humid, moderate. The Pied and Black-capped are warm, fairly wet-climate specialists, while the White-throated and Grey-headed Kingfishers occupy a more Central position on the humidity gradient (Appendix 1).

Habitat

The kingfishers are forest birds that are often, but not always, associated with streams, rivers and other water bodies. The dacelonine kingfishers are less restricted to forest than the other kingfishers, and are frequent in light woodland and savanna. The five Palearctic kingfishers are associated with standing water but are not confined to forest habitats. The ability to break away from forests would have been advantageous in the Plio-Pliestocene world of shrinking forests in the Palearctic.

Migratory Behaviour

Most kingfishers are sedentary. All 92 species have sedentary populations while only 5 have migrants. The Common Kingfisher is among the few migratory kingfishers. South-western and southern populations of the Common Kingfisher are resident and northern, continental populations are highly migratory (Cramp, 1985).

Fossil kingfishers

Fossil kingfishers are extremely rare in the Pleistocene record of Europe (Appendix 2) and only appear in the Late Pleistocene (Tyrberg, 1998, 2008). This apparently late appearance may simply reflect the paucity of the record as a whole, or it may point to a severe restriction in Europe during cold glacial periods when kingfisher habitats may have become extremely limited.

Taxa in the Palearctic

Alcedo

One species breeds in the Palearctic – the Common Kingfisher, a bioclimatic generalist with a discontinuous pan-palearctic multi-latitude (Latitude Category F) range. It provides a further example of a 'single species generalist in a genus' trend, with this one species dominating the entire Palearctic. Like the rollers, and to some extent the bee-eaters, it is the only migratory species in a genus of tropical, mainly southern Asian, resident species. The migratory kingfishers are those from northern and continental populations. Those of the Western Palearctic are sedentary or short-distance migrants, as are the extralimital populations of southern Asia.

Halcyon

Two species are marginal along the subtropical (Latitude Category E) belt of the Western Palearctic – White-throated Kingfisher and Grey-headed Kingfisher. Both are resident bioclimatic semi-specialists. Two other species – Ruddy *H. coromanda* (bioclimatic moderate migrant) and Black-capped *H. pileata* (bioclimatic specialist migrant) are also Latitude Category E specialists of the Eastern Palearctic.

Ceryle

This single-species genus is represented marginally in the southern and eastern fringe of the Western Palearctic by the Pied Kingfisher, a sedentary semi-specialist of the subtropical (Latitude Category E) belt.

CONCLUSIONS

(a) **Bioclimatic generalisation.** The five bioclimatically tolerant species (20%) in this group come from the hoopoes, woodpeckers and kingfishers. The bee-eaters and rollers have no bioclimatically tolerant species. Three – Eurasian Hoopoe, Great Spotted Woodpecker and Common Kingfisher – have multi-latitude (Latitude Category F) ranges, and the other two – Grey-headed and White-backed Woodpeckers – are temperate (Latitude Category C). They therefore conform with the pattern observed in other groups, that of a single bioclimatically tolerant species per genus.

(b) **Mid-latitude belt representation.** The representation of mid-latitude species is low (4 species, 16%, to which we may add the split-category (B/D) Three-toed Woodpecker). This is not surprising as the species are either temperate, multi-latitude species (woodpeckers in forest, kingfisher, hoopoe, roller) or birds of the subtropical southern fringe (kingfishers, bee-eaters and some woodpeckers).

(c) **Families.** Most families have few genera and species. The woodpeckers provide 15 (60%) of the species. The genus *Dendrocopos* has the most species (8).

(d) **Geographical nature of the genera.** The major influence of the Himalayan-Tibetan massif as a platform for species to spread across the Palearctic is evident in the woodpecker genera *Picus, Dryocopus* and *Dendrocopos.* The remaining families and genera are tropical Afro-Asian in which a limited number of species have managed to survive in the Palearctic, most of them on the southern fringes and with the aid of migratory behaviour.

(e) **Representation in the Western Palearctic.** 19 species (76%) of the species in this group are found in the Western Palearctic, a remarkably high proportion compared to other groups. All families and genera are represented in the Western Palearctic.

CHAPTER 11
Owls

Group 1, subgroup e

The barn owls are an ancient lineage that diverged from the rest of the owls early in the Oligocene (Ericson *et al.,* 2006). The true barn owls diverged from the bay owls, the other subfamily of the Tytonidae, in the mid-Miocene, over 10 mya (Wink and Heidrich, 1999; Wink *et al.,* 2009). The radiation of the different species of *Tyto* followed shortly after and its centre of origin appears to be South-east Asia and New Guinea, where most species are found today. There is evidence of two separate settlements of the Australasian region (Wink *et al.,* 2009). The next stage of divergence involves the Common Barn Owl in the Old World, from which a later branch leads to the New World taxon (now sometimes considered a discrete species). The New World was presumably colonised from Asia at a point where the climate was relatively mild, as barn owls are essentially species of warm climates. Barn owls are absent from large areas of the Palearctic, especially the dry interior, the more humid Western Palearctic being exceptional in this respect. They must have once been present in northern Asia (from where the New World would have been colonised) but have since become extinct during Plio-Pleistocene aridification.

The Striginae

The separation of the large Striginae subfamily from the hawk and pygmy owls (the Surninae) predates the split of the barn and bay owls (Wink *et al.,* 2009) which, as we have seen, dates to over 10 mya. The group of owls that include the *Otus* scops owls also separated within the Striginae prior to 10 mya. The other three major groups in the subfamily, the *Strix* wood owls, the *Bubo* eagle owls and the *Asio* eared owls, diverged some time later, probably during the mid-Miocene. Within the wood owls, Hume's Owl *S. butleri* is most closely related to the African Wood Owl *S. woodfordi*, with these taxa separating at some point between the split of the New and Old World *Strix* owls (5-6 mya) and the separation of the Old World species (Tawny *S. aluco* and Ural *S. uralensis* Owls, around 4 mya). The separation of the Holarctic Great Grey Owl, *S. nebulosa,* would seem to predate that of Tawny from Ural Owl (Wink *et al.,* 2009).

Hume's Owl is a geographical counterpart of the African Wood Owl in the Middle East, where it occupies desert and open country. The Tawny and Ural Owls are also geographical counterparts of each other; the Tawny in the oceanic west and the warm south of the Palearctic and the Ural in the continental north, with an area of overlap in southern Scandinavia and eastern Europe. The larger Great Grey Owl occurs to the north of the Ural Owl, with which it overlaps across a large belt of the northern Palearctic. The size difference between the two presumably permits coexistence.

The eagle owls, *Bubo,* constitute a sister group to the *Strix* owls, with a divergence date of around 10 mya (Wink *et al.,* 2009). Within this group, a number of species separated at an early stage to form a loose cluster that includes African and south-east Asian forest and fish-eating species. Within this group, the fish owls (formerly *Ketupa*) include the Brown Fish Owl *B. zeylonensis*, which reaches the Western Palearctic in the Middle East, and Blakiston's Fish Owl *B. blakistoni* in the Eastern Palearctic. The remaining species separate out into an American cluster, which includes the Holarctic Snowy Owl *B. scandiacus,* a southern African cluster, and a Palearctic–North and East African group. This latter group includes the widespread Eurasian Eagle Owl *B. bubo,* and its geographical counterparts, the Cape Eagle Owl *B.capensis,* which occupies the Rift Valley down to South Africa, and the Pharaoh Eagle Owl *B. ascalaphus* of North Africa, the Sahara and Arabia.

The eared owls, *Asio,* are a small and homogeneous group within the Striginae. They separated from the rest of the group around 6–8 mya. Eight species make up this group, with three occurring in the Western Palearctic. The eared owls separate into two groups. The long-eared owls (Eurasian *A. otus,* African *A. abyssinicus,* Madagascar *A. madagascariensis* and Stygian *A. stygius*) are geographical counterparts that between them span the Palearctic, the New World and Africa. The Neotropical Striped Owl, *A. clamator,* is closely related. The second group (Short-eared Owl *A. flammeus,* Marsh Owl *A. capensis* and Galápagos Owl *A. galapagoensis*) are also geographical counterparts, with the first species widespread across the Holarctic and large parts of South America, and *capensis* replacing it in Africa. This second group is characterised by preferring open ground and marshland, and is absent from the Neotropical and African rain forests. The eared owls are absent from South-east Asia and Australia.

The fourth owl group, the scops owls *Otus,* separated from the North American *Megascops* screech owls around 6–8 mya. The split must have followed a period of contact when warm forests would have connected the two worlds. Many *Otus* are South-east Asian or Indian Ocean island endemics. The four Western Palearctic species (Eurasian Scops Owl *O. scops*, Pallid Scops Owl *O. brucei*, Arabian Scops Owl *O. pamelae* and Socotra Scops Owl *O. socotranus*) are closely related and geographical counterparts– Eurasian Scops across a wide area of the southern and Central Palearctic, *O. brucei* within a warm, dry region of the south-central Palearctic, and the other two localised within the Arabian peninsula. This group is closely-related to the Oriental Scops Owl *O. sunia* of southern and eastern Asia and the African Scops Owl *O. senegalensis* of sub-Saharan Africa.

The Surninae

Three major clusters make up the Surninae. The first group that separates from the rest are South-east Asian and Australasian hawk owls of the genus *Ninox.* There are no Western Palearctic species in this group (though there is one in the Eastern Palearctic).

The second major cluster is composed of the four species in the genus *Aegolius*. All occur in the New World, where they must have originated, and the most northerly species, Tengmalm's Owl *A. funereus,* has a wide distribution across the Holarctic west to Scandinavia and the Pyrenees. The separation between this species and the Nearctic Northern Saw-whet Owl *A. acadicus* is older than 6 mya (Wink *et al.*, 2009). The third major cluster is subdivided into two subgroups, one containing the genus *Athene* and the other the genera *Taenioglaux, Surnia* and *Glaucidium* (Wink *et al.*, 2009). *Athene* consists of six species that are geographical counterparts: Burrowing Owl *A. cunicularia* in the New World, Little Owl *A. noctua* across much of the mid-latitude belt, Lilith Owlet *A. lilith* in south-eastern Europe and the Middle East, Ethiopian Little Owl *A. spilogaster* in north-east Africa, Spotted Owlet *A. brama* in south and South-east Asia, and Forest Owlet *A. blewitti* in central India. In the second subgroup, the New and Old World pygmy owls separated from each other around 6–8 mya (Wink and Heidrich, 1999). The three Old World species are geographical counterparts, with the Western Palearctic species, the Eurasian Pygmy Owl *Glaucidium passerinum*, occupying a wide belt of forest north of the mid-latitude belt from Scandinavia to eastern Russia. The other two species are African and are separated from that of the Palearctic by desert, steppe and savanna and must have diverged from a common ancestor with a larger distribution in the Miocene, when forests were continuous between Africa and the Palearctic. The Northern Hawk Owl *Surnia ulula* has a widespread Holarctic distribution across the boreal forests, from north-east Canada to Scandinavia, generally north of *Glaucidium,* from which the genus must have diverged in the mid-Miocene.

Climate

The barn owls are overwhelmingly species of warm, wet/humid climates, with a small number of species having managed to occupy the dry end of the warm climatic spectrum (Figures 11.1–11.2 combine Tytonidae and Strigidae). Occupation of cooler climates is rare and limited to a single species in each climatic type. The Western Palearctic is occupied by the Common Barn Owl *Tyto alba,* the barn owl with the widest bioclimatic tolerance, occupying a more central position on the temperature and humidity gradients than any other barn owl (Appendix 1).

Like the barn owls, the striginine owls are largely species of warm, wet/humid climates (Figures 11.1–11.2). A significant number of species have become specialists of mountain climates. Otherwise, small numbers have occupied temperate and cool bioclimates, dry as well as humid. Palearctic species include four with the widest bioclimatic tolerances in the group, and comparable to the most bioclimatically tolerant of all birds: Eurasian Eagle, Tawny, Eurasian Long-eared and Short-eared Owls. A second group includes three species with relatively broad tolerances and tendencies towards cool and wet bioclimates (Ural Owl), cool and dry (Great Grey Owl), and warm and dry (Eurasian Scops Owl). An exceptional species in the group is the Snowy Owl, which has become a specialist of the coldest and driest bioclimates. The remaining Palearctic species are all specialists of warm climates with restricted distributions in the south of the region. Two are in dry climates, in south-west and Central Asia: Hume's and Pallid Scops Owls. Two others are species of warm climates but occupy a wetter part of the humidity gradient: Brown Fish Owl and Marsh Owl.

The surninine owls are, as all other owls, species of warm, wet/humid, climates. It is the four genera that have reached the Palearctic that have the widest climatic tolerance and have managed to occupy temperate and cool climates (Appendix 1). Even so, the surninine owls do not reach the bioclimatic breadth of the most generalised striginine owls. The Little Owl stands out as the species with the broadest tolerance, occupying a central position in the temperature and humidity gradients. The Northern Hawk, Eurasian Pygmy and Tengmalm's Owls are species with fairly broad tolerance to climates at the cool and dry end of the spectrum. The Lilith Owlet is a specialist of warm, dry climates.

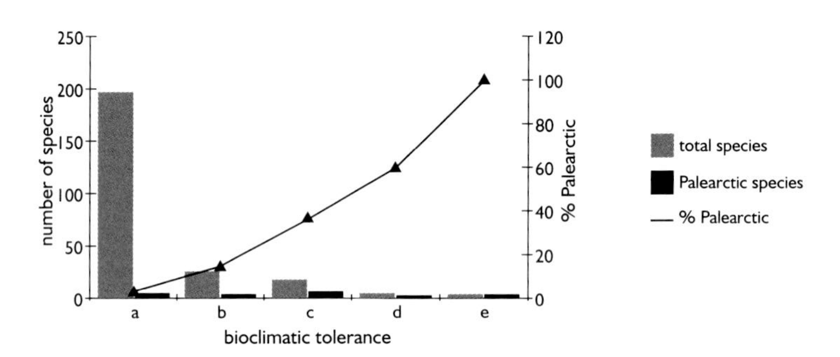

Figure 11.1 *Bioclimatic tolerance of owls. Definitions as on Figure 4.1 (see p. 38).*

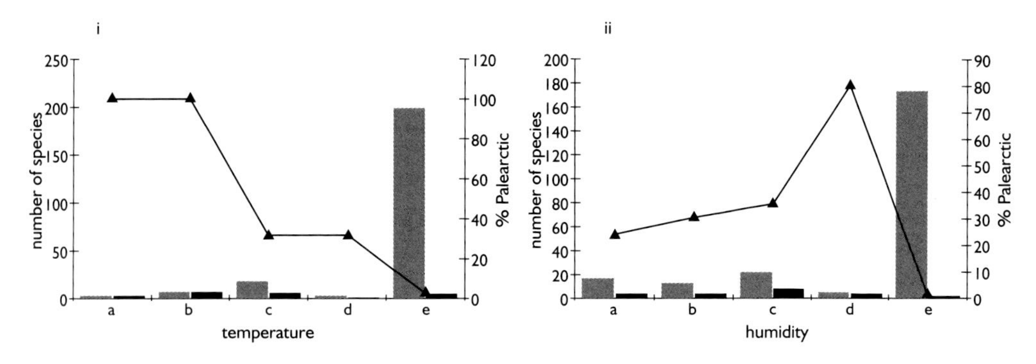

Figure 11.2 *Temperature (i) and humidity (ii) tolerance of owls. Definitions as in Figure 4.2 (see p. 38). Key as in Figure 11.1.*

Habitat

Barn owls are birds of forest, with a few species occupying more open habitats, including grasslands (Figure 11.3). The striginine owls are predominantly species of forest (Figure 11.3). Very few species have abandoned forest habitats to occupy savanna, shrubland and wetland, with even fewer being successful in treeless habitats, including tundra and desert. The 15 Western Palearctic species deviate from this pattern. Even though seven species are typical in that they are forest inhabitants, a wide range of habitats is occupied: savanna (5), shrubland (4), treeless (2), tundra (2), desert (6), rocky areas (6) and wetlands (3). The forest dwellers include those of the temperate and boreal forests that lie north of the mid-latitude belt, and these species have uninterrupted distributions either across the Palearctic (Ural Owl) or across the Holarctic (Great Grey, Eurasian Long-eared and Short-eared Owls). A notable subset is of species that occupy open habitats in the warm and arid part along the south of the mid-latitude belt (Pharaoh's Eagle, Hume's, Pallid Scops, Arabian Scops and Socotra Scops Owls).

The surninine owls are predominantly birds of forests, with a significant number of species in savanna. The *Athene* owls appear to have managed to break away from the forest theme and have become at home in savanna and treeless habitats (including desert and rocky). Palearctic species follow this trend of forest-dwelling, with *Athene* in more open habitats.

Although owls are heavily dependent on forests, the occupation of rocky habitats (Figure 11.3) is a feature that characterises some Palearctic owls. This ability to occupy rocky habitats has ensured the success of some species where forests were absent.

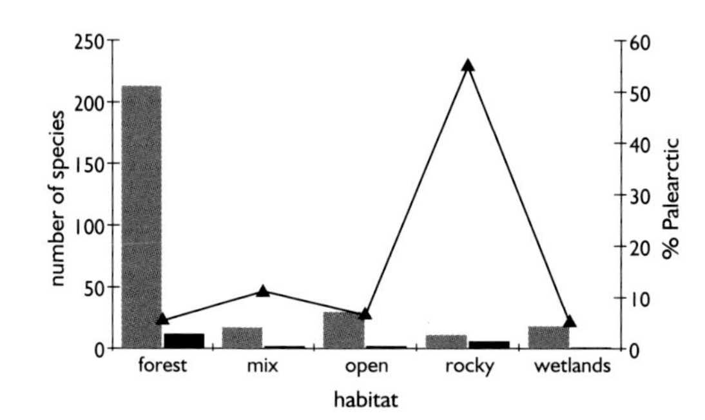

Figure 11.3 *Habitat occupation by owls. Definitions as in Figure 4.3 (see p. 39). Key as in Figure 11.1.*

Migratory behaviour

Barn owls are strictly resident species. There are no migratory species of barn owl and such a limitation may have been a major contributory factor to their absence from the more continental environments of the Palearctic.

The striginine owls include the few species of owl that exhibit migratory behaviour. All seven migratory species occur in the Palearctic: Snowy, Eurasian Scops, Oriental Scops, Pallid Scops, Brown Hawk, Eurasian Long-eared and Short-eared Owls. In addition, a number of other species show irruptive behaviour related to breeding output which is, in turn, related to prey abundance. The Snowy and Eurasian Scops Owls are fully migratory, the former moving to areas immediately to the south and the latter to tropical Africa. Pallid Scops Owls in the south of their limited range are resident, with populations in the continental interiors being migratory. Oriental Scops and Brown Hawk Owls are migratory in the Eastern Palearctic with sedentary populations to the south. Short-eared Owls are almost wholly migratory but populations in the extreme south and west of the breeding range are resident. The Eurasian Long-eared Owl, by contrast, is mainly sedentary with only populations in the northernmost continental areas being migratory. This difference between two closely related owls would suggest that forest habitats (occupied by Long-eared) are possible to occupy throughout the year, whereas open ones (used by Short-eared, and incidentally also Snowy) are not. This suggestion is reinforced by the behaviour of owls of the temperate and boreal forests, all of which are resident. Their success, without having to migrate south for the winter, lies in their ability to exploit small mammals during the long, dark northern winter.

Migratory behaviour is scarce in this group and is virtually absent from Western Palearctic species which are, at most, irruptive (the exceptions are the striginine owls discussed above). Overall, 247 of 250 species have sedentary populations, contrasting with only 10 with migrants, 7 of which are Palearctic.

Fossil owls

Barn owls first appear in the fossil record of Europe in the Oligocene, and *Tyto* is present from the Early Miocene (Mlíkovský, 2002). This observation is in keeping with an early split of the Tytonidae from the Strigidae during the Oligocene, which must have been followed by a rapid radiation and geographical expansion.

Fossil striginine owls are certainly present in Europe from the Early Miocene (*Asio, Strix*) and modern species from at least the Early Pliocene (*B. bubo, B. zeylonensis*; Mlíkovský, 2002). Eleven species are all present from the Early Pleistocene (Appendix 2). Noteworthy among these records is the presence of northern species well to the south of the present range. Snowy Owls are common in France and many other central and western European countries, reaching south to Gibraltar. Ural Owls also reach Italy, France and Spain, while Great Grey Owls occur in central and eastern Europe, west to Italy (Tyrberg, 1998, 2008). These observations indicate a significant southward and westward extension of boreal forest and tundra during the Pleistocene glaciations.

Compared to other owls, surninine owls appear rather later in the fossil record of Europe: *Aegolius* in the Late Miocene, and *Surnia, Glaucidium* and *Athene* in the Late Pliocene. All Western Palearctic species are present in the fossil record throughout the Pleistocene (Appendix 2), the Little Owl being the most frequently recorded. As with the previous group, the more northerly surninine owls are recorded south and west of their present range during the Pleistocene. The Northern Hawk Owl has been found in deposits from Germany, Italy, Austria, Switzerland, Hungary, Ukraine, France and the United Kingdom; the Eurasian Pygmy Owl in France, Italy, Greece and Spain; and Tengmalm's Owl in the United Kingdom, Italy and Spain.

Taxa in the Palearctic

TYTONIDAE

Tyto

The Common Barn Owl *T. alba* is the sole Palearctic representative of this genus of tropical forest owls. The Barn Owl itself has a wide tropical distribution, in the Americas, Africa, southern Asia and Australia, but outside the tropics it has only succeeded in North America and the Western Palearctic. It is absent from the Central and Eastern Palearctic. It is a bioclimatic multi-latitude (Latitude Category F) semi-generalist but avoids cold and arid conditions so it would have been prevented from surviving in the harsh conditions of the Central Palearctic. Its absence from the Eastern Palearctic which, like North America, once had connections with southern tropical forests, is harder to understand. It may be that conditions in the oceanic Western Palearctic permit the survival of this sedentary owl but the harshness of Eastern Palearctic winters does not. If so, the answer would lie in the bird's inability to migrate.

STRIGIDAE

Otus

Five (9.8%) of the world's scops owls breed in the Western Palearctic. *Otus* are overwhelmingly bioclimatic specialists of warm and wet forests. The Palearctic species include the two most bioclimatically generalised of the genus – European Scops (moderate, Latitude Category D) of the Western and Central Palearctic, and its Eastern Palearctic counterpart, the Oriental Scops Owl (moderate, multi-latitude, Category F).

These two species have the largest ranges of the Palearctic scops owls; they are successful because they are migratory.

The only other migratory scops owl is Pallid Scops Owl, which has a restricted range in the Central Palearctic and the eastern part of the Western. To the south resident populations occur. This is a bioclimatic specialist of the mid-latitude belt (Latitude Category D) so, in spite of its migratory habits, its bioclimatic intolerance has restricted it much more than the two previously discussed species. The other species are resident Eastern Palearctic bioclimatic specialists – Collared Scops *O. lettia* (multi-latitude, Category F) and Japanese Scops *O. semitorques* (mid-latitude, Category D).

Strix

Four species (16.6%) of *Strix* owls breed in the Palearctic. They are mainly tropical forest owls with two major clusters, one in southern Asia and the Himalayas, which may have been the centre of radiation of the genus, and the second in the New World. The colonisation of the Palearctic may have followed the 'launch-pad model' that we have described for other, particularly forest, genera (see p. 41). Entry into the Americas would have been during a mild climatic phase, when forests provided habitat continuity.

Three of the four species have pan-palearctic ranges. The Tawny Owl is a generalist with a continuous, multi-latitude (Latitude Category F), range in the Western Palearctic and discontinuous, as we would expect with the fragmentation of forests, in the Central and Eastern Palearctic. In this respect the Tawny Owl resembles the Barn Owl in having a wide distribution in the west and a restricted one in the east. Great Grey and Ural Owls are bioclimatic moderates that owe their success to having occupied the Boreal (Latitude Category B) forest belt. They have continuous pan-palearctic ranges except in the west, where they are marginal. The Great Grey Owl's range continues in the Nearctic. A limited southern area (Latitude Category D) in the Western Palearctic range of the Ural Owl indicates a larger range that became disconnected with global warming. Finally, Hume's Owl *S. butleri* is a specialist of the subtropical belt of the Western Palearctic, which has adapted to arid treeless environments by adopting a rocky habitat lifestyle. Its range is consequently restricted. This genus conforms with the 'single generalist per genus' pattern (see p. 46).

Bubo

Four species (16%) of largely Afro-Asian tropical forest eagle owls breed in the Palearctic. Eurasian Eagle Owl is a bioclimatic generalist with a multi-latitude (Latitude Category F) pan-palearctic range that is discontinuous in the west. In spite of its large range, this owl never managed to colonise the Nearctic. Its success has been due in part to its ability to breed in rocky, as well as forested, habitats. The Snowy Owl is a semi-specialist of open Arctic (latitude Category A) habitats and has a Holarctic distribution. It is the *Bubo* solution to life on the tundra, and involves migration; the Snowy Owl is the only migratory owl of the genus. Two bioclimatic semi-specialists, with the unusual habit of catching fish, occupy limited areas of the Western (Brown Fish Owl – Latitude Category E) and Eastern (Blakiston's Fish Owl – Latitude Category C) Palearctic.

Asio

Three species (37.5% of the species in this genus) breed in the Palearctic. The genus is globally widespread. The Palearctic species include two generalists, which would appear to form a break from the 'single generalist per genus' rule. But the two species belong to the two major lineages within *Asio* and differ hugely in behaviour. The Eurasian Long-eared Owl is a forest bird with a multi-latitude (Latitude Category F) pan-palearctic range. It is sedentary across this range but northernmost populations are migratory; the Short-eared Owl *A. flammeus* is a Boreal (Latitude Category B) pan-palearctic species and, like the Long-eared, its range is fragmented in the west. It is a species of open, treeless habitats and is highly migratory. These major ecological and behavioural differences, in species that belong to different lineages within the genus, would seem sufficient to permit two widespread generalist congeners to overlap widely in range. The third species – Marsh Owl – is African and reaches the southern fringe of the Western Palearctic where it is on the northern tip of its range. It is a sedentary semi-specialist of the subtropical (Latitude Category E) belt, and is a species of wetlands.

Ninox

The Brown Hawk Owl *N. scutulata* is the only species (4%) of this large south-east Asian and Australasian genus to occur in the Palearctic. It is the most bioclimatically tolerant (moderate) of all the species in its genus and occupies a multi-latitude (Latitude Category F) in the Eastern Palearctic. It is a migratory forest insectivore that appears to have been confined to the east, as is the case for a number of other forest insectivores.

Aegolius

This is a New World genus of four species, one of which – Tengmalm's (or Boreal) Owl – has a Holarctic range. It is the most bioclimatically tolerant (semi-generalist) in the genus and it has a split-category (B/D) pan-palearctic range that is discontinuous in the west. It follows the pattern observed in a number of other forest species that must have had a wider latitudinal spread during colder climatic phases.

Athene

Two (33.33%) of the little owls breed in the Palearctic. The Little Owl *A. noctua* is a semi-generalist with a pan-palearctic range across the mid-latitude belt (Latitude Category D). An inhabitant of rocky habitats, it has spread right across this belt as other species have done (e.g. choughs, wheatears). The Lilith Owlet *A. lilith* is confined to south-eastern Europe and the Middle East where it must have speciated during a period of climatic isolation. It is a specialist of warm and dry habitats of the mid-latitude belt (Latitude Category D). The little owls seem good candidates for a 'launch-pad range expansion' from south or south-east Asia via the Himalayas, with a gradual abandonment of forest for open and rocky habitats. Little Owl spread across the belt and two species emerged in isolation within this range, Lilith and Ethiopian Little Owl in north-east Africa. There must have been a wider range during a warmer past, which allowed the genus to enter the New World. It is represented there today by the widespread Burrowing Owl.

Glaucidium

The Eurasian Pygmy Owl is the only Palearctic representative (4%) of this large New World genus of warm and wet climate forest owls. Along with the Nearctic Northern Pygmy Owl *G. californicum,* it is the most bioclimatically tolerant (moderate) of the genus. Its range is Boreal (Latitude Category B) pan-palearctic, discontinuous in the west.

Surnia

This single-species Holarctic genus is a Miocene offshoot of *Glaucidium*. The Northern Hawk Owl *S. ulula* is a bioclimatic moderate with a continuous pan-palearctic range, split between Boreal (B) and mid-latitude (D), as in a number of other forest birds (including Tengmalm's and Ural Owls).

CONCLUSIONS

(a) **Bioclimatic generalisation**. Bioclimatic generalists and semi-generalists make up a large proportion of the Palearctic owls – 7 species, 29.17%. It appears to have been the solution among birds that have not developed migratory strategies: 16 (66.67%) of the Palearctic owls are sedentary. Four species have multi-latitude (Latitude Category F) ranges, three of them pan-palearctic. The other bioclimatically tolerant species have Boreal (Latitude Category B, 1 species), mid-latitude (Latitude Category D, 1 species) and split (Latitude Category B/D, 21 species) ranges.

(b) **Mid-latitude belt representation**. Five species have ranges centred on 40°N, and they are all little and scops owls. To these we may add three with split boreal/warm (B/D) ranges. In fact, other than the multi-latitude species (6) most owls are either warm (Category D), split boreal/warm or Boreal (B, 4 species). In contrast there is only one Temperate species (Category C), which seems to indicate that ideal conditions for owls, other than multi-latitude ones, lie in the 40th and 60th parallel belts.

(c) **Geographical nature of the genera**. Several genera fit into the 'launch-pad model' of South-east Asian species penetrating the Palearctic via Himalayan altitude belts. This would seem to apply to *Strix* and

Athene, and possibly also to *Bubo* and *Otus*. Two, possibly three, genera reveal a different kind of 'launch-pad' spread, one we have come across occasionally so far (e.g. in *Troglodytes*); the dispersal of a single species in a genus (usually bioclimatically tolerant) from the Nearctic. The genera in question are *Aegolius, Glaucidium* and, possibly, *Surnia*. In such cases the Palearctic is dominated by one species, a situation akin to the south-east Asian–Himalayan model. In both cases the sources of species have connections with the tropical south.

(d) **Habitat and diet modification**. A feature among the owls has been the ability to move away from forest habitats and exploit a range of other habitats available to them in the Palearctic. Rocky habitats are exploited for breeding in three genera – *Athene* (Little Owl, Lilith Owlet), *Strix* (Hume's Owl) and *Bubo* (Eurasian Eagle Owl). There is also adaptation to the open habitats of the north in *Bubo* (Snowy) and *Asio* (Short-eared), and movement into wetland habitats in *Asio* (Marsh Owl) and among the fish-eating eagle owls (Brown and Blakiston's Fish Owls).

(e) **Representation in the Western Palearctic**. The sub-region with fewest owls is, not unsurprisingly given the great loss of forests there, the Central Palearctic (13). There are 15 species in the Eastern Palearctic, but the sub-region with most species is the Western Palearctic with 18. The reason for this may be that the Western Palearctic contains the greatest range of owl habitats (forest and rocky), more than the forested Eastern Palearctic or the forest-sparse Central Palearctic.

CHAPTER 12
Raptors

Group 1, subgroup f

PANDIONIDAE

The osprey *Pandion haliaetus* represents an ancient lineage that separated from other raptor families long ago (Lerner and Mindell, 2005; Griffiths *et al.,* 2007). Mitochondrial DNA analysis of the four osprey subspecies, *P. h. haliaetus* (Palearctic, largely migratory), *P. h. carolinensis* (North America, migratory), *P. h. ridgwayi* (Caribbean, sedentary) and *P. h. cristatus* (Sulawesi, Java, Australia, New Caldeonia, sedentary) seems to suggest that these geographical forms separated from each other some time between 1 and 2 mya (Wink *et al.,* 2004c), within the broad time frame when climatic variability was intensifying and significant cooling was underway. The degree of genetic distance between the osprey taxa is such that there are grounds to consider specific separation.

Ospreys are widespread, including populations that are resident on islands and along the coasts on the western, northern and eastern ends of the Sahara (Cramp, 1980). This fragmented distribution is the product of significant decline in historical times due to human intervention (e.g. Irby, 1895) so that we may expect ospreys to have been much more widespread along many stretches of coast and islands in the Mediterranean, the Atlantic coasts of Iberia, Morocco and Mauritania and the Red Sea before the intense occupation of the coasts by humans. This circum-Saharan coastal distribution was probably a relict of a much more widespread distribution of ospreys across the width of the Sahara when it was a vast network of wetlands. Osprey remains have been excavated in Late Palaeolithic contexts in Kom Ombo in the Nile Valley, dating to between 11,500 and 17,000 years ago; these ospreys were occupying fish-rich wetlands and riverine habitats, on a grassland plain, alongside many species of aquatic birds (Churcher and Smith, 1972). Ospreys neither breed nor winter in this arid region today. Osprey remains have also been recovered from the 7,600–6,000 year-old site of Tell Mureybet on the upper Euphrates in Syria, where the species does not breed or winter today either, alongside other raptors and many waterbirds (Tyrberg, 1998). To the south osprey remains have been recovered in similar waterbird assemblages in Natufian sites dating to between 13,000 and 10,000 years ago, at Hayonim and Mallaha, both in Galilee, Israel (Tyrberg, 1998). These meagre observations support the view that ospreys were more widespread across the Middle East and Sahara during wetter times, reaching well into the Holocene, and that there must have been a significant fragmentation of southern populations due to climate aridification followed by human intervention in historical times. The fragmentation of the southern populations would have been part of a longer term process that would have separated other osprey populations too.

Climate and habitat

The ospreys are species of warm wet/humid climates that are also at home in warm and dry climates. Temperate and cool climates have also been colonised. They are essentially birds of wetlands.

Migratory behaviour

Ospreys exhibit behaviours from sedentary to fully migratory. Eurasian birds are fully migratory across the range (including the Western Palearctic), except in the extreme south-west where they are sedentary.

Fossil ospreys

The earliest fossil ospreys are from the Oligocene formation of the Jebel Qatrani in the Fayum, Egypt, dating to around 30 mya (Rasmussen *et al.* 1987; Seiffert, 2006). Qatrani was a fish-rich swampy region with overgrown river banks, floating vegetation, open riverine habitat and a small lake (Murray, 2004), ideal for fish-eating birds which included an early fish eagle, a rail, several species of jacana, storks, herons and cormorants (Rasmussen *et al.*, 1987). Western Palearctic fossil ospreys are fairly scarce in comparison to other raptors, and they are all from sites within the present geographical distribution (Appendix 2).

Taxa in the Palearctic

Pandion

The Eurasian Osprey *P. haliaetus* is the only one of four ospreys to breed in the Palearctic (note that these lineages are usually considered subspecies, but Wink *et al.* (2004c) suggest that the taxa represent full species since they exceed the genetic threshold of species definition, even though morphological differences are slight. For this reason, and because it helps to clarify the bioclimatic position of the different lineages, I have followed Wink *et al.*'s split). The Eurasian Osprey is a bioclimatic generalist with a pan-palearctic distribution, fragmented in the west in part at least because of human disturbance. A fish-eating raptor, it is highly migratory except in the extreme south-west where it is sedentary. The Eurasian Osprey therefore represents an extreme example of the single generalist per genus, taken to the family level. It occupies a central position in the temperature and humidity gradients, tending towards humid and cool (Appendix 1).

ACCIPITRIDAE

These raptors are a diverse group of ancient lineage and deserve detailed treatment of their inter-relationships.

Elanine and pernine kites

The elanine kites form the basal split within the Accipitridae, and are therefore a sister group of the rest of the family (Lerner *et al.*, 2005; Griffiths *et al.*, 2007). This group is composed of three genera and six species. They form an ancient group that is represented in western North America, parts of South America, Africa, south and South-east Asia and Australia. One, the Black-winged Kite *Elanus caeruleus* just reaches the Western Palearctic along its southern rim. The connection between New and Old World taxa must have been ancient and severed by the cooling climate of the Plio-Pleistocene

The pernine kites form a group of 16 species divided into eight genera. Six of these have traditionally been grouped together: *Henicopernis, Chondrohierax, Leptodon, Pernis, Elanoides* and *Aviceda.* Two former milvine kite genera, *Lophoictinia* and *Hamirostra*, are now added on the basis of genetic analyses (Lerner and Mindell, 2005; Griffiths *et al.*, 2007). The pernine kites are species of tropical regions of the New World, Africa, south and South-east Asia and Australia. The separation of New and Old World genera, as with the elanine kites, must reach back to an early part of the Tertiary, when tropical climates reached well to the north and connected the two worlds. This group is closest to the bearded vultures. The genus *Pernis* includes the only Western Palearctic species, the European Honey-buzzard *P. apivorus.* This species appears to have been the first to split within the genus, and the separation had probably already taken place by the Pliocene (Gamauf and Haring, 2004). The radiation of the eastern and South-east Asian populations of *Pernis*, which may necessitate a taxonomic re-assessment of certain island forms, appears to have been caused by the sea-level changes and contraction of rain-forest habitats that occurred during the Pleistocene. The distribution of the Palearctic honey-buzzards follows the bands of forest that would have shifted latitudinally with changing climates during glaciations. These species, at the northern limit of distribution of the pernine kites, are the only migratory populations of the subfamily.

Bearded vultures

Although usually placed with the Old World vultures, recent genetic studies show that the bearded vultures are a separate group that diverged earlier than the Old World vultures (Lerner and Mindell, 2005). They share a distant common ancestry with the pernine kites, which are closer to them than the Old World vultures. Together, the bearded vultures and pernine kites form a distinct and ancient group that is separate from other raptors (Griffiths *et al.,* 2007).The bearded vultures form a small group of distinctive species that are nevertheless related, though less so than the Old World vultures among themselves. The four species include, unexpectedly, the Madagascar Serpent Eagle *Eutriorchis astur* and, less surprisingly, the Palmnut Vulture *Gypohierax angolensis.* The remaining two species, and those most closely related to each other within this group, are the Bearded Vulture *Gypaetus barbatus* and the Egyptian Vulture *Neophron percnopterus.* These latter two are found in the Western Palearctic.

As with the solitary, tree-nesting Eurasian Black Vulture *Aegypius monachus,* which has a similar distribution across the mid-latitude belt of the southern Palearctic, the Bearded Vulture has remained sedentary. Populations of Bearded Vulture have apparently undergone severe isolation during glacials with subsequent recolonisation, to the degree that African and Asian populations show distinctive genetic features and differ from Western European birds, with both genetic lineages mixing equally in the Alps (Godoy *et al.,* 2004). Bearded and Egyptian Vultures co-occur over large tracts of their geographical ranges, including the Western Palearctic. A major difference in size, and in food taken, effectively isolates them ecologically.

Snake-eagles

The snake-eagles are an exclusively Old World group (Lerner and Mindell, 2005). The six species in the genus *Spilornis* are inhabitants of warm, wet forests and are a distinct lineage, separate from the other snake-eagles. They are the counterpart of *Circaetus* in south and South-east Asia and represent an early divergence among the snake-eagles. The Philippine Eagle *Pithecophaga jefferyi* represents a divergent lineage within south-east Asia which has become a large rain forest species that hunts monkeys, hornbills and other large species. The remaining snake-eagles today occupy Africa, with one species reaching the Palearctic. The Bateleur *Terathopius caudatus* is another unusual species and is closely related to the *Circaetus* eagles, occupying savannas and plains across much of sub-Saharan Africa. The remaining snake eagles, six species in all, are in the genus *Circaetus.* The Short-toed Snake-eagle *C. gallicus* and the closely related Black-chested Snake-eagle *C. pectoralis* may be geographical counterparts. The Brown Snake-eagle *C. cinereus* overlaps geographically with the sub-Saharan populations of the previous two species, with which it is very closely-related, but occupies a denser part of the woodland gradient (Brown, 1971). The Congo Serpent Eagle *C. spectabilis* is the next closest relative and is separated by habitat, living in deep rain forest. The two remaining species, Banded Snake-eagle *C. cinerascens* and Southern Banded Snake-eagle *C. fasciolatus,* are closely related to each other and occupy the denser end of the woodland spectrum, between Congo Serpent Eagle and Brown Snake-eagle (Brown, 1971). The ecological separation within *Circaetus* is largely by habitat, in contrast to the booted and fish eagles.

One species from this group occupies the Western Palearctic, and it is the one with the widest bioclimatic tolerance of all the snake-eagles (Table 12.1). It is also the species that occupies the most open part of the spectrum of habitats occupied by the snake-eagles – the Short-toed Snake-eagle. This species has breeding populations just south of the Sahara Desert (race *beaudouini*) and in India, indicating a wider former breeding distribution that has been severed by the advance of the Sahara and Arabian Deserts. The present distribution of this species across the Palearctic need not therefore represent an expansion from Africa during interglacials. As with many other raptors it is likely that the common ancestor of the *Circaetus* eagles had a wide geographical distribution across Africa, India and the Palearctic. With the opening up of habitats in the Plio-Pleistocene, Palearctic eagles survived by increasing their bioclimatic tolerance, living in open habitats or by becoming migratory.

Table 12.1 *The species of diurnal raptor by bioclimate. Range estimates vary from extreme generalist (E) to extreme specialist (A). Temperature estimates range from warm (E) to cold (A) and humidity from wet (E) to dry (A). * indicates Palearctic species.*

Species	Global range	Global temperature	Global humidity	Montane
Eurasian Osprey*	E	B	D	
American Osprey	D	C	B	
Caribbean Osprey	A	E	E	
Eastern Osprey	B	E	C	
Black-winged Kite*	B	E	C	
Black-shouldered Kite	A	E	A	
White-tailed Kite	B	E	B	
Letter-winged Kite	A	E	A	
Pearl Kite	B	E	C	
Scissor-tailed Kite	A	E	C	
Long-tailed Honey-buzzard	A	E	E	
Black Honey-buzzard	A	E	E	
Hook-billed Kite	A	E	E	
Grey-headed Kite	A	E	E	
White-collared Kite	A	E	E	
European Honey-buzzard *	C	C	B	
Oriental Honey-buzzard *	C	C	B	
Barred Honey-buzzard	A	E	E	
Swallow-tailed Kite	A	E	E	
African Baza	A	E	E	
Madagascar Baza	A	E	E	
Jerdon's Baza	A	E	E	
Pacific Baza	A	E	E	
Black Baza	A	E	E	
Square-tailed Kite	A	E	A	
Black-breasted Buzzard	B	E	C	
Palm-nut Vulture	A	E	E	
Bearded Vulture*	C	D	B	x
Egyptian Vulture*	C	D	B	
Madagascar Serpent Eagle	A	E	E	
Crested Serpent Eagle	A	E	E	
South Nicobar Serpent Eagle	A	E	E	
Mountain Serpent Eagle	A	E	E	
Sulawesi Serpent Eagle	A	E	E	
Philippine Serpent Eagle	A	E	E	
Andaman Serpent Eagle	A	E	E	
Philippine Eagle	A	E	E	
Bateleur	B	E	B	
Banded Snake Eagle	B	E	B	
Southern Banded Snake Eagle	A	E	E	
Congo Serpent Eagle	A	E	E	
Black-chested Snake Eagle	B	E	C	
Brown Snake Eagle	B	E	B	
Short-toed Snake Eagle*	C	D	C	
Indian Black Vulture	A	E	E	
White-headed Vulture	B	E	B	
Eurasian Black Vulture*	C	D	B	x
Lappet-faced Vulture*	B	E	B	
Hooded Vulture	B	E	B	

Species	Global range	Global temperature	Global humidity	Montane
Indian White-rumped Vulture	C	E	C	x
Himalayan Griffon Vulture	A			x
African White-backed Vulture	B	E	B	
Ruppell's Vulture	B	E	C	
Eurasian Griffon Vulture*	C	D	B	
Cape Griffon Vulture	A	E	A	
Slender-billed Vulture	A	E	E	
Indian Vulture	A	E	E	
Lesser Spotted Eagle*	B	D	A	
Indian Spotted Eagle	A	E	E	
Greater Spotted Eagle*	A	C	A	
Long-crested Eagle	B	E	B	
Black Eagle	A	E	E	
Wahlberg's Eagle	B	E	B	
Rufous-bellied Eagle	B	E	E	x
Mountain Hawk Eagle	C	C	E	x
Booted Eagle*	C	D	B	x
Little Eagle	B	E	C	
Ayres's Hawk Eagle	B	E	C	
New Guinea Hawk Eagle	A	E	E	
African Tawny Eagle*	A	E	A	
Asian Tawny Eagle	B	E	C	
Steppe Eagle*	A	C	A	
Spanish Imperial Eagle*	A	E	A	
Imperial Eagle*	B	D	A	
Gurney's Eagle	A	E	E	
Golden Eagle*	E	B	B	x
Wedge-tailed Eagle	B	E	C	
Verreaux's Eagle*	C	E	B	x
Bonelli's Eagle*	D	D	C	x
African Hawk Eagle	B	E	B	
Cassin's Hawk Eagle	A	E	E	
Dark Chanting Goshawk*	B	E	B	
Eastern Chanting Goshawk	B	E	C	
Pale Chanting Goshawk	A	E	A	
Gabar Goshawk	B	E	B	
Grey-bellied Goshawk	A	E	E	
Crested Goshawk	A	E	E	
Sulawesi Goshawk	A	E	E	
Red-chested Goshawk	A	E	E	
African Goshawk	B	E	C	
Chestnut-flanked Sparrowhawk	A	E	E	
Shikra*	C	D	B	
Nicobar Sparrowhawk	A	E	E	
Levant Sparrowhawk*	A	E	A	
Chinese Goshawk *	B	D	E	
Frances's Goshawk	A	E	E	
Spot-tailed Goshawk	A	E	E	
Grey Goshawk	B	E	C	
Brown Goshawk	B	E	C	
Black-mantled Goshawk	A	E	E	
Pied Goshawk	A	E	E	
White-bellied Goshawk	A	E	E	

Species	Global range	Global temperature	Global humidity	Montane
Fiji Goshawk	A	E	E	
Moluccan Goshawk	A	E	E	
Slaty-mantled Sparrowhawk	A	E	E	
Imitator Sparrowhawk	A	E	E	
Grey-headed Goshawk	A	E	E	
New Britain Goshawk	A	E	E	
Tiny Hawk	A	E	E	
Semi-collared Hawk	A	E	E	
Red-thighed Sparrowhawk	A	E	E	
Little Sparrowhawk	B	E	C	
Japanese Sparrowhawk*	C	B	D	
Besra	B	E	E	x
Small Sparrowhawk	A	E	E	
Rufous-necked Sparrowhawk	A	E	E	
Collared Sparrowhawk	B	E	C	
New Britain Sparrowhawk	A	E	E	
Vinous-breasted Sparrowhawk	A	E	E	
Madagascar Sparrowhawk	A	E	E	
Ovampo Sparrowhawk	B	E	C	
Eurasian Sparrowhawk*	E	C	D	x
Rufous-chested Sparrowhawk	C	E	C	x
Sharp-shinned Hawk	D	C	B	
White-breasted Hawk	A			x
Plain-breasted Hawk	A			x
Rufous-thighed Hawk	A	E	E	
Cooper's Hawk	C	C	B	
Gundlach's Hawk	A	E	E	
Bicolored Hawk	A	E	E	
Black Goshawk	A	E	E	
Henst's Goshawk	A	E	E	
Northern Goshawk*	E	B	B	
Meyer's Goshawk	A	E	E	
Western Marsh Harrier*	C	C	C	
African Marsh Harrier	B	E	C	
Eastern Marsh Harrier*	C	B	B	
Papuan Harrier	A	E	E	
Swamp Harrier	B	E	C	
Madagascar Marsh Harrier	A	E	E	
Long-winged Harrier	B	C	C	
Spotted Harrier	B	E	C	
Black Harrier	A	E	A	
Hen Harrier*	E	B	B	
Cinereous Harrier	C	C	A	x
Pallid Harrier*	A	C	A	
Pied Harrier*	C	B	B	
Montagu's Harrier*	C	C	B	
Red Kite*	C	C	D	
Black Kite*	E	C	C	
Yellow-billed Kite*	B	E	C	
Whistling Kite	B	E	C	
Brahminy Kite	B	E	D	
Lesser Fish Eagle	A	E	E	
Grey-headed Fish Eagle	A	E	E	

Species	Global range	Global temperature	Global humidity	Montane
African Fish Eagle	B	E	B	
Madagascar Fish Eagle	A	E	E	
White-bellied Sea Eagle	B	E	D	
Sanford's Sea Eagle	A	E	E	
White-tailed Eagle*	D	B	B	
Bald Eagle	D	B	B	
Pallas's Fish Eagle*	B	D	B	
Steller's Sea Eagle*	B	A	C	
Grasshopper Buzzard	B	E	C	
White-eyed Buzzard	B	E	C	
Rufous-winged Buzzard	A	E	E	
Grey-faced Buzzard*	B	A	C	
Mississippi Kite	B	E	C	
Plumbeous Kite	A	E	E	
Snail Kite	A	E	E	
Crane Hawk	A	E	E	
Plumbeous Hawk	A	E	E	
Barred Hawk	A	E	E	
Semiplumbeous Hawk	A	E	E	
Black-faced Hawk	A	E	E	
White-browed Hawk	A	E	E	
White-necked Hawk	A	E	E	
White Hawk	A	E	E	
Grey-backed Hawk	A	E	E	
Savanna Hawk	A	E	E	
Harris's Hawk	B	E	C	
White-rumped Hawk	B	E	E	x
Black-chested Buzzard-Eagle	B	E	E	x
White-tailed Hawk	B	E	C	
Red-backed Hawk	C	C	C	x
Great Black-hawk	A	E	E	
Solitary Eagle	A			x
Crowned Eagle	A	E	E	
Roadside Hawk	B	E	E	x
Slate-coloured Hawk	A	E	E	
Rufous Crab-Hawk	A	E	E	
Common Black-Hawk	B	E	C	
Mangrove Black-Hawk	A	E	E	
Mantled Hawk	A	E	E	
Grey Hawk	B	E	C	
Grey-lined Hawk	A	E	E	
Red-shouldered Hawk	C	C	B	
Ridgway's Hawk	A	E	E	
Broad-winged Hawk	B	C	E	
Short-tailed Hawk	B	E	E	x
White-throated Hawk	A			x
Swainson's Hawk	C	B	B	
Galápagos Hawk	A	E	E	
Puna Hawk	A			x
Zone-tailed Hawk	B	E	C	
Hawaiian Hawk	A	E	E	
Red-tailed Hawk	C	B	B	
Rufous-tailed Hawk	A	A	A	

Species	Global range	Global temperature	Global humidity	Montane
Common Buzzard*	E	B	C	
Eastern Buzzard	C	B	C	
Himalayan Buzzard	A			x
Mountain Buzzard	A			x
Madagascar Buzzard	A	E	E	
Long-legged Buzzard*	B	D	A	
Upland Buzzard	A	C	A	
Ferruginous Hawk	B	D	A	
Rough-legged Buzzard*	C	B	B	
Red-necked Buzzard	B	E	C	
Augur Buzzard	A			x
Archer's Buzzard	A			x
Jackal Buzzard	A	E	A	

Old World vultures

The six genera and 13 species of Old World vultures are derived from recent genetic studies (Seibold and Helbig, 1995; Lerner and Mindell, 2005; Johnson *et al.,* 2006; Griffiths *et al.,* 2007). They are closest genetically to the snake-eagles. The large, solitary vultures are related to each other and form a separate group from the more sociable griffons of the genus *Gyps.* The Red-headed Vulture *Sarcogyps calvus,* Eurasian Black Vulture and Lappet-faced Vulture *Torgos tracheliotus* are geographical replacements, in south and South-east Asia, the mid-latitude belt and sub-Saharan Africa respectively. These species may feed alone but will also attend gatherings with the social griffons. A fourth species, smaller than the rest, is the White-headed Vulture *Trigonoceps occipitalis.* It occurs over much of the geographical area and habitat of the Lappet-faced Vulture.

The large vultures are specialist scavengers that consume sinews and other tough body parts with the aid of powerful beaks. The two African species are known to kill prey (Brown, 1971; Houston, 1979). All four species have a preference for nesting in trees, which may keep them away from the griffons in areas where there are no cliffs for the latter to nest on. Of the two species that reach the Western Palearctic, the Lappet-faced Vulture is restricted by climatic tolerance, being a species of warm climates. The Eurasian Black Vulture, on the other hand, is among the most bioclimatically tolerant of all Old World vultures (Table 12.1). The large solitary vultures would appear to represent geographical separation by isolation from a common, but distant, ancestor. The northernmost species, the Eurasian Black Vulture, is largely sedentary. Although bioclimatically tolerant for a vulture, it is restricted to southern parts of the Palearctic and is primarily a species of warm–temperate climates.

The Hooded Vulture *Necrosyrtes monachus* is a genetically distinctive vulture that is nevertheless closest to the *Gyps* vultures and not, as often considered, to the Egyptian Vulture *Neophron percnopterus* (these two species are an example of convergent evolution).

The eight *Gyps* vultures may represent four distinct evolutionary splits, and are not all geographical or ecological equivalents. The most distinctive, and the species that first separated within the genus, is the Indian White-rumped Vulture, *G. bengalensis.* Next is the Himalayan Griffon *G. himalayensis,* which is very close to the easternmost populations usually attributed to the Eurasian Griffon, *G. fulvus fulvescens*; it is therefore possible that *fulvescens* belongs with *G. himalayensis* and not with *G. fulvus.* The next to separate out is the African White-backed Vulture *G. africanus.* These species represent an early geographical expansion of *Gyps* across Africa and southern Asia, including the Himalayas, with significant subsequent geographical isolation.

The remaining species represent a more recent radiation. The Eurasian Griffon Vulture *G. fulvus* and Rüppell's Vulture *G. rueppellii* are very closely related and are likely geographical counterparts. The remaining species, though less closely related, appear to be geographical equivalents: the Cape Griffon *G. coprotheres* in South Africa, the Indian Vulture *G. indicus* in India, and the Slender-billed Vulture *G. tenuirostris* in northern India and South-east Asia.

Booted eagles

The phylogenetic relationships of the booted eagles are complicated and do not follow the classic subdivisions of the main genera (*Aquila, Hieraaetus, Spizaetus*) that were based on morphology and plumage (Helbig *et al.*, 2005; Lerner and Mindell, 2005). Here I adopt a classification based on these recent studies. The single-species genera *Morphnus* (Crested Eagle), *Harpia* (Harpy Eagle) and *Harpyopsis* (New Guinea Eagle) are distinct from the rest of the booted eagles. The Asian Hawk-eagles, formerly *Spizaetus*, are best considered a separate genus – *Nisaetus*. The New World Hawk-eagles (*Spizaetus, Oroaetus, Spizastur*) are a distinct and separate cluster of eagles from the Asian group. The Western Palearctic booted eagles are members of five distinct evolutionary-ecological lineages. Within the closely related genera *Aquila* and *Hieraaetus* three of the four main lineages are separated by size, which indicates underlying differences in prey taken, and the fourth by bioclimate and habitat. Within the lineages themselves habitat and bioclimatic tolerance separate the different species.

(a) **Large generalist *Aquila* group** The ecological and geographical space occupied by the largest eagles is divided among species and genera, suggesting an ancient radiation from a common ancestor that would have lived in warm, wet forest. This group is made up of five large species that replace each other geographically and ecologically: Gurney's Eagle *Aquila gurneyi* of New Guinea and South-east Asia; Wedge-tailed Eagle *A. audax* of Australia; Verraux's *A.verrauxi* of the Rift Valley; Golden Eagle *A. chrysaetos*, which has a Holarctic distribution; and Cassin's Hawk Eagle *A. africana*, the smallest species in the group, of the African rain forest. The absence of *Aquila* from Central and South America may be related to an early occupation of this region by other large eagles (e.g. Crested *Morphnus guianensis,* Harpy *Harpia harpya*). Their absence from woodland and savanna in sub-Saharan Africa may be for similar reasons, these habitats being occupied by Crowned *Stephanoaetus coronatus* and Martial *Polemaetus bellicosus* Eagles; and the limited distribution in New Guinea and south-east Asia by the presence of New Guinea Eagle *Harpyopsis novaeguineae* and the Philippine Eagle *Pithecophaga jefferyi.*

The Golden Eagle, the species in the subfamily with the widest bioclimatic tolerance (Table 12.1) and capable of hunting over a wide range of open terrain, evolved in response to the seasonal and open environments that came to cover huge areas of the Holarctic in the Plio-Pleistocene. It is the only species in this group, because of the high latitudes occupied, that has some migratory populations. These are from the most northerly and continental parts of the range. At the same time, its high bioclimatic tolerance makes it less migratory than other eagles occupying similar latitudes. Its ability to nest in rocky habitats, a feature that is unusual among a largely tree-nesting subfamily, would have permitted the exploitation of treeless habitats. The second Western Palearctic species in this group only reaches the far south. Verreaux's Eagle is the Golden Eagle's ecological counterpart along the mountainous ground of the Rift Valley down to southern Africa and, like the Golden Eagle, also nests in rocky terrain. It is bioclimatically less tolerant than the Golden Eagle, occupying the warm part of the temperature gradient (Table 12.1).

(b) **Medium-sized generalist *Aquila* group** Two very closely related species, until recently regarded as subspecies, take up the medium-sized booted eagle ecological niche. Bonelli's Eagle *A. fasciata* and African Hawk Eagle *A. spilogaster* occupy the southern part of the mid-latitude belt and savanna habitats in sub-Saharan Africa respectively. The Sahara and Arabian Deserts separate the two species today. Part of the Bonelli's Eagle's success in the mid-latitude belt may have to do with its habit of nesting in rocky habitat, which has permitted its survival in many parts where trees are absent or thin on the ground. Even so, Bonelli's Eagles do occasionally nest in trees (pers. obs.). The African Hawk Eagle nests in trees, and this habit has allowed it to spread across large areas of savanna where there is no rocky ground.

These medium-sized eagles represent a later radiation from a common ancestor of the large eagles, which were successful hunters of birds and small mammals over savanna and open terrain. With the loss of savanna with suitable trees for nesting across large areas of the northern and eastern parts of the joint geographical range during the Plio-Pleistocene, the populations split into tree and rocky-ground nesters. Bonelli's Eagle developed a wider bioclimatic tolerance than the African Hawk Eagle (Table 12.1), which aided its survival away from the tropics, although never as far as the more climatically tolerant Golden Eagle. Bonelli's Eagle's

▲ Among the earliest lineages to diversify were the nightjars, which had already split from the swift-hummingbird line in the Cretaceous, before 65 million years ago – a lineage that predates the K/T event. Red-necked Nightjar *Caprimulgus ruficollis* by Clive Finlayson.

▼ The storks were among the waterbird lineages that started to emerge in the Palaeocene (65–55.8 mya). The Ciconiidae emerged in the Early Miocene (*c.* 20 mya), and the genus *Ciconia* in the Middle Miocene (*c.* 17 mya). Black Stork *Ciconia nigra* by Clive Finlayson.

▲ Herons, pelicans and ibises diverged during the Early Eocene as part of the diversification of waterbirds. The Ardeidae first appeared in the Late Oligocene–Early Miocene (*c.* 25–20 mya) and the genus *Ardea* in the Early Miocene. Purple Heron *Ardea purpurea* by Clive Finlayson.

▲ The tern lineage separated from the gulls very early in the Tertiary, in the Late Palaeocene (*c.* 60 mya). Tern radiation started in the Eocene but accelerated in the Middle Miocene (*c.* 17 mya) and continued into the Pliocene. Whiskered Tern by Clive Finlayson.

▼ The Otididae emerged during the aridification of the northern hemisphere in the Middle Miocene (*c.* 17 mya). *Otis* appeared in the Middle Miocene with other bustard genera emerging later. Great Bustards *Otis tarda* by Clive Finlayson.

▲ Among the lineages that diversified during the Miocene aridification were the owls. *Strix*, *Bubo* and *Asio* split and radiated around 10 mya. Short-eared Owls *Asio flammeus* by Clive Finlayson.

▼ The greatest evolutionary activity in the Miocene took place among the recently evolved passerines. *Sylvia* warblers, rock thrushes, trumpeter finches, pipits and larks diversified into the new open habitats that were expanding at the expense of the forests. Rufous-tailed Rock Thrush *Monticola saxatilis* by Clive Finlayson.

▲ The Pliocene (5.33-2.6 mya) saw the continuation of the Miocene aridification and further adaptation of passerines to the opening up of the forests. These radiations included *Locustella* warblers, tits, *Turdus* thrushes, dippers, sparrows and shrikes. Woodchat Shrike *Lanius senator* by Clive Finlayson.

▼ The Pleistocene (2.6–0.01 mya) saw the tail-end of the Miocene–Pliocene explosion of avian diversity. Lineage splits continued among closely-related species. The glaciations of this period separated geographical populations of species like the Lesser Kestrel *Falco naumanni*. Clive Finlayson.

▲ The Common Raven *Corvus corax* is one of only fifteen Palaearctic species that have been able to adjust to the entire range of available bioclimates while remaining largely resident. It is an intelligent and versatile mixed-strategy carnivore with a multi-latitude, pan-palearctic breeding range. Once taken up by one species, this strategy allows it to become widespread, and effectively closes off the option to others in the genus. Clive Finlayson.

▼ The Black Kite *Milvus migrans* exhibits a successful strategy that has been adopted by 19 other species in the Palearctic, combining the highest level of bioclimatic tolerance with migratory behaviour. Most such species are mixed-strategy carnivores, omnivores or insectivores with large pan-palearctic ranges. Clive Finlayson.

▲ The White-fronted Goose *Anser albifrons* represents a specialised strategy that is found in 42 Palearctic species. These are species with the narrowest bioclimatic tolerances and breeding ranges that are confined to the Arctic. All species are fully migratory. Species with these characteristics are dominated by shorebirds and geese and tend to be omnivorous or herbivorous (the latter a rare strategy among Palearctic birds). Clive Finlayson.

▼ The Cream-coloured Courser *Cursorius cursor* is typical of the 65 Palearctic species that have the narrowest bioclimatic tolerances and breeding ranges, limited to the extreme south of the region. With a few exceptions, these birds are largely resident and are derived from a diversity of lineages. Spanish Nature/Peter Jones.

▲ Of 134 Palearctic species that partly or wholly breed in mountains, 88 do so along the chain of mountains that make up the mid-latitude belt. The White-winged Snowfinch *Montifringilla nivalis* is a typical example. It has a narrow bioclimatic tolerance and an omnivorous diet that permits it to occupy harsh, rocky mountain peaks and slopes throughout the year. Such species find suitable habitat across the mid-latitude belt, from Iberia to the Himalayas, and so have pan-palearctic ranges within a narrow latitude corridor. Roberto Ragno.

▼ The Ortolan Bunting *Emberiza hortulana* is one of 23 species that have split latitudinal ranges between the boreal and mid-latitude belts. A further five species have a similar distribution but in the Arctic belt. The populations that breed in the mid-latitude belt tend to occupy high mountain zones, bioclimatically equivalent to latitude bands to the north. These are species that must have had larger ranges during glacial periods. With post-glacial warming their ranges retreated northwards and up mountain ranges. Like the Ortolan, a good proportion of these species are migratory omnivores of open habitats that move in to exploit spring and summer abundance of food in these extremely seasonal environments. Clive Finlayson.

▲ Throughout this book I have emphasised the importance of the heterogeneous mid-latitude belt as a reservoir of avian species diversity. The main species occupying this belt are those that, like the Rock Sparrow *Petronia petronia*, inhabit open habitats. They are invariably species of limited bioclimatic tolerance with a range of strategies, among which resident omnivores, among them the Rock Sparrow, predominate. Clive Finlayson.

▼ A second important group of the mid-latitude belt are those of mixed habitats, usually open ground with shrub cover. The *Alectoris* partridges form a genus that occupies the length of the belt, with regional speciation. The Barbary Partridge *Alectoris barbara* is the north-west African representative. Like many in this habitat group, it is a sedentary omnivore. Clive Finlayson.

▲ The third most important habitat category among the mid-latitude belt species is wetland. The Black-winged Stilt *Himantopus himantopus* typifies a group of largely migratory mixed-strategy carnivores and omnivores. This species is also typical in that its heavily fragmented range reflects the present-day disconnection between wetlands across this belt. Clive Finlayson.

▲ Woodland and savanna once dominated the mid-latitude belt between Iberia and the Far East. The increasing seasonality and aridity that affected central parts of the belt, largely due to the uplift of the Tibetan Plateau, fragmented this habitat. Vast areas were taken up by steppe and desert. The Iberian Azure-winged Magpie *Cyanopica cooki* (above) and its Far-eastern sister species, the Asian Azure-winged Magpie *C. cyanus*, represent the extremes of a once-continuous range that became severed in the Pliocene. Clive Finlayson.

▼ Though fewest in number of species (45), many birds of the rocky habitats of the mid-latitude belt are distributed from the Iberian Peninsula to the Himalayas. In some cases, like the Bearded Vulture *Gypaetus barbatus* (below), these birds also occupy the southern extension of rocky habitats down to southern Africa. Clive Finlayson.

▲ The Cattle Egret *Bubulcus ibis* colonised South Africa and the Americas from the Palearctic in the 20th century. This expansion hints at the powers of dispersal of the herons, which may account for the global ranges of many species. The Cattle Egret is a bioclimatic semi-generalist with a fragmented mid-latitude belt range. Though mainly sedentary, populations in the continental interior of the Palearctic are migratory. A combination of bioclimatic tolerance and a capacity for long-distance movement may be the key to the Cattle Egret's recent success. Stewart Finlayson.

▼ The Snowy Owl *Bubo scandiacus* has adapted to living in the tundra, with its success dependent on migration. A bioclimatic semi-specialist, the Snowy Owl is the only regular migratory species in its genus. It is among the birds that has a clear signal of southward range displacement during glacial periods, with subfossil remains having been excavated as far south as Gibraltar. Stefan McElwee.

▲ The fish eagles fall into two groups, the tropical and the non-tropical species. The latter include three Palearctic species – White-tailed Eagle *Haliaeetus albicilla*, Pallas's Fish Eagle *H. leucophrys* and Steller's Sea Eagle *H. pelagicus*. The White-tailed Eagle is a bioclimatic semi-generalist with a pan-palearctic range that has relied on a combination of mixed-strategy carnivory and migratory behaviour for its success. Its absence from many parts of Western Europe is historical and due to human causes. Pallas's Fish Eagle is a bioclimatic semi-specialist of warm, dry, regions of the Central Palearctic while Steller's Sea Eagle is a bioclimatic semi-specialist of the boreal belt in the Eastern Palearctic. Ian Fisher.

▼ The *Sitta* nuthatches are a largely southern Asian genus with many species clustered around the Himalayas. One group of closely related species includes the Corsican Nuthatch *Sitta whiteheadi* (below). Six have localised ranges within the Palearctic and southern Asia while a seventh (Red-breasted Nuthatch *S. canadensis*) is widespread across North America. The Red-breasted Nuthatch was able to spread widely in the absence of competitors, while the nuthatches of the mid-latitude belt became isolated from each other in a world in which forests were patchily distributed. Stephen Daly.

▲ *Sylvia* is among the few predominantly Palearctic species-rich genera. It has a complex history of differentiation, which started in the Miocene. Later periods of aridification in the mid-latitude belt appear to have contributed to isolation and speciation. One such period, around 3.4–3.1 million years ago, led to a lineage in the central Mediterranean that gave rise to five species, including the Subalpine Warbler *Sylvia cantillans* (above). Clive Finlayson.

▼ A small number of seabird genera include species that are endemic to the temperate and warm waters of the western Palearctic. The genus *Calonectris* has three species, all Palearctic, including Cory's Shearwater *C. diomedea*, which is exclusive to the Western Palearctic. These birds have succeeded by performing long migrations, returning to the breeding grounds to capitalise on spring and summer marine productivity. Clive Finlayson.

▲ A recurring theme in this book is the success achieved by species from tree-nesting lineages that have been able to switch to nesting on rock faces, habitats that are relatively widespread in the Palearctic. The *Aquila* eagles are a perfect example. Most are tree-nesters but three, among them Golden *A. chrysaetos* and Bonelli's *A. fasciata* (above), are predominantly cliff-nesters, and have colonised large areas of the Palearctic. Bonelli's Eagle's sister species – the African Hawk Eagle *A. spilogaster* – remains a tree-nester on savanna habitats. Clive Finlayson.

▼ Auks are an ancient lineage of birds with origins in the Palaeocene (around 60 million years ago). The lineage radiated in the Atlantic and Pacific within the northern hemisphere. The Atlantic Puffin *Fratercula arctica* (below) is one of three species in a genus that has its origins in the Pacific low Arctic around 50 million years ago. Wide-ranging and migratory behaviour has often accompanied success in the auks, which exploit seasonal concentrations of pelagic fish in otherwise vast spaces of open ocean. Clive Finlayson.

▲ Sylvioids, muscicapoids, passeroids and a number of terrestrial non-passerine lineages have succeeded in the Palearctic with a combination of low bioclimatic tolerance and high migratory activity. Today the smaller passerines outnumber the larger non-passerines, such as the European Bee-eater *Merops apiaster*, which were the dominant terrestrial birds of the pre-Miocene Palearctic. With the arrival and diversification of passerines in the Palearctic in the Miocene, these non-passerines seem to have been pushed to the ecological margins. Clive Finlayson.

generalised diet allowed it to remain a resident of these places, as many southern populations of Golden Eagle were also able to do.

(c) **Large savanna-steppe *Aquila* group** This is a sister group to the other four booted eagle groups. It consists of five closely related large eagles, all larger than group (b) species and overlapping in size with group (a), though not reaching the dimensions of the largest species in that group (Golden, Wedge-tailed). They are separated from these other large eagles by habitat. These species have radiated within warm to temperate arid bioclimates, and are at home in open savannas, steppe and even desert (Table 12.1). They represent another solution to the opening of habitats in the Plio-Pleistocene. The Steppe Eagle *A. nipalensis* is the most divergent lineage within the group. This species occupies large areas of steppe across Eurasia today and may have reached west as far as France in the Late Pleistocene. It is therefore separated ecologically from the other species, some of which it overlaps geographically with, in the group by habitat.

The remaining species are geographical replacements of each other: the Imperial Eagle *A. heliaca* across a large part of the central and eastern mid-latitude belt, the Spanish Imperial Eagle *A. adalberti* in the extreme west of this belt, the African Tawny Eagle *A. rapax* to the south across much of Africa, and the Asian Tawny Eagle *A. vindhiana* in South Asia. The Imperial Eagle reached west to Italy with the expansion of open habitats during cold and dry episodes of the Pleistocene. The response to surviving in the most open and seasonal environments of the steppes and savannas of the mid-latitude belt has been migration. Steppe Eagle is fully migratory and most Imperial Eagles, excepting some populations in the south and west of the range, are also migratory. In contrast, the southern and western species (Spanish Imperial, African and Asian Tawny) are sedentary. Thus the migratory response is seen between and within species following a similar pattern. The ability to occupy open, seasonal environments has made this group highly successful in the Palearctic and they must have been more abundant in the Western Palearctic during cold and dry episodes of the glacials, when forests were reduced at the expense of savanna and steppe.

(d) **Small savanna *Hieraaetus* group** The five species in this group constitute the genus *Hieraaetus*. They are the smallest of the booted eagles and occupy the warm end of the climatic gradient, avoiding extreme aridity and humidity. All are species of open woodland and savanna and are geographical replacements of each other, with the exception of the two larger species, Wahlberg's Eagle *H. wahlbergi* and Ayres's Hawk-eagle *H. ayresii,* which are separated by size and therefore prey taken. The Booted Eagle *H. pennatus* occupies a large area of the mid-latitude belt. It is, ecologically, the small counterpart of the Imperial and Spanish Imperial Eagles. This northernmost species has the widest bioclimatic tolerance in the group (Table 12.1), as we would expect, and is highly migratory, with only the south-westernmost populations (in Iberia and Morocco) having individuals that remain in the breeding area all year. The New Guinea Hawk-eagle *H. wesikei* and the Little Eagle *H. morphnoides* occupy similar ecological niches in New Guinea and Australia (the African and South-east Asian rain forests are not occupied).

(e) **Small to medium-sized wet forest *Lophaetus* group** The *Lophaetus* eagles are a sister group of the other booted eagles (Helbig *et al.*, 2005; Lerner *et al.*, 2005; Griffiths *et al.*, 2007). They are a group of forest species that are at home along edges, close to glades, wet meadows, rivers and mangroves but also occasionally in dry mountain and hillside forest. The five species are largely geographical replacements although there is overlap in breeding range between the most recently separated pair, the Greater Spotted *L. clanga* and Lesser Spotted *L. pomarina* Eagles (Cramp, 1980). This pair may have split around 1 mya (Seibold *et al.*, 1996) and, where they overlap, the Greater Spotted Eagle tends to hunt in wetter and more open habitats than the Lesser, although both species nest within well-structured forests (Lõhmus and Väli, 2005).

The first species to split within this group was the Long-crested Eagle *L. occipitalis* of sub-Saharan Africa, which is also the smallest species, being comparable in size to the *Hieraaetus* eagles. The south and south-east Asian Black Eagle *L. malayensis* was the next species to branch out, while the Indian Spotted Eagle *L. hastata* is thought to have split from the *L. clanga/pomarina* pair around 1.8 mya (Väli, 2006). Thus, the final separation of the tropical taxa from the Palearctic ones seems to have occurred at around the start of the Pleistocene.

The absence of *Lophaetus* across huge tracts of the mid-latitude belt would seem to indicate that they were ecologically replaced in the more open environments of the belt by other booted eagles, most notably those of groups (c) and (d). These eagles have narrow bioclimatic ranges of tolerance (Table 12.1) which, together with their rather specific habitat preferences, may have promoted their geographical separation. Unlike other authors who have suggested a southern origin for this group with subsequent range expansion northwards (Helbig *et al.*, 2005), I prefer to view this group as another example of an ancestral forest dweller with a large geographical range covering much of Eurasia and Africa that became progressively split into regional populations, eventually species, by the cooling and drying climate and consequent loss of forest cover across much of the mid-latitude belt during the Plio-Pleistocene.

Not surprisingly, with life in the northern forests came the need to migrate south in the winter. The Lesser Spotted Eagle is wholly migratory, wintering in tropical East Africa while the Greater Spotted Eagle has fully migratory and partially migratory populations, some reaching the northern parts of east Africa. In Africa the wintering *Lophaetus* eagles are potentially in contact with the Long-crested Eagle. They appear to be separated by habitat, the local resident in forest and the migrants in savanna and open woodland (Brown, 1971).

Chanting goshawks

Four species of chanting goshawks comprise this subfamily (Lerner and Mindell, 2005). They are African species of warm bioclimates and restricted tolerance (Table 12.1), at home in savanna, shrubland and grassland country. The Dark Chanting Goshawk *Melierax metabates* reaches the extreme south-west of the Western Palearctic, where a population survives north of the Sahara Desert in Morocco (Thévenot *et al.*, 2003). All chanting goshawks are resident.

Accipiters and harriers

Here I follow recent phylogenetic analyses that indicate that the harriers form part of a larger grouping that includes *Accipiter* (Lerner *et al.*, 2008). This large group, most closely related to the milvine kites and the fish eagles (Griffiths *et al.*, 2007), is therefore divided into two genera: *Circus* with 14 species and *Accipiter* with 49 species. The relationships of the various species of *Accipiter* are poorly understood so any interpretation must be tentative. Recent studies (Wink and Sauer-Gurth, 2004; Griffiths *et al.*, 2007; Lerner *et al.*, 2008) would seem to point to an early separation of the east Asian-New Guinean species (Brown Goshawk *A. fasciatus,* Collared Sparrowhawk *A. cirrocephalus,* Japanese Sparrowhawk *A. gularis,* Imitator Sparrowhawk *A. imitator,* Grey Goshawk *A. novaehollandiae*) from the other species. The closest relatives may be the African species (African Goshawk *A. tachiro,* Rufous-chested Sparrowhawk *A. rufiventris*) which would provide a link with a Eurasian-North American group including the Eurasian Sparrowhawk *A. nisus* and the North American Sharp-shinned Hawk *A. striatus.*

The group of large goshawks (Northern Goshawk *A. gentilis,* Meyer's Goshawk *A. meyerianus,* Henst's Goshawk *A. henstii,* Black Goshawk *A. melanoleucus*) form a separate cluster with the Neotropical Cooper's *A. cooperii* and Bicolored Hawks *A. bicolor,* which may suggest an ancestry in that continent with subsequent radiation to Africa, Asia and New Guinea. The large goshawk body appears to have evolved by a process of heterochrony, that is by a change in the timing or rate of development, from a smaller hawk (Cubo and Mañosa, 1999).

Recent studies show that the *Circus* harriers are nested within the accipiters and are therefore part of this group. One such study places *Circus* between the east Asian-New Guinean accipiters and the Eurasian Sparrowhawk (Griffiths *et al.*, 2007). This observation suggests that the harriers evolved from short-winged forest accipiters to become long-winged open-country species. The Spotted Harrier *C. assimilis* of Australia and South-east Asia separates out from other *Circus* harriers first. The Hen Harrier *C. cyaneus* splits next from the remaining harriers; the North American and Eurasian races of this widespread harrier are considered separate species, on account of degree of genetic difference, by a number of authors (Wink and Sauer-Gurth, 2004).

The remaining harriers cluster into two groups: the marsh harriers share a common ancestry and represent a radiation across the Eurasia, Africa and New Guinea (Wink and Sauer-Gurth, 2004). The other group is composed of three species: the Palearctic Pallid Harrier *C. macrourus,* the African Black Harrier *C. maurus* and Neotropical Cinerous Harrier *C. cinereus,* suggesting a separate and widespread geographical radiation. Montagu's Harrier *C. pygargus* and the Long-winged Harrier *C. buffoni* occupy intermediate positions between the two major clusters (Wink and Sauer-Gurth, 2004). In sum, the four Western Palearctic species (Hen, Pallid, Montagu's and Western Marsh *C. aeruginosus*) are not part of a single group of closely-related harriers but are instead members of groups that represent separate radiations.

Milvine kites

The milvine kites, comprising the genera *Milvus* and *Haliastur*, share a common ancestry with the fish eagles of the subfamily Haliaeetinae (Lerner and Mindell, 2005; Griffiths *et al.*, 2007). Of interest to us is the genus *Milvus,* two of the three species occurring in the Western Palearctic. The evolutionary relationships and taxonomy of *Milvus* require work, especially as the different species and subspecies are very close to each other genetically (Schreiber *et al.,* 2000; Scheider *et al.*, 2004; Johnson *et al.,* 2005).

The Yellow-billed Kite *Milvus aegypticus* of central, west and north-eastern Africa is the most distinct of the *Milvus* kites. The Red Kite *M. milvus* would seem to be the product of a relatively recent speciation, almost certainly in the Iberian Peninsula (Schreiber *et al.,* 2000) or possibly north-west Africa. The colonisation of north-western Europe would have been a post-glacial expansion. Intriguingly, Red Kites (including the Cape Verde form) appear closest genetically to southern African and Madagascar Black Kites (*M. migrans parasitus*) and then to other Black Kite lineages. One possible, and tentative, explanation might be that there was an early split in the lineage that separated the ancestor of the Yellow-billed Kite from the common ancestor of the Red and Black Kites. The Red Kite emerged from a split in the southern, resident Black Kite populations that may have separated those to the north (which evolved into Red) from those to the south (Black) of the Sahara. The Cape Verde Red Kites represent the southernmost populations of the northern group.

Fish eagles

The classification of the fish eagles is less complicated than that of the booted eagles. Here, I include the Lesser Fish Eagle *humilis* and the Grey-headed Fish Eagle *ichthyaetus* within the genus *Haliaeetus* (formerly *Ichthyophaga*), in keeping with recent genetic studies (Lerner and Mindell, 2005). Three species breed in the Palearctic, White-tailed Eagle *H. albicilla,* Pallas's Fish Eagle *H. leucoryphus* and Steller's Sea Eagle *H. pelagicus.* Only the first of these breeds in the Western Palearctic and it is genetically closest to the American Bald Eagle *H. leucocephalus,* with a relatively more recent split between these two species in comparison to the other northern species.

The fish eagles constitute a group of species that has colonised all continents except South America. The different species are geographical replacements of each other and there is a deep lineage split between tropical species and northern ones (Lerner *et al.,* 2005). Within the tropical subgroup, the African Fish Eagle *H. vocifer* and the Madagascar Fish Eagle *H. vociferoides* are very close. Their nearest relatives are two south and south-east Asian species, the Lesser Fish Eagle and the Grey-headed Fish Eagle, which live close to wetlands within forests. These four species cluster with the White-bellied Sea Eagle *H. leucogaster*, which has a wide distribution from coastal India to Australia, and Sanford's Sea Eagle *H. sanfordi*, which is endemic to the Solomon Islands. The northern group consists of two very closely-related species, the Bald Eagle of North America and the White-tailed Eagle of the Palearctic. Pallas's Fish Eagle, with a Central and Eastern Palearctic distribution south of the White-tailed, and Steller's Sea Eagle of the north-east Asian Pacific coast complete the northern group. This latter species overlaps significantly geographically with White-tailed Eagle, but its larger size, one of the largest of all raptors, separates it ecologically.

The distribution of the fish eagles is consonant with a widespread common ancestor of the group, which may have lived in the Pliocene or even the Middle Miocene (Wink *et al.,* 1996). The closest species pairs, Bald/White-tailed and White-bellied/Sanford's may have diverged 1 mya and 150,000 years ago respectively. The 1 mya estimate depends heavily on the calibration of the mitochondrial DNA clock; the separation may have occured as long ago as 3–4 mya. Either way, the Bald Eagle must be the descendant of a population that colonised North America from Asia, where we find the greatest diversity of species of this genus.

As was the case with the booted eagles, the split between the groups would have started with the climatic deterioration during the Plio-Pleistocene. Northern species survived in part by becoming migratory. Coastal (Bald, White-tailed, Steller's) and southern/western (Bald, White-tailed, Pallas's) populations were able to remain sedentary. In addition, in a group of species with low bioclimatic tolerance, the Bald and the White-tailed Eagles occupy a broad bioclimatic range that is comparable to the Golden Eagle. The fish eagles should be viewed as complementary to the large *Aquila* eagles, taking up the fish predator ecological niche in wetlands and coastal areas. The Western Palearctic fell within the domain of one species, the White-tailed Eagle.

Buzzards

Progress in unravelling the complex group of the buzzards, most closely related the fish eagles, has been made in recent years even though the relationships of certain species may still not be clear. I have followed the most recent genetic studies (Riesing *et al.*, 2003; Kruckenhauser *et al.*, 2004; Lerner and Mindell, 2005; Griffiths *et al.*, 2007; Lerner *et al.*, 2008; Amaral *et al.*, 2009), allocating the species into 17 genera and 59 species. The genus with most species is *Buteo* with 28, including the three species that occupy the Western Palearctic. This genus originated in South America (Riesing *et al.*, 2003) and colonised North America at least once, some time between 7.7 and 3.3 mya (Amaral *et al.*, 2009). From here *Buteo* colonized the Old World, presumably in moments when suitable habitat became available in Beringia (now submerged land between Alaska and Chukotka). Modern Old World buzzards seem to have evolved relatively rapidly, as populations were separated and reconnected by glacial cycles (Kruckenhauser *et al.*, 2004), making taxonomic separation difficult (Amaral *et al.*, 2009). The differences between races of the Common Buzzard *B. buteo* and the Long-legged Buzzard *B. rufinus*, for example, are not clear, and these buzzards should probably be considered members of a superspecies (Kruckenhauser *et al.*, 2004).

Across the Palearctic (island forms aside), we may consider the situation to include a group of species with broad latitude ranges (Common Buzzard in the west, Steppe Buzzard *B. b. vulpinus* to the east as far as the Yenisei River, and the Eastern Buzzard *B. japonicus* east of Lake Baikal); a second group, represented by the Long-legged Buzzard, occupies arid areas approximately to the south of these forms; a third, represented by the Rough-legged Buzzard *B. lagopus*, is to the north of the other two groups. The Himalayan Buzzard *B. refectus* and the Upland Buzzard *B. hemilasius* are mountain species.

Climate

The diurnal raptors (Pandionidae is included here) are bioclimatic specialists of warm and mainly humid climates (Figures 12.1–12.2). Those that have been successful in the Palearctic form a clear subset of this group and comprise the most bioclimatically tolerant species. These are spread across the temperature and humidity gradients reaching the coldest and driest climates (Figure 12.2). Bioclimatic tolerance is a key factor in their success in the Palearctic.

Elanine kites are species of warm climates outside which they do not venture. *Elanus* is particularly at home in dry conditions. All species show very narrow bioclimatic ranges. The Black-winged Kite is among

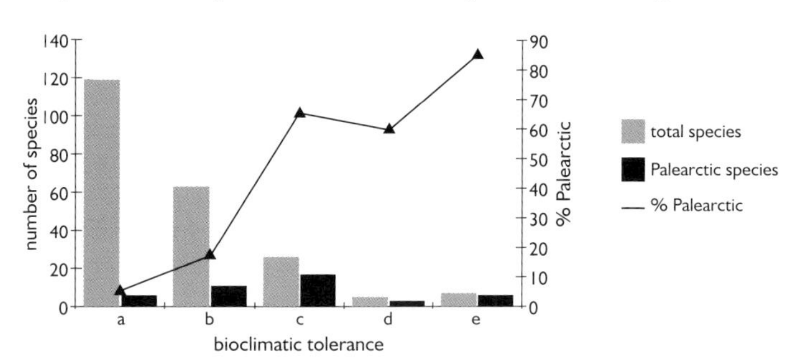

Figure 12.1 *Bioclimatic tolerance of raptors. Definitions as on Figure 4.1 (see p. 38).*

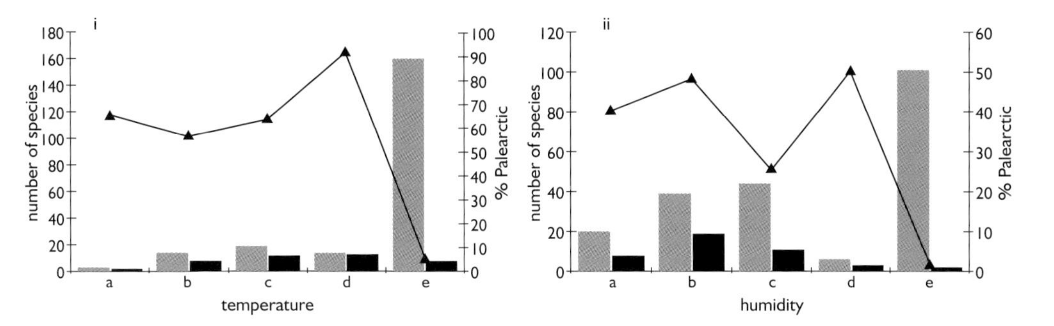

Figure 12.2 *Temperature (i) and humidity (ii) tolerance of raptors. Definitions as in Figure 4.2 (see p. 38). Key as in Figure 12.1.*

the more bioclimatically tolerant of the elanine kites and is less restricted to dry conditions than other elanines. It reaches the southern fringe of the Western Palearctic, strictly within the warmest bioclimates (Table 12.1). The pernine kites are species of warm/wet-humid climates (Table 12.1). Only three genera depart from this restricted pattern: *Lophoictinia* in warm/dry climates, *Hamirostra* from warm wet to dry, and *Pernis.* The latter genus has been the most successful at living in a wide bioclimatic range that has included temperate and cool climates (Table 12.1). The two Palearctic species, *P. apivorus* and *P. ptilorhynchus*, occupy the most central positions on the temperature and humidity gradients of all pernine kites, and they are the species with the widest global bioclimatic tolerances. The European Honey-buzzard, the only Western Palearctic pernine kite, occupies a fairly central bioclimatic position tending towards the dry part of the gradient (Table 12.1). The European Honey-buzzard has therefore succeeded away from tropical regions, which typify the pernine kites, by shifting its climatic tolerance from the warm/wet preference of the group towards a more central one.

There is an absence of pattern among the disparate Gypaetinae species other than a general avoidance of cool/cold climates (Table 12.1). The two Western Palearctic species are tolerant of a wider range of bioclimates than the other two species. They resemble the Eurasian Black and Griffon Vultures in range and bioclimatic position, with a tendency towards the warm and dry end of the spectrum. Snake-eagles are primarily species of warm/wet-humid climates (Table 12.1). Other than *Circaetus,* the Bateleur represents the only species to have broken away from this climatic regime, extending it to warm and dry climates. Four of the six *Circaetus* species also occupy a large portion of the humidity gradient within warm climates. The Short-toed Snake-eagle is the only snake-eagle to have colonised temperate climates (Table 12.1). Within the Western Palearctic the Short-toed Snake-eagle is tolerant of a range of climates, being more restricted than the White-tailed or Golden Eagles and comparable to the Booted Eagle (Table 12.1). The Short-toed Snake-eagle occupies the warm part of the climatic gradient and is close to the centre of the humidity gradient, tending towards humid. These observations are in accord with a species within an essentially warmth-loving group, which has managed to live away from the tropics. A diet based predominantly on snakes and other reptiles presumably limits the climatic tolerance of all snake-eagles.

Old World vultures belong to genera that predominate in warm climates (Table 12.1). Two genera, *Gyps* and *Aegypius*, have penetrated temperate and montane climates and it is from these that the widespread Palearctic species are derived. The Eurasian Black Vulture is, along with the Eurasian Griffon and the Indian White-rumped Vulture, the most bioclimatically tolerant of all the Old World vultures; it is a species of warm and dry bioclimates and also reaches into the montane zones (Table 12.1). The Eurasian Griffon has identical bioclimatic requirements to the Eurasian Black Vulture (Table 12.1). Within the relative bioclimatic limits of the Old World vultures, those that occupy the Western Palearctic are thus among the species with the broadest climatic range of tolerances.

There are 40 species of booted eagles in the world and they fall into 10 genera (Table 12.1). The majority are species of warm-wet/humid climates. The genera that have managed to occupy the widest climatic regimes, *Lophaetus, Hieraaetus* and *Aquila,* are those that are represented in the Palearctic Region. The

Palearctic eagles are therefore the most climatically tolerant of all the booted eagles. The main trend away from the warm-wet/humid climates is shown by genera that occupy the rainfall gradient, within warm climates, towards warm-dry climates; these climatically tolerant genera appear to have also had a degree of success in living in temperate-dry environments. Very few species appear to have been successful in cooler and wetter climates: the Mountain Hawk-eagle *Nisaetus nipalensis* which reaches Japan and Korea in the Eastern Palearctic; the Booted Eagle *Hieraaetus pennatus,* penetrating western Europe; and the widespread Golden Eagle, which occupies much of the Holarctic. A number of species have also locally occupied mountain climates, notably the three species just referred to, along with Bonelli's Eagle, and the African Verreaux's Eagle that penetrates the south-Western Palearctic. Five of the eight montane booted eagles thus occur in some part of the Palearctic, suggesting that climatic tolerance to mountain habitats has allowed them to reach further north than less tolerant congeners.

Ten of the 40 species of booted eagles breed within the Western Palearctic. Some species, for example the Golden Eagle, are widespread while others, like Verreaux's Eagle, are highly localised. The Western Palearctic booted eagles are birds of hot to warm climates (Table 12.1). Three of the 10 species appear to have very specific climatic tolerances that limit their distribution to the warmest areas of the Western Palearctic: the Spanish Imperial Eagle in the south-west and Verreaux's and Tawny Eagles in the south-east and south. Six species have broader climatic tolerances than these three species: Booted and Bonelli's Eagles are in warm climates; Greater Spotted, Lesser Spotted, Imperial and, especially, Steppe Eagles in drier climates than the former species. Finally, the Golden Eagle stands out by having the widest climatic range and occupying a central position along the temperature and climatic ranges (Table 12.1). This feature has permitted it to become widespread within the Western Palearctic and beyond, with this bird being the most cosmopolitan of all the booted eagles.

The overall picture from Table 12.1 is that the booted eagles that occupy the Western Palearctic are, in global terms, the most climatically generalised of the booted eagles (Golden Eagle, Bonelli's Eagle, Booted Eagle and Verreaux's Eagle). The first three are also widespread within the Western Palearctic. The remaining species that occupy the Western Palearctic are climatic specialists and largely occupy the warm/temperate-dry part of the climatic gradient.

Accipiters are predominantly species of warm and wet climates (Table 12.1). This diverse genus has managed to occupy all available bioclimates, however, although they become decidedly fewer away from warm climates. The harriers, in contrast, are less frequent in warm and wet climates and appear predominantly in warm and temperate dry bioclimates. Thus the radiation of the harriers from the accipiters is related to a shift along the humidity gradient, from wet to dry. The breadth of bioclimates tolerated accounts for the global spread of the two genera, and possibly explains the repeated radiations and colonisations that we have observed.

The three most bioclimatically tolerant accipiters/harriers have Western Palearctic populations: Eurasian Sparrowhawk, Northern Goshawk and Hen Harrier (Table 12.1). Three other Western Palearctic species also have relatively broad tolerances: Shikra *Accipiter badius,* Western Marsh Harrier *Circus aeruginosus* and Montagu's Harrier. Finally, there are two bioclimatic specialists that are typical of dry bioclimates: Levant Sparrowhawk *A. brevipes* and Pallid Harrier. (Table 12.1). The overall picture is of generalised species capable of living across a range of bioclimates, and a few specialists that are typical of dry bioclimates.

The milvine kites are primarily birds of warm climates (Table 12.1). The genus *Milvus* emerges as more tolerant of a wider range of conditions than *Haliastur*, occupying temperate and even cold climates. The two species that have been successful away from warm climates, Black and Red Kites, have the widest climatic tolerances of the Milvinae (Table 12.1). They have achieved this by shifting their bioclimatic position away from the warm end of the spectrum towards its centre. The Black Kite has a wider climatic tolerance than the Red Kite within the Western Palearctic (Table 12.1). Both occupy a similar position along the warm part of the gradient, but the Black Kite appears more tolerant of dry conditions than the Red Kite, which stands out as restricted to the wettest part of the gradient. This observation would appear to be in agreement with the putative origins of the Red Kite along the Atlantic rim of the continent.

Most fish eagles are species of warm-wet/humid climates (Table 12.1). Two species, White-tailed and Bald Eagle, have been able to exploit all available bioclimates except montane, and together occupy large areas of the Holarctic. The other Palearctic species, Pallas's Fish Eagle and Steller's Sea Eagle, have succeeded in colder or drier climates than the rest, having apparently become climate specialists in the process.The African Fish Eagle occupies a drier part of the gradient than the remaining warm climate species. Within the Western Palearctic the White-tailed Eagle shows a wide bioclimatic tolerance (Table 12.1). It occupies a cool and dry position in the climatic gradients.

The buzzards are birds primarily of warm and wet climates (Table 12.1), consistent with their tropical and equatorial South American origins. The genus *Buteo* stands out as having the broadest climatic tolerance, having occupied all bioclimates. Half of the species in this genus remain faithful to warm/wet climates but a significant number have moved along the temperature and humidity gradients, occupying warm and temperate dry climates, cool wet/humid and montane climates. It is this ability to occupy a wide range of bioclimates, including cold and dry, that permitted an expansion from the New to the Old World across high latitudes. The Common Buzzard has the broadest climatic range of all the buzzards, which must account for its widespread distribution across the Western and Central Palearctic. Within the Western Palearctic it occupies a central, tending towards cool, position on the bioclimatic gradient and it has the widest bioclimatic tolerance of the three *Buteo* species, with a range comparable with the most tolerant Western Palearctic raptors. The Long-legged Buzzard is more restricted than the Common Buzzard, occupying a warmer and drier position on the bioclimatic gradient, while the Rough-legged Buzzard is in the cold and humid part of the gradient.

Habitat

Diurnal raptors are species of forest and open woodland with a significant number also in open and wetland habitats (Figure 12.3). Palearctic species are noteworthy because of their association with rocky habitats, an unusual habit among these birds. In this respect they are very similar to falcons and owls. This suggests that the ability to nest in rocky habitats has been a significant feature that has allowed groups that otherwise nest in trees to colonise large areas of the Palearctic.

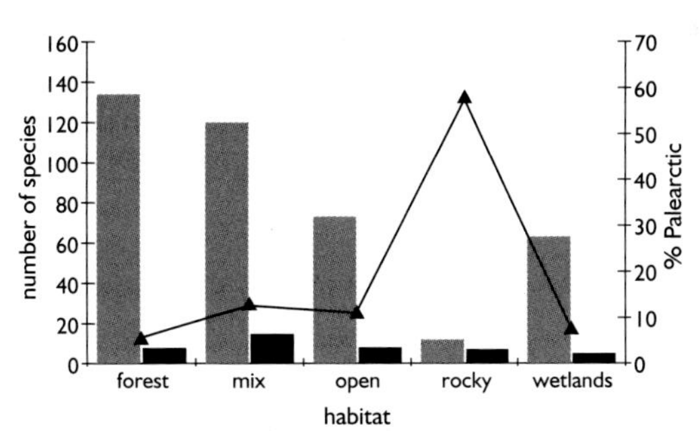

Figure 12.3 *Habitat occupation by raptors. Definitions as in Figure 4.3 (see p. 39). Key as in Figure 12.1.*

The elanine kites are species of open habitats, occupying predominantly savanna but also treeless habitats, desert and wetlands (Table 12.2). The Black-winged Kite is a typical bird of savanna habitat. In the Western Palearctic it has become particularly at home in the dehesas (grazed parkland) of south-western Iberia, which are structurally similar to wooded savanna (Finlayson, 2006). This species has benefited from the growth of this habitat through the clearing of Holm Oak *Quercus rotundifolia* woods (Molinero and Ferrero Cantisán, 1985; Balbontín *et al.*, 2008), indicating that geographical restriction within the warm bioclimates of south-western Iberia was due to lack of habitat availability.

The pernine kites are essentially birds of forest (Table 12.2). Some species have managed to colonise more open habitats and three genera have incorporated, presumably independently, wetlands into their

Subfamily	Species	Forest	Savanna	Shrubland	Treeless	Tundra	Desert	Rocky	Wetlands	Marine
Pandioninae	4	0	0	0	0	0	0	0	4	0
Elaninae	6	1	6	0	3	0	1	0	2	0
Perninae	16	14	5	3	1	0	0	1	3	0
Gypaetinae	4	2	1	0	1	0	1	2	1	0
Circaetinae	14	9	5	4	2	0	1	0	1	0
Aegypinae	13	2	11	2	9	0	4	4	0	0
Aquilinae	40	26	21	6	12	1	4	3	5	0
Melieraxinae	4	0	3	4	2	0	0	0	0	0
Accipitrinae	63	42	19	10	14	0	0	0	17	0
Milvinae	5	1	4	0	3	0	2	0	5	0
Haliaeetinae	10	1	0	0	0	0	0	0	10	0
Buteoninae	59	28	35	7	20	1	8	2	14	0

Table 12.2 *The subfamilies of raptors by habitat.*

repertoire. The species in *Pernis* have remained forest birds, indicating that the ability to live outside the tropics has been due to a change in climatic tolerance and not in habitat requirements. The European Honey-buzzard is a forest bird and is even restricted to this habitat when roosting on migration (Finlayson, 1992).

There is no pattern of habitat occupation among the four species of bearded vultures other than an association with trees in the two exclusively tropical African species and a dependence on rocky habitats for nesting in the two that have reached the Western Palearctic (Table 12.2). The Egyptian and Bearded Vultures breed in caverns, ledges and fissures within cliffs but they are not necessarily mountain species. The Bearded Vulture once nested on coastal cliffs, as at Gibraltar, but is now largely confined to mountains having been driven away from many areas by human pressure. The range of feeding habitat is governed by access from nesting or roosting sites and typically covers open ground, where mammalian herbivores, their main but not only source of food, are abundant.

Like the booted eagles, the snake-eagles are forest raptors (Table 12.2). Only *Circaetus* and *Terathopius* species have adapted to life in wooded savannas, shrublands and grasslands. Unlike some booted eagles, snake-eagles have not managed to occupy rocky habitats and all remain tree-nesting species. The Short-toed Snake-eagle has the broadest habitat range of all snake-eagles, occupying savannas, shrublands, grasslands, steppe and even desert. This feature may have enabled it to successfully exploit a wide range of ecological opportunities away from the tropics,

The Old World vultures are species of open country, inhabiting predominantly savanna and treeless country, where they largely feed on the carcasses of mammalian herbivores (Table 12.2). They are primarily tree-nesters but four species in the genus *Gyps* nest on cliffs. Availability of suitable nest-sites must therefore be the main limiting factor in the distribution of these vultures, although their ability to range widely in search of food may have compensated. The two Western Palearctic species are typical of the open habitats that characterise the Old World vultures. The Eurasian Griffon is largely a colonial cliff-nester and the larger Eurasian Black is a solitary tree-nesting species. They coincide in areas with trees and cliffs, over many parts of the Mediterranean region, for example, where limestone cliffs are often in close proximity to trees and open landscapes.

Booted Eagles are predominantly forest and savanna raptors (Table 12.2). Several genera tend towards open habitats, and it is noteworthy that the genera that reach the Palearctic seem to be those that have most successfully moved away from dense forest. Among these the *Aquila* eagles appear particularly well suited for savanna and treeless (e.g. steppe) habitats. Species in the genera *Aquila* and *Lophaetus* are also typical of wetlands. The number of habitats occupied by Western Palearctic eagles varies between one and four per species (Golden and Verreaux's Eagles). Nine of the 10 Western Palearctic eagles live in savanna habitats and then across a range of open or semi-open habitats, from treeless steppe and desert to shrublands, but only two species, Lesser Spotted Eagle and Booted Eagle, can be considered regular occupiers of forest. This observation confirms the view that the occupation of non-tropical areas by booted eagles has been successful in species that have prospered away from dense forest. The Golden Eagle is the only species to live on the tundra. Three species, Golden, Bonelli's and Verreaux's, are at home in rocky habitats. All three are in the genus *Aquila* and they are the only species that occupy such habitats, even outside the Palearctic.

There is a striking contrast between the largely forest-dwelling accipiters and the harriers that dominate in treeless habitats and wetlands (Table 12.2). Between them almost every habitat, excepting tundra, desert and rocky areas, has been occupied. The extremely specialised accipiter morphology, with short wings suitable for flight in dense forest vegetation, would have required greater modification in species colonising open habitats. Such changes would have been more drastic than in many of the less morphologically specialised groups of raptors that we have examined. The outcome, the harriers, therefore appear as departing more significantly from the ancestral forest accipiters than other open habitat raptors do from forest forms. The accipters of the Western Palearctic include forest species (Eurasian Sparrowhawk, Northern Goshawk) and those of more open, wooded savannah, country (Shikra, Levant Sparrowhawk). The four harriers are characteristic of treeless habitats and wetlands. Overall the Western Palearctic accipiters and harriers largely occupy the less-forested end of the habitat spectrum, Eurasian Sparrowhawk and Northern Goshawk being the exceptions, having colonised a broad range of temperate and boreal forests. In this regard it is interesting that such occupation of northern forests has involved only two species and that these differ significantly in size, indicating partitioning of food resources by size (van Beusekom, 1972).

Milvine kites are largely birds of open woodland, savanna and treeless habitats. Wetlands are habitats common to all species and some have also been able to live in arid habitats (Table 12.2). These habitat requirements appear to have predisposed the milvine kites to success in a wide range of environments. The Black Kite and the Yellow-billed Kite have the broadest habitat ranges of all the milvine kites, closely followed by the Red Kite, which differs by its absence from arid habitats. These kites occupy the greatest range of habitats of all the diurnal raptors.

The fish eagles are restricted in habitat to wetlands, including coastal habitats (Table 12.2). Two species, Sanford's Eagle and Lesser Fish Eagle, are restricted to such habitats within forest, but the rest do not appear to have such limitations. The distribution of the fish eagles is therefore governed largely, within their climatic tolerances, by the location of wetlands. Those species that have managed to exploit coastal habitats have been able spread along coastal corridors, while others presumably used rivers and other watercourses. The White-tailed Eagle's success has been the ability to occupy inland and coastal wetlands across a wide bioclimatic range. As wetlands have opened and closed in response to climate change and local geological factors (e.g. changes in the flow of major rivers), so the size of the Western Palearctic population must have fluctuated. The presence of huge inland lakes and seas in Siberia during the last glaciation, as the outflow of many rivers was blocked to the north by advancing ice (Finlayson and Carrión, 2007), must have marked a zenith for the White-tailed Eagle population.

The buzzards are primarily a group of open woodland and wooded savanna species, with a significant number also occupying forest, followed by treeless habitats and wetlands (Table 12.2). All habitats, except marine, have been occupied by buzzards and the genus *Buteo* stands out as having the widest habitat range of all buzzard genera. *Buteo* are primarily wooded savanna raptors that have expanded towards either end of the habitat gradient, equally into forest and treeless habitats (Table 12.2). The Common Buzzard has the broadest habitat range of the buzzards, occupying a wide range of wooded habitats, venturing into shrublands and treeless environments. The most restricted of the three is the Long-legged Buzzard, which is a species of open, treeless habitats, including arid areas.

	Global	Palearctic	Global %	Palearctic %	Palearctic of global %
Migratory	56	22	22.49	51.16	39.28
Sedentary	234	31	93.97	72.09	13.25
Total Species	249	43			

Table 12.3 *Migratory behaviour in diurnal raptors.*

Migratory behaviour

Diurnal raptors are largely sedentary, but the Palearctic species include a significantly higher proportion of migratory species than elsewhere (Table 12.3). This is remarkable given the energetic constraints that raptors face on migration and seems to have been an added factor, along with bioclimatic tolerance and ability to nest on rocky surfaces, in their success in the Palearctic.

The elanine kites are sedentary species (Table 12.4). The only species that performs migratory movements is the Black-winged Kite. In this species Western Palearctic populations, which are at the extreme northern edge of the geographical range, are partially migratory (Molinero and Ferrero Cantisán, 1985). Two genera of pernine kites have migratory populations (Table 12.4). One of the five *Aviceda*, the Black Baza *A. leuphotos*, is partially migratory with substantial numbers gathering at raptor crossing points (Decandido *et al.,* 2004). The other genus is *Pernis.* Two of the three species have migratory populations and these are the ones that breed in the Palearctic. Palearctic races of the Oriental Honey-buzzard are fully migratory, wintering in India, while tropical races are resident (Gamauf and Haring, 2004). The other is the strongly migratory European Honey-buzzard. Tropical *Pernis* are fully resident. Migratory behaviour in *Pernis* (at both subspecies and species levels) is therefore related with latitudinal location of the breeding grounds. The entire population of European Honey-buzzard winters in tropical Africa (Moreau, 1972). It is one of the more spectacular migrant raptors, with birds gathering at key crossing points in vast numbers (Finlayson, 1992; Bildstein, 2006). The European Honey-Buzzard is the only *Pernis* that has no resident populations.

Subfamily	Number of species	Resident	Partially migratory	Fully migratory
Pandion	4	4	2	2
Elaninae	6	6	1	0
Perninae	16	14	2	2
Gypaetinae	4	4	1	1
Circaetinae	14	14	1	1
Aegypinae	13	13	2	0
Aquilinae	40	37	6	5
Melierax	4	4	0	0
Accipitrinae	63	55	8	13
Milvinae	5	5	3	1
Haliaeetus	10	6	4	0
Buteoninae	59	51	10	9

Table 12.4 *The subfamilies of diurnal raptors by migratory behaviour.*

All four species of Palearctic vultures have resident populations and only the Egyptian Vulture has migratory and partially migratory populations (Table 12.4); these are invariably those along the northern edge of the breeding range, in Europe and eastwards across Central Asia (Cramp, 1980).

Immature Bearded Vultures are sometimes seen in association with other raptors on migration (pers. obs.) but observations are rare and suggest that migration is not usual in this species. Their scarcity may account in part for the paucity of records in traditional migration watch points. The Egyptian Vultures of the Western Palearctic are fully migratory, wintering in tropical Africa close to the southern edge of the Sahara (Moreau, 1972), although isolated individuals may remain year-round in countries such as Spain and Morocco (Finlayson, 1992). Arrival in the European breeding grounds is early in the year, some in January and most during February and early March, with a second passage of non-breeding immature birds between April and June (Finlayson, 1992).

The Short-toed Snake-eagle is the only migratory snake-eagle (Table 12.4). Even the African populations perform long-distance migratory movements in response to seasonal changes in the productivity of the savannas (Curry-Lindahl, 1981b). Migratory behaviour in this species is therefore linked to seasonality of food supplies, avoiding the cool northern winters in high latitudes, and responding to rainfall seasonality in the tropics. Western Palearctic Short-toed Snake-eagles spend the non-breeding season within the northern part of the range of the Brown Snake-Eagle and within that of the southern, presumed resident, populations of their own species in Africa. Some individuals remain in the extreme south-west of the Western Palearctic in winter, as with some booted eagles, but these are few in number (Finlayson, 1992). Birds from the eastern part of the Palearctic range winter in India among residents of the same species (Cramp, 1980).

All twelve species of Old World vultures have resident populations, and it is the two Western Palearctic species that have partially migratory populations (Table 12.4). Migratory behaviour in these vultures therefore has a strong latitudinal component. A few Black Vultures cross the Strait of Gibraltar each year but numbers are very low compared to Griffon Vultures, which migrate in very large numbers, totalling over a thousand in a single day during late October and November. These are mostly juvenile birds from northern Spanish colonies that return late in the spring, between April and June. The passage has increased significantly in the last two decades as conservation has permitted the population to increase. The large passage thus reflects an increase in numbers and not a change in migratory behaviour. These birds spend the winter months across a vast area on both sides of the Sahara, as far south as Niger (Finlayson, 1992). There is also a small passage of Griffons along the eastern Mediterranean, of birds that presumably stay in north-east Africa.

The ability to succeed away from warm-wet/humid forests appears to mark the success of a number of booted eagles, especially species in the genera *Lophaetus, Hieraaetus* and *Aquila*, and to a lesser degree *Nisaetus*, away from strictly tropical regions. These attributes have enabled these eagles to survive in the diverse climates and vegetation types of the Palearctic, at least seasonally. Booted Eagles are largely sedentary, and migratory behaviour is only typical of a few species in the genera that occupy the Palearctic (Table 12.4). In the case of the booted eagles, then, migratory behaviour is exclusive to Palearctic populations. Of the Western Palearctic eagles only the African Tawny and Bonelli's Eagles can be considered strict residents: they only have resident populations. The Spanish Imperial Eagle could be added to this short list but young birds are known to move away from nesting areas to regular winter haunts, and even cross the Strait of Gibraltar in spring and autumn, so some populations may be partially migratory. Another four species have resident populations, making a total of seven out of 10 species with resident populations somewhere in the Western Palearctic. Another seven species have partially or fully migratory populations within the same range. The degree of migratory behaviour among the Western Palearctic booted eagles is dependent on each species and its requirements and, within species, on geographical location. Migratory behaviour is most intense among populations that live furthest into continental parts of the Western Palearctic. The spotted eagles are largely migratory and spend the non-breeding season in tropical Africa and Asia. In the case of the Greater Spotted Eagle, part of the population remains north of the tropics, even within Europe, and these are at the south-western end of the geographical range where climatic extremes are less pronounced. The only representative of the genus *Hieraaetus* in the region is the Booted Eagle. This species is almost entirely migratory within the Palearctic, spending the non-breeding season in tropical Africa and India. A few remain within the breeding area in the winter, as with the Greater Spotted Eagle, in the extreme south-west, in this case in south-western Iberia.

Among the closely related steppe-savanna group of eagles of the genus *Aquila,* the Steppe Eagle, which breeds in the extremely seasonal environment of the Eurasian steppe, is fully migratory, spending the non-breeding season in east Africa down to southern Africa, with some individuals remaining to the north, in Arabia, Iran and Iraq. The Imperial Eagle occupies wooded savanna countryside and migrates south to north-east Africa, the Middle East, Arabia and also south Asia. As with the spotted eagles and Booted Eagle, the extreme south-west of the breeding range is occupied in the winter. The Spanish Imperial Eagle breeds in the extreme south-west of the Palearctic and is largely resident. The Tawny Eagle, which breeds in the south-westernmost part of the range occupied by the steppe-savanna eagles, occupies open habitats with trees and is resident. In this group, then, the pattern of migratory behaviour between species, recently separated species, subspecies and populations, follows a gradient from full residency (Tawny) to fully migratory (Steppe) along a south-west to north-east gradient. This pattern is replicated within a single species in the Golden Eagle, in which the north-easternmost populations are migratory and the southernmost are resident. The remaining species, Bonelli's and Verreaux's Eagles, have southern distributions and are, predictably, non-migratory.

The accipiters and harriers are predominantly sedentary with a small proportion of migratory species. Overall, the harriers have significantly more migratory species than the accipiters (Table 12.4), not surprising given the arid and open environments that they occupy, which are likely to be more seasonal than the warm and wet forests in which the accipiters predominate. All the Western Palearctic accipters and harriers have migratory populations. Of these, Shikra, Levant Sparrowhawk, Pallid and Montagu's Harriers, all species of dry savanna and steppe, are fully migratory. At the other end of the spectrum, the forest-dwelling Eurasian Sparrowhawk and Northern Goshawk have resident populations within a spectrum that includes partially migratory and migratory populations. In all cases where species have migratory and resident populations, the resident ones are in the south and west of the range with the migratory ones to the north and east: Western Marsh Harrier, Hen Harrier, Eurasian Sparrowhawk and Northern Goshawk (Cramp, 1980). In the case of these four species, the open ground harriers exhibit a greater proportion of migratory populations than the forest-dwelling accipiters.

All Milvinae have resident populations but only *Milvus* has migratory ones (Table 12.4). These include long-distance migrants, intra-tropical migrants and partially migratory populations (Curry-Lindahl, 1981a). It is therefore not surprising that it is from this genus that the Western Palearctic populations have emerged. Within the Western Palearctic, the Black Kite is fully migratory, wintering in tropical Africa, although individuals occasionally remain in the south-western extreme of the breeding range in the winter. It is not easy to separate such birds from early northward migrants that are frequent in January, the main period of absence from Iberia being during the months of summer drought (July-September) (Finlayson, 1992). This species is, along with the European Honey-buzzard, one of the most numerous at traditional raptor crossing points over narrow bodies of water (Finlayson, 1992; Bildstein, 2006). The Red Kite is partially migratory within the Western Palearctic and, as with other raptors, it is the populations from the more continental, north-eastern, parts of the range that are migratory (Cramp, 1980).

The six tropical species of fish eagles are resident and it is only the four non-tropical species that have migratory populations (Table 12.4). Migratory behaviour in fish eagles is thus related to climatic seasonality and the freezing-up of waters in the most northerly populations. The pattern in all cases is for the most northerly populations to be fully migratory, wintering within the breeding area of more southerly, resident populations or even further south. In the case of the White-tailed Eagle, Western Palearctic populations tend to move south-west, Central Palearctic ones south towards Iraq and India, and Eastern Palearctic birds south-east towards the Chinese coast. Migration is thus in all cases away from continental interiors. This pattern is repeated in the Central Asian population of Pallas's Fish Eagle, which winters in South Asia, among local residents of the same species, and in the Persian Gulf. Populations of White-tailed Eagles living in the south-western parts of the geographical range are resident but those to the north and east are migratory, moving south-west into western Europe and areas around the Black Sea. Populations that bred in south-western Europe in the past may have been resident along the western Mediterranean and nearby Atlantic coasts, but this must remain conjecture in the absence of concrete evidence.

Migratory behaviour appears to have evolved on more than one occasion among the buzzards but only once in *Buteo* (Amaral *et al.*, 2009). Migratory behaviour appears to have been favoured among species living in open habitats or forest edge, where tracking of variable food resources would have been at a premium, while residency was the rule in forest species. In Holarctic *Buteo,* migratory behaviour may have been linked to life in seasonal environments. This genus has the greatest number of migratory species of the buzzards (Table 12.4). The three Western Palearctic species vary in migratory behaviour, which seems heavily dependent on climate seasonality. The Rough-legged Buzzard of Arctic environments is fully migratory. The Common Buzzard has resident, partially migratory and migratory populations depending on the degree of continentality of the environments occupied, those of the westernmost Atlantic rim, including island forms, being resident and those of the continental interiors being highly migratory. The Steppe Buzzard *B. b. vulpinus* is fully migratory and performs spectacular migrations (Finlayson, 1992) to African winter quarters between Sudan and South Africa (Moreau, 1972). The Long-legged Buzzard also exhibits a wide range of migratory behaviour, and it is the continental populations that are migratory whereas the most southerly and westerly are resident (Cramp, 1980).

Fossil raptors

There are no reliable fossil records of elanine kites in Europe. Given its restriction to forest habitats, it is not surprising that fossil sites with European Honey-buzzard remains are scarce: one Early Pleistocene, one Middle Pleistocene site and six Late Pleistocene sites altogether (Appendix 2; Mlíkovský, 2002; Tyrberg, 1998; 2008). All are within the present breeding or migratory range.

The Bearded Vulture is common in the Pleistocene fossil record of the Western Palearctic, though not as frequent as the Eurasian Black and Griffon Vultures. Bearded Vultures are present from the Early Pleistocene, distributed more widely than today across suitable rocky habitat from the Iberian Peninsula and France in the west through Italy, Hungary and Romania to Georgia and Iraq in the east (Tyrberg, 1998). It does not appear as far north as the Eurasian Black or Griffon Vultures, however, seemingly restricted to the heterogeneous topography of middle and southern Eurasia. The Egyptian Vulture is surprisingly scarce in the fossil record, occurring in sites within the present breeding range during the Middle and Late Pleistocene. This paucity is hard to explain in a large bird associated with caves and rocky cavities where many paleontological sites are found.

Western Palearctic fossil Short-toed Snake-eagles are uncommon; as this is not a cliff-nesting species, this absence may reflect habitat preference, as many fossil sites are in caves and rock shelters. A parallel would seem to be the tree-nesting booted eagles that are similarly scarce in the fossil record. The species is reported from the Early Pleistocene of Italy (1 site) and northern Spain (1) and from the Late Pleistocene of France (3), Italy (2), Spain (1), Libya (1), Iraq (1) and Syria (1), all within the present-day range.

Griffon and Black Vultures are common in the Western Palearctic fossil record (Appendix 2). A large form of Griffon, *Gyps melitensis,* is often described from the Middle and Late Pleistocene of Europe but I consider this to be a large form of *G. fulvus* that suffered a size diminution towards present-day sizes (Sánchez Marco, 2007a). All other fossil forms may easily be subsumed into either *G. fulvus* or *A. monachus.* The disproportionate representation of fossils by time periods is similar to others that we have already observed and which reflects preservation rather than actual population size. The pattern for these vultures resembles that of the Golden and White-tailed Eagles and it is interesting that it includes the tree-nesting Black Vulture, all the other species being cliff-nesters that might have been expected to occur most frequently in cave sites. The geographical distribution of fossil sites reflects a wider range in the past, both species reaching further north in Western Europe than today, into Germany (as far as 51°N), France and across Central Europe (Hungary, Austria, Bulgaria and Romania) and Italy. The disappearance of these species from many of these places seems to have taken place as recently as the 19th Century (Cramp, 1980). Other than these two species, there is only a single, Late Pleistocene record of African White-backed Vulture *G. africanus* from Egypt (Bir Tarfawi in the Western Desert), much further north than its present African range and within the Western Palearctic (Tyrberg, 1998).

Booted eagles of uncertain affinity appear in the European fossil record during the Eocene, and they become increasingly frequent in the Miocene. The genus *Aquila* is present from the Late Miocene. At least two species, which we can relate to living ones, appear in the pre-Pleistocene European fossil record: *Hieraaetus edwardsi*, ancestor of Booted Eagle *H. pennatus*, in the Middle Miocene of France and Spain; and Bonelli's Eagle in the Late Pliocene of Bulgaria (Mlíkovský, 2002).

In Appendix 2, I summarise 234 records of fossil booted eagles for the Western Palearctic in the last two million years from the information in Tyrberg (1998; 2008). As expected for species that would have been in existence at the start of the Pleistocene, a number of these are present in Early Pleistocene deposits and the rest are certainly around by the Middle Pleistocene. The only exception is the Spanish Imperial Eagle, for which records are not unsurprisingly scarce given the species' limited geographical distribution, and which only appears in the Late Pleistocene. This observation is consistent with a recent speciation event following isolation of a small (judging from its exceptionally low genetic variability; Negro and Hiraldo, 1994), Iberian population from the main *A. heliaca* population (Helbig *et al.*, 2005).

There is little by way of surprise in the distribution of these fossils by countries as the majority are from sites where we would expect these birds today, either breeding or outside the breeding season. There are some exceptions, which we must treat with some caution because of their scarcity. There are fossil Greater Spotted Eagles from Germany, France and Italy that would suggest a westerly extension of its range during the Middle and Late Pleistocene. Some Greater Spotted Eagles do winter west, reaching the Iberian Peninsula, but they are scarce so it is somewhat surprising to find fossils of them this far west. A similar situation, though less pronounced, occurs with the Imperial Eagle, which appears in at least two Late Pleistocene Italian sites, west of the present breeding range. This is not altogether unexpected either, as there must have been a link between Imperial Eagle populations across the range, including the Iberian Peninsula, with isolation and speciation of the latter population possibly in the Late Pleistocene. The presence of a possible Tawny Eagle in the Early Pleistocene of France is tantalising as it may represent a once wider, and more northerly, geographical range of this African species. Alternatively, if the fossil is closer to Steppe Eagle, it may represent a wider western extension of the range of this steppe species in the past. Finally, the presence of Booted or Bonelli's Eagle in the Late Pleistocene-Holocene site of Ightham Fissure in Kent is unusual as neither species breeds in the UK. Taken together this meagre evidence could be interpreted to point to the existence of more widespread extensions of the open woodland, and even steppe, habitats suitable for Imperial, Greater Spotted and other booted eagles which may have been severed by the advance of the ice sheets of the last glaciation. The warm and humid conditions that followed would have favoured denser woodland in western Europe, unsuitable for these species.

Fossil accipiters first appear in the European fossil record in the Early Miocene but the first harriers are found in the early Pleistocene (Mlíkovský, 2002). Accipiters and harriers are well-represented in the European fossil record (Appendix 2). It is surprising that woodland species should be so well represented, as other woodland raptors are not. Of particular interest is the greater frequency of Pallid Harrier compared to Montagu's Harrier, a situation that is reversed in present-day Europe (Cramp, 1980). The appearance of the steppe-dwelling Pallid Harrier in the Middle Pleistocene of France and the Late Pleistocene of the Czech Republic, Hungary, France and Italy (Tyrberg, 1998; 2008) would seem to coincide with similar westward expansions of other steppe raptors that we have observed at this time.

Considering their abundance and ubiquity, milvine kites are remarkably scarce in the Western Palearctic fossil record. Just as striking is the relative abundance of the Red Kite compared to the Black Kite (Appendix 2) which, today at least, is approximately four times more numerous as a breeding Western Palearctic raptor (Hagemeijer and Blair, 1997) and occupies a wider geographical area. The overall scarcity of milvine kites in the fossil record resembles other tree-nesting species and may reflect the locations where fossils accumulate, which tend to be rocky. The fossil sites of the two species are overwhelmingly from circum-Mediterranean countries, lending support to the view that during periods of glaciation, these warmth-loving species were largely absent from central Europe, which could only have been colonised during post-glacial warming. One intriguing possibility is that kites expanded over areas of Europe only with the advent of agriculture during the Neolithic (around 7,000 years ago), and even later with the forest clearance that took place in the Middle Ages (Schreiber *et al.*, 2000).

The origin of the fish eagles may date back to the Oligocene, judging from the remains of a large eagle, with close similarities to *Haliaeetus*, in El Fayum, Egypt (Rasmussen, 1987). As discussed above, the origins of the various extant species of *Haliaeetus* may be older than previously supposed and may go back to the Miocene. If so *Haliaeetus* is an ancient genus with an early geographical separation of species, around the ancient Tethys Ocean, where most species remain today. One species managed to break away from this hub and gave rise to the four species that colonised Palearctic and Nearctic wetlands. The subsequent glaciations, after 2 mya, may have played a part in isolating the various Holarctic populations but not in the speciation process itself, as the current species would already have existed by then. It is my view that the speciation of the Holarctic *Haliaeetus* probably occurred during periods of desiccation in the Pliocene, when connectivity between wetlands would have been severed, and not during the glaciations of the Pleistocene that need not have kept populations apart. The distribution of Western Palearctic fossils covers two sites in the Early Pleistocene, nine in the Middle Pleistocene and 36 in the Late Pleistocene (Appendix 2; Tyrberg,1998; 2008). Many of the sites are within the current breeding or wintering range of this species and are therefore uninformative. The presence of a number of White-tailed Eagle sites outside the current range, in England, Wales, France, Italy and the Iberian Peninsula during the Middle and Late Pleistocene indicates a wider distribution that was probably largely severed by persecution since 1800 (Hagemeijer and Blair, 1997). Of interest is a fossil record of a Late Pleistocene African Fish Eagle *H. vocifer* in the Sinai Peninsula of Egypt (Tyrberg, 1998) indicating a northerly extension of this species' range, presumably during wet periods, in the past.

The earliest *Buteo* dates to the Upper Oligocene of North America (Cracraft, 1969). *Buteo* first appears in the European record during the Middle Miocene (Mlíkovský, 2002). These buzzards represent extinct lineages with extant species deriving from a later radiation from the New World. Buzzards are well-represented in the Western Palearctic fossil record throughout the Pleistocene, becoming most frequent in the Late Pleistocene (Appendix 2). The Common Buzzard, the most frequent, is present from the Early Pleistocene. The Long-legged Buzzard first appears in the Middle Pleistocene and the Rough-legged Buzzard in the Late Pleistocene, which may reflect the relatively recent radiation of the Old World Buzzards. Part of the reason may, however, have to do with expansion south (of Rough-legged) and west (of Long-legged) during the coldest and driest moments of the Pleistocene, coinciding with the long period leading up to the Last Glacial Maximum. The presence of Rough-legged Buzzard, presumably in the winter, in areas well south of the present wintering range (Italy, France, Spain and Gibraltar) would bear this out (Tyrberg, 1998; 2008). Likewise, the appearance of the Long-legged Buzzard west in Italy, Spain, France and Luxembourg may reflect the westward expansion of the steppe at this time (Finlayson and Carrión, 2007).

Taxa in the Palearctic

Elanus

Only one (25%) of the species in this genus – the Black-winged Kite *E. caeruleus* – breeds in the Palearctic. It is a bioclimatic semi-specialist of the subtropical (Latitude Category E) belt and is restricted to a marginal range in the south-western Palearctic. Like other Palearctic species which just reach the Palearctic (mainly Western) from Africa (e.g. Rufous Bush Chat *Cercotrichas galactotes*, some of the larks and gallinules) they are severely limited in their geographical expansion, and the south-west simply represents the northern edge of the range. The large expanse of the Sahara must hinder the establishment of these and other potential candidates for Palearctic colonisation, and they risk isolation and extinction north of the desert in times of cold and aridity. This pattern is in clear contrast with the situation in the Eastern Palearctic, in which tropical and subtropical species always had an escape route south and others had the benefit of the Himalayas as a 'launch-pad' for colonisation of the Palearctic. For this reason the Palearctic fauna is largely tropical Asian in origin, despite the presence of the large African land mass much closer to the Western and parts of the Central Palearctic.

Pernis

Two of three species in this genus occupy the Palearctic with non-overlapping ranges, the European Honey-buzzard in the Western and Central Palearctic and the Oriental Honey-buzzard *P. ptyliorhynchus* in the east. The two species are bioclimatic moderate forest raptors that rely heavily on insects, and are therefore highly migratory. The third species, the south-east Asian Barred Honey-buzzard *P. celebensis* is a sedentary specialist. This small genus could provide another example of expansion across the forests of the Palearctic, with subsequent severance of eastern and western populations by aridity and loss of forest habitats, from a south-eastern tropical source. In this case, these warmth-loving species must have connected with the Palearctic during a mild period. Alternatively, honey-buzzards could have been widely distributed across the Palearctic-Asian area in the Miocene, with northern populations surviving by increasing their bioclimatic tolerance and becoming migratory.

Gypaetus

The Bearded Vulture is the sole species of its genus. A bioclimatic moderate, this sedentary species of rocky habitats has a pan-palearctic range across the mid-latitude belt (Latitude Category D). Its distribution is continuous in the Central Palearctic but fragmented in the west and marginal in the east. Like other mid-latitude belt genera (e.g. *Monticola, Oenanthe*), the range extends southwards along the north-east African mountains down to southern Africa.

Neophron

The Egyptian Vulture is the sole species of its genus. It is a bioclimatic semi-generalist of the mid-latitude belt (Latitude Category D) of the Western and Central Palearctic. This range is the northern fringe of a wider range that encompasses Africa and India, where the species is sedentary. Palearctic populations, on the other hand, are highly migratory. The Egyptian Vulture appears to be an example of a bioclimatically tolerant wide-ranging Old World species that was able to survive the seasonality of the southern Palearctic by becoming migratory.

Circaetus

One (20%) of five snake-eagles – Short-toed – breeds in the Palearctic. The remaining species are African but the Short-toed Snake-eagle also breeds in India, where it is a resident. It is a bioclimatic moderate with a multi-latitude (Latitude Category F) range across the Western and Central Palearctic, where it is highly migratory. This is another case of a genus with a wide Old World distribution containing one species that is more bioclimatically tolerant than the rest having the Palearctic as the northern part of the genus range. Migration, as in the Egyptian Vulture, is the solution to survival in these edge-regions. As with the Egyptian Vulture we can interpret this as *Circaetus* having had an Old World range in the Tertiary, which became increasingly seasonal; the populations experiencing the rising seasonality became migratory, thus surviving in an increasingly hostile world. In the case of the Egyptian Vulture the patterns developed within a single species, while in the snake-eagles it was within a multi-species genus.

Gyps

The Afro-Asian *Gyps* vultures appear to have undergone several radiations, with the oldest lineages in south-east Asia and the Himalayas. Early expansions may have followed the mid-latitude belt, with a subsequent spread into Africa followed by later radiations. One species (12.5%) breeds in the Western (discontinuous) and Central Palearctic, the Eurasian Griffon, a bioclimatic moderate of the mid-latitude belt (Latitude Category D). It is one of two moderates (the other is the Indian White-rumped Vulture) in a genus of specialists and semi-specialists. All griffons are sedentary except for the most continental and northerly populations of Eurasian Griffon, which can be strongly migratory, reaching sub-Saharan Africa. The griffons are comparable to the snake-eagles, with continental populations of the northernmost species having become migratory.

Aegypius

The Eurasian Black Vulture is the sole species of its genus and has a mid-latitude belt (Latitude Category D) range across forests of the Western (discontinuous) and Central Palearctic. Other single-species genera perform the Black Vulture's function in Africa (*Torgos*) and southern Asia (*Sarcogyps*). Like the Eurasian Griffon and Bearded Vultures, the Eurasian Black Vulture is a bioclimatic moderate, but it remains sedentary across most of the range. In this case the Black Vulture represents the northern part of the range of a functional vulture unit (the three components of which are not particularly closely related). So among the vultures we have species in the Palearctic that are at the northern limit of vulture survival: two are populations within a species, another is a species within a genus and yet another is a genus within a functional group (large vultures) within a subfamily (Aegypinae).

Torgos

The Lappet-faced Vulture is the only species in its genus. It is a large, bioclimatic semi-specialist African resident that just reaches the southernmost Western Palearctic (subtropical, Latitude Category E). It therefore represents an extreme version of the single-species pattern that we have observed in the Egyptian Vulture, except that its northernmost range is much further south than the mid-latitude belt; the southerly position of the northern edge of the range does not require the species to be migratory.

Nisaetus

The Mountain Hawk Eagle *N. nipalensis* is the sole representative (12.5%) of this South-east Asian genus in the Palearctic. The other species in the genus are specialist or semi-specialist residents, but this species is a partially migratory bioclimatic moderate. It is restricted in range to the Eastern Palearctic, a pattern typical of many tropical South-east Asian forest genera of bioclimatic specialists.

Hieraaetus

This is a group of small hawk eagles that occupy a broad range in Africa, the Palearctic, southern Asia and into Australasia. They are bioclimatic specialist and semi-specialist residents of woodland and savanna, with the exception of the Booted Eagle, the sole Palearctic representative (16.67%), which is a migratory bioclimatic moderate. Only in the extreme south-west of the range are a few Booted Eagles sedentary. The Booted Eagle is another example of a genus with a single species, at the northern edge of the genus range, which is bioclimatically more tolerant than its congeners and is also migratory.

Aquila

Seven species (58.33%) in this genus breed in the Palearctic. The Golden Eagle is the only generalist. It has a continuous, multi-latitude (Latitude Category F) range where it is largely sedentary except in the extreme north. Bonelli's Eagle is a sedentary semi-generalist with a discontinuous pan-palearctic range across the mid-latitude belt (Latitude Category D). Golden and Bonelli's Eagles are just two of three species in the genus that have adapted to nesting in rocky habitats, instead of the tree-nesting behaviour that typifies the genus. Both species have spread right across the mid-latitude belt where, given their bioclimatic tolerance, they have remained sedentary. The third species to have adapted to rocky habitats is Verreaux's Eagle. It is the counterpart of the previous species along the rocky habitats that range along the Rift Valley down to southern Africa. It is a sedentary bioclimatic moderate that reaches the southern edge of the Western Palearctic. The range occupied by the Bearded Vulture is shared in this genus by three species.

The remaining eagles in this genus are tree-nesting specialists or semi-specialists and three are mid-latitude belt (Latitude Category D) species: Imperial Eagle is a bioclimatic semi-specialist of the savannas of the Central Palearctic with marginal occupation of areas of the Western and Eastern Palearctic; Steppe Eagle is a bioclimatic specialist with a similar range to the Imperial Eagle, which it replaces in more open steppe and semi-desert habitats; Spanish Imperial Eagle is the Imperial Eagle's specialist counterpart in the savannas of the extreme west of the Western Palearctic, in the Iberian Peninsula and (formerly) North Africa (Thévenot *et al.* 2003). The African Tawny Eagle is the Steppe Eagle's counterpart in the desert and semi-desert habitats of the south-Western Palearctic; it is a sedentary bioclimatic specialist of the subtropical belt (Latitude Category E). Imperial Eagles are highly migratory, though those from the south-west of the range and Spanish Imperial Eagles are sedentary. A similar pattern describes the Steppe (highly migratory) Eagle in the north-east and the Tawny (sedentary) Eagle in the south-west of the joint range of these two closely related eagles.

Lophaetus

This is an unusual genus, which has two (40%) Palearctic species, the others occupying African and southern Asian ranges. The Lesser Spotted Eagle is a highly migratory bioclimatic semi-generalist with an exclusively Western Palearctic temperate (Latitude Category C) range. The larger Greater Spotted Eagle is a part-migratory bioclimatic specialist with a pan-palearctic temperate (Latitude Category C) range. They are the only migratory species in this genus of forest and woodland eagles. The spotted eagles may represent an ancient radiation and widespread distribution that has been significantly severed through time, the component species having diverged significantly in habits. The Palearctic species would, in this scheme, be solutions to a changing world of increasing seasonality in habitats not occupied by the *Aquila* eagles.

Melierax

The Dark Chanting Goshawk is the only species, of four tropical African bioclimatic sedentary specialists and semi-specialists, to reach the Palearctic. It has a subtropical (Latitude Category E) range in the Western Palearctic and resembles a number of other African species and genera that incorporate the southernmost Palearctic as the northern edge of their range.

Accipiter

Six species (12.5%) of this large genus breed in the Palearctic. The Northern Goshawk is a Holarctic bioclimatic resident generalist of temperate (Latitude Category C) forests. Its range in the Palearctic is continuous throughout. The smaller Eurasian Sparrowhawk, which may be derived from an Asian group, is also a pan-palearctic generalist with continuous distribution throughout, but its multi-latitude (Latitude Category F) range is not restricted to the temperate belt. It is a partial migrant, many northern and continental populations being highly migratory. The presence of two generalists in a genus is unusual; in this case it may reflect different lineages and significant differences in body size, effectively making the two species functionally distinct within the genus.

The Shikra represents a very different strategy. It is essentially a bioclimatically moderate tropical Afro-Asian woodland resident species. Palearctic populations, in the centre, are on the northern limits of the species' range and contrast with those to the south by being highly migratory. In distributional terms it could be regarded as intermediate between Verreaux's Eagle and Egyptian Vulture. The Levant Sparrowhawk is a Western and Central Palearctic solution to the savannas that spread across these regions with the loss of forests. In this respect it resembles the Imperial Eagle and, like that species, is highly migratory.

Two other accipiters in the region, Japanese Sparrowhawk and Chinese Sparrowhawk *A. soloensis* are moderate and semi-specialist respectively. Highly migratory, they occur in the Eastern Palearctic.

Circus

Six harriers (42.86%) breed in the Palearctic. The Hen Harrier, which constitutes an early lineage that is separate from other harriers, is a Holarctic generalist with a pan-palearctic distribution that is discontinuous in the west. It is a migrant of the Boreal (Latitude Category B) belt, a range conducive to pan-palearctic distributions. The two marsh harriers – Western *C. aeruginosus* and Eastern *C. splionotus* – are bioclimatic moderates of the temperate (Latitude Category C) belt and are geographical complements, Western in the Western and Central Palearctic and Eastern in the east. Eastern Marsh Harrier is migratory, as are continental and northern populations of Western Marsh, but south-western populations are sedentary or short-distance migrants.

Of the remaining species, the Pallid Harrier, which is related to the Hen Harrier but more distant to the apparently similar Montagu's Harrier, is a bioclimatic specialist of the temperate belt of the Central Palearctic with a marginal presence in the Western Palearctic. As with other Central Palearctic steppe species (e.g. Steppe Eagle), the Pallid Harrier is highly migratory. Montagu's Harrier is a bioclimatic moderate with a multi-latitude (Latitude Category F) range and a Western and Central Palearctic distribution. The Pied Harrier *C. melanoleucos* is also a bioclimatic moderate, of the temperate belt of the Eastern Palearctic. Montagu's and Pied are both highly migratory.

Milvus

The three kites of this genus breed in the Palearctic. The Black Kite is a migratory biolcimatic generalist with a continuous, multi-latitude (Latitude Category F) pan-palearctic range. This genus appears to fit with the 'single generalist species per genus' rule as the two other species are not so tolerant. The Red Kite is a bioclimatic moderate of the temperate (Latitude Category C) belt that is exclusive to the Western Palearctic. The north-eastern, continental populations of this species are migratory but south-western ones are sedentary. The third species – Yellow-billed Kite – is an African resident that occupies the subtropical (Latitude Category E) belt in the Western Palearctic, this being the northern edge of the species' range (as in other African species, such as Dark Chanting Goshawk).

Haliaeetus

Three fish eagles (37.5%) breed in the Palearctic. The White-tailed Eagle is a bioclimatic semi-generalist with a pan-palearctic multi-latitude (Latitude Category F) range that is fragmented in the west in large measure, like the Eurasian Osprey, because of human activity. Only its Nearctic counterpart – the Bald Eagle – has a comparable bioclimatic tolerance. The other Palearctic species are semi-specialists: Pallas's Fish Eagle in the Central Palearctic (Warm, Latitude Category D) and Steller's Sea Eagle in the Eastern Palearctic (Boreal, Latitude Category B). All three Palearctic fish eagles are largely migratory.

Butastur

The Grey-faced Buzzard *B. indicus* is the only species in a genus of four Afro-Asian savanna species which reaches the Palearctic. It is a bioclimatic semi-specialist of the temperate belt (Latitude Category C) of the Eastern Palearctic and is highly migratory. Like other species that have their northern outposts in the Eastern Palearctic, the Grey-faced Buzzard's range, centred in the temperate belt, contrasts with many Western Palearctic fringe species whose ranges are centred on the subtropical belt. It is a reflection of the latitudinal habitat continuity in the east that is missing in the west.

Buteo

Four (14.28%) of the species in this largely American genus breed in the Palearctic. The Common Buzzard is a multi-latitude (Latitude Category F) generalist with a continuous, pan-palearctic, range. Its northern and continental populations are highly migratory while those in the mild south-west are short distance migrants or residents. The Steppe Buzzard, race *vulpinus* of Common Buzzard, is (like Imperial Eagle and Pallid Harrier) a full, long-distance migrant. The Rough-legged Buzzard is a bioclimatic moderate with a Holarctic range. It has a continuous, pan-palearctic range centred on the Boreal (Latitude Category B) belt. It is highly migratory. The third species – Long-legged Buzzard – is a mid-latitude belt semi-specialist. Its continental populations are highly migratory, while those of the west are sedentary. Finally, the Upland Buzzard *B. hemilasius* is a migratory specialist at the Eastern Palearctic end of the mid-latitude belt. In the Palearctic the buzzards conform to the rule of a single bioclimatically tolerant species per genus.

CONCLUSIONS

(a) **Bioclimatic generalisation**. Bioclimatic generalists and semi-generalists make up a large proportion of the Palearctic diurnal raptors – 10 species, 22.22% of the total. Six species have multi-latitude ranges, all of them being pan-palearctic – Eurasian Osprey, Golden Eagle, Black Kite, Eurasian Sparrowhawk, White-tailed Eagle and Common Buzzard. The other four are spread out between one Boreal (Hen Harrier), one temperate (Northern Goshawk) and two mid-latitude belt (Bonelli's Eagle and Egyptian Vulture) species.

(b) Mid-latitude belt representation. Fifteen (33.33%) species have ranges centred on 40°N. Curiously, there are no split-category species among the diurnal raptors. The vultures (4) and booted eagles (6) seem to have been particularly suited to this belt. Species of rocky (4), open (especially steppe, 3) and mixed (especially savanna, 5) habitats dominate in the mid-latitude belt.

(c) Geographical nature of the genera. Two genera – *Pernis* and *Gyps* – seem to fit into the 'launch-pad model', starting with South-east Asian-Himalayan species that expanded across temperate forests and rocky mid-latitude habitats. A large number of genera and species appear to be at the northern fringe, in the Western Palearctic, of an otherwise tropical distribution: Black-winged Kite, Egyptian Vulture, Short-toed Snake-eagle (of genus *Circaetus*), Lappet-faced Vulture, Booted Eagle (of genus *Hieraaetus*), Dark Chanting Goshawk, Shikra, Yellow-billed Kite, African Tawny Eagle and Verreaux's Eagle. They are mainly species of the subtropical (Latitude Category E) belt (6 species) or the mid-latitude (Latitude Category D) belt (3 species). This is an unusually large proportion of species. In contrast, only two species – Mountain Hawk Eagle and Grey-faced Buzzard – appear as northern fringe species in the Eastern Palearctic. One is a mid-latitude species and the other a temperate species, suggesting that northern fringe species extend further north in the Eastern than the Western Palearctic. Two genera in particular – *Buteo* and *Accipiter* – seem to have fed the Palearctic from the Nearctic.

(d) Representation in the Western Palearctic. The diurnal raptors are best represented in the Western Palearctic (34 species), followed by Central (27) and Eastern (24) Palearctic. Part of the reason for this is the number of fringe species from African populations in the west. There are three uniquely Western Palearctic raptors – Red Kite, Lesser Spotted Eagle and Spanish Imperial Eagle.

CHAPTER 13
Gulls, terns, auks and waders

Group 2

In this book I follow the three broad divisions (suborders) of these birds identified by Ericson *et al.* (2003), Paton *et al.* (2003), Paton and Baker (2006) and Baker *et al.* (2007). The first, Lari, includes the gulls (Laridae), terns (Sternidae), skuas (Stercorariidae), skimmers (Rynchopidae) and auks (Alcidae) as well as the Crab Plover *Dromas ardeola* (Dromadidae), the pratincoles and coursers (Glareolidae) and, remotely, the buttonquails (Turnicidae). The second group, Scolopaci, incorporates the jacanas (Jacanidae), painted snipes (Rostratulidae), seed snipes (Thinocoridae), Plains Wanderer *Pedionomus torquatus* (Pedionomidae) and the large wader family Scolopacidae. The third group, Charadrii, includes the Magellanic Plover *Pluvianellus socialis* (Pluvianellidae), sheathbills (Chionidae), thick-knees (Burhinidae), oystercatchers (Haematopodidae), avocets (Recurvirostridae) and plovers (Charadriidae).

These are ancient lineages, the three suborders having originated in the Late Cretaceous between 79 and 102 mya with at least 14 lineages having survived the K/T mass extinction (Baker *et al.*, 2007). It is not unsurprising that many of the families in this group have worldwide distributions, as they would have existed prior to the present-day configuration of the continents.

Climate

In this section I have subdivided the group into ecologically similar units for the sake of clarity. The first unit includes gulls, terns and skuas, which are global bioclimatic specialists (Figure 13.1i). The Palearctic species deviate from the pattern and include a high proportion of bioclimatically tolerant species, including all generalists. The second unit – the auks – are all at the specialist end of the tolerance gradient, the most tolerant being moderates (Figure 13.1ii). The Palearctic pattern matches the global one, as most species live in the Palearctic. The shorebirds are also global bioclimatic specialists but resemble the first unit in that Palearctic species include the most tolerant and all the generalists (Figure 13.1iii). Finally, the buttonquails are largely specialists (11 of 17 species) and semi-specialists (5 of 17).

The larids occupy a wide range of bioclimates, which is unusual among birds. They tend towards warm climate but are equally at home in the coldest and wettest climates (Figures 13.2). The Palearctic species have wide bioclimatic ranges but tend away from the warmest and wettest.

The auks are concentrated on the cool part of the gradient and tend towards the arid while occupying the entire humidity range (Figure 13.3). Palearctic species are more evenly spread across the humidity range than the entire set of species. The shorebirds are similarly widely distributed across the temperature and humidity gradients (Figure 13.4). The Palearctic species appear to form two subgroups – one in the coldest end of the gradient and the other towards the warm, but not hot. Most are also at the dry end of the humidity range.

The low levels of bioclimatic tolerance and the wide ranges of temperature and humidity occupied indicate that this group is made up of many species, each taking up a portion of the gradients. The buttonquails are exceptional, with most species occupying warm and humid bioclimates.

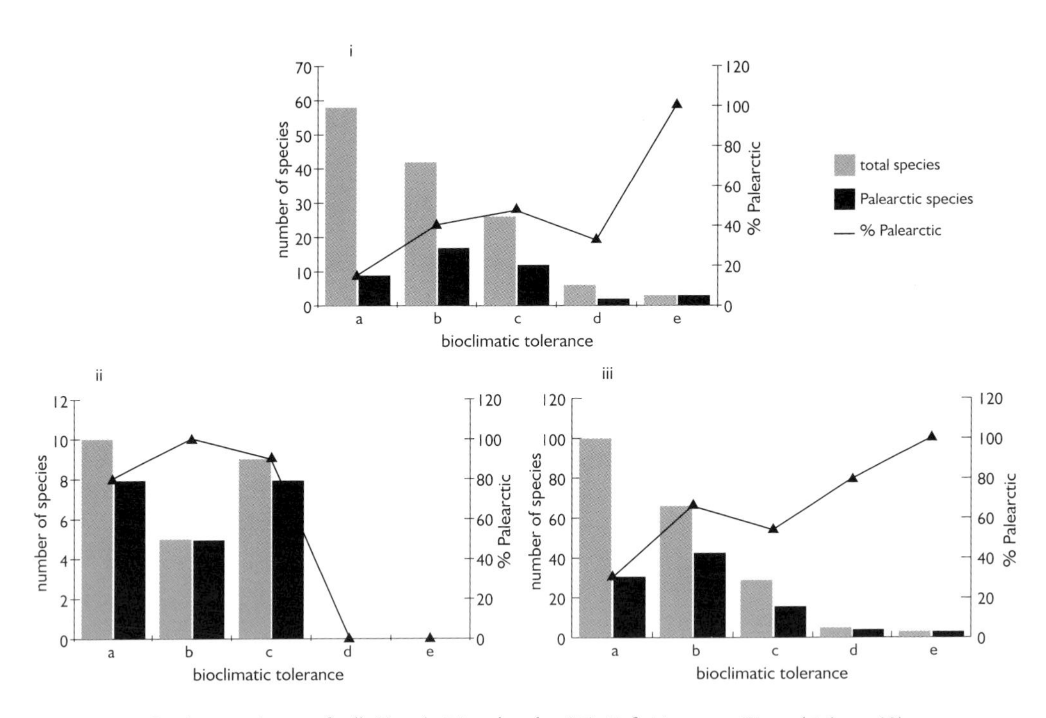

Figure 13.1 *Bioclimatic tolerance of gulls (i), auks (ii) and waders (iii). Definitions as on Figure 4.1 (see p. 38).*

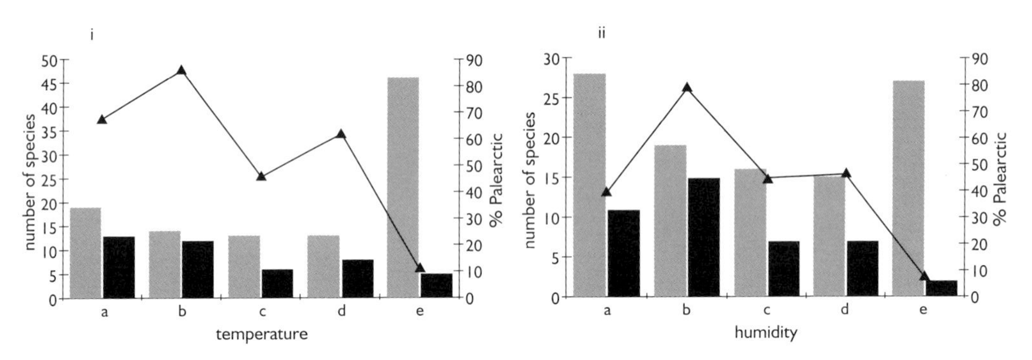

Figure 13.2 *Temperature (i) and humidity (ii) tolerance of gulls. Definitions as in Figure 4.2 (see p. 38). Key as in Figure 13.1.*

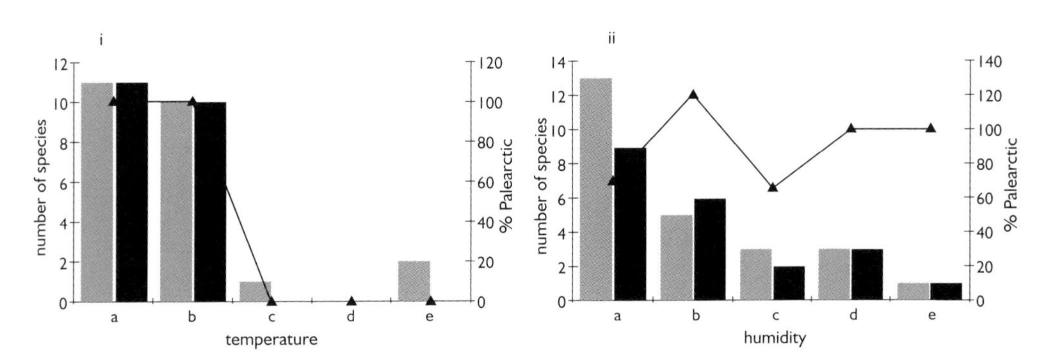

Figure 13.3 *Temperature (i) and humidity (ii) tolerance of auks. Definitions as in Figure 4.2 (see p. 38). Key as in Figure 13.1.*

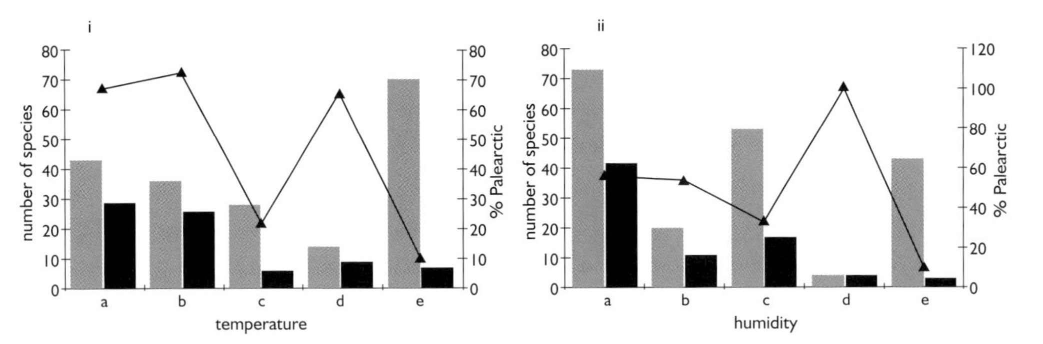

Figure 13.4 *Temperature (i) and humidity (ii) tolerance of waders. Definitions as in Figure 4.2 (see p. 38). Key as in Figure 13.1.*

Habitat

The larids are largely wetland and marine but are also well represented on rocky habitats, where many breed, in the Palearctic (Figure 13.5i). Alcids are strongly dependent on rocks and exploit the marine environment (Figure 13.5ii), while shorebirds occupy wetland and open habitats, both globally and within the Palearctic (Figure 13.5iii). Between them, the species in this group sort themselves out across the marine-wetland-open habitat range of habitats, many breeding on rocky habitats. In this latter way, they recall the nocturnal and diurnal birds of prey. The buttonquails mostly occur in open and mixed habitats.

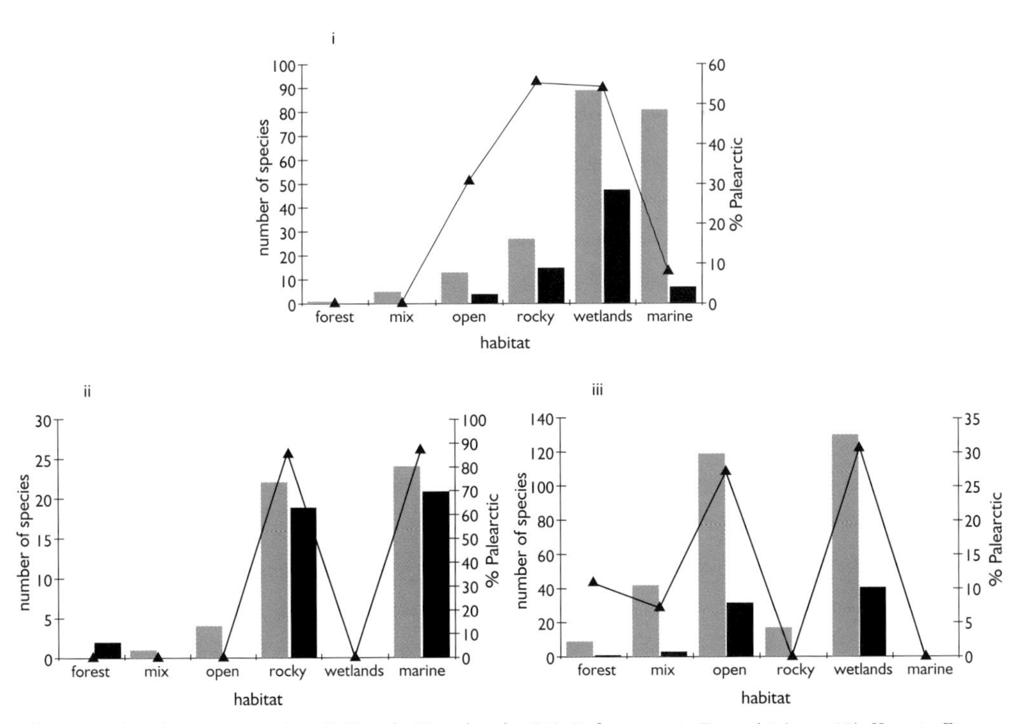

Figure 13.5 *Habitat occupation by gulls (i), auks (ii) and waders (iii). Definitions as in Figure 4.3 (see p. 39). Key as in Figure 13.1.*

Shorebirds	Global	Palearctic	Global %	Palearctic %	Palearctic of global %
Migratory	119	73	58.33	93.59	61.34
Sedentary	115	15	56.37	19.48	13.04
Total Species	204	78			

Gulls	Global	Palearctic	Global %	Palearctic %	Palearctic of global %
Migratory	90	42	84.12	97.67	46.67
Sedentary	59	15	55.14	34.88	25.42
Total Species	107	43			

Auks	Global	Palearctic	Global %	Palearctic %	Palearctic of global %
Migratory	23	21	95.83	100	91.3
Sedentary	18	17	75	80.95	94.44
Total Species	24	21			

Buttonquails	Global	Palearctic	Global %	Palearctic %	Palearctic of global %
Migratory	1	1	5.88	50	100
Sedentary	17	1	100	50	5.88
Total Species	17	2			

Table 13.1 *Migratory behaviour in Palearctic and global Charadriiformes.*

Migratory behaviour

The species in this group, excepting the buttonquails, excel as migrants. This behaviour is particularly noteworthy among Palearctic species (Table 13.1). Migratory behaviour, along with occupation of open habitats and exploitation of wetlands and the sea, seem to be the main factors in the success of these birds in the Palearctic.

Fossil species

In addition to a range of species already present in the early Pleistocene, Common Redshank *Tringa totanus*, Eurasian Dotterel *Charadrius morinellus* and Common Guillemot *Uria aalge* are present in the Late Pliocene (Mlíkovský, 2002). The species (Appendix 2) indicate little, especially as many are highly migratory and cover huge geographical areas. Nevertheless, the presence of Grey Plover *Pluvialis squatarola* and Glaucous Gull *Larus hyperboreus* in a number of sites may indicate a more southerly distribution than today. This seems certain for the Little Auk *Alle alle*, which is found in sites down to Gibraltar, where it is not found today even in winter. The presence of Pallas's Gull *Ichthyaetus ichthyaetus* in Italy, if correct, suggests a westward expansion that may have been during a dry period, when the steppe fauna moved west.

Taxa in the Palearctic: Suborder Lari

LARIDAE

The common ancestor of gulls and skimmers diverged in the Early Oligocene, and the radiation of modern gull genera commenced much later, during the late Oligocene, when the ancestor of the genera *Creargus* (Swallow-tailed Gull *C. furcatus*) and *Rhodostethia* (Ross's Gull, *R. rosea*) diverged from the other gulls, followed by *Xema*, *Pagophila* and *Rissa* successively during the Early Miocene (Baker *et al.*, 2007). The modern gull genera are therefore Tertiary lineages that have survived to the present day.

Larus

Twelve species (57.14%) of the species in this genus breed in the Palearctic. Ten of them (71.43%) are part of a Holarctic white-headed or *argentatus* group, which dominates the genus by species. These species are all closely related and interconnected with a glacial history in two refugia, one in north-west Europe and the other in the Aral-Caspian region, the populations from the first group spreading into the Mediterranean and Black Sea area (including *michahellis* and *armenicus*) and the other (*cachinnans*) north-east and then along the Arctic-Atlantic coasts, where they overlap with the first group (Liebers *et al.*, 2004). The *fuscus* group is derived from this stock (Liebers and Helbig, 2002). According to this scenario, the Great black-backed Gull *L. marinus* seems to have emerged in north-eastern North America during a period of isolation before making contact, in north-west Europe, with *fuscus* and *argentatus*.

The species derived from the *cachinnans* group include the Lesser Black-backed Gull *L. fuscus*, a bioclimatic moderate with a Boreal (Latitude Category B), Western and Central Palearctic range; Glaucous-winged Gull *L. glaucescens*, a bioclimatic semi-specialist with a Boreal Eastern Palearctic range; Iceland Gull *L. glaucoides*, a semi-specialist with an Arctic (Latitude Category A) Western Palearctic distribution; and Caspian Gull *L. cachinnans*, a semi-specialist of the mid-latitude belt (Latitude Category D). Together, this group encompasses a pan-palearctic multi-latitude range. One feature that all the species have in common is that they are highly migratory.

The North Atlantic species (including *michahellis* and *armenicus*) include two Boreal bioclimatic moderates – Herring Gull *L. argentatus* with a pan-palearctic range and Great Black-backed Gull, which is confined to the Western Palearctic; the two species have Nearctic counterparts. Also in this group is the pan-palearctic Arctic semi-specialist, Glaucous Gull, and two mid-latitude belt (Latitude Category D) species – Yellow-legged Gull *L. michahellis* (mid-latitude Category D semi-specialist) and Armenian Gull *L. armenicus* (mid-latitude Category D specialist). Curiously, these gulls tend to have migratory and sedentary populations, bar the Armenian Gull which, nevertheless, is a short-distance migrant.

The Slaty-backed Gull *L. schistisagus* is a partially migratory semi-specialist with a multi-latitude (Latitude Category F) range in the Eastern Palearctic. The Common Gull *L. canus* is a pan-palearctic bioclimatic moderate with a Boreal (Latitude Category B) range and is highly migratory except in the oceanic west. The last Palearctic *Larus* belongs to a separate lineage, the band-tailed gulls, which have Asian, Australasian and South American representatives. It is the part-migratory Black-tailed Gull *L. crassirostris*, a mid-latitude (D) bioclimatic moderate of the Eastern Palearctic.

Ichthyaetus

Four of the six species in this genus, closely linked to *Larus*, breed in the Palearctic and the remaining two – Sooty Gull *I. hemprichii* and White-eyed Gull *I. leucopthalmus* – occur just outside the southern fringe of the Western Palearctic, in the Red Sea. *Ichthyaetus* gulls have ranges focused on the mid-latitude belt and adjacent areas. Audouin's Gull *I. audouinii* is a mid-latitude belt (Latitude Category D) specialist of the Western Palearctic; Pallas's Gull *I. ichthyaetus* is a semi-specialist of the mid-latitude belt in the Western and Central Palearctic; Relict Gull *I. relictus* is an Eastern Palearctic specialist with a range just to the north of the mid-latitude belt (making it Latitude Category C). Finally, Mediterranean Gull *I. melanocephalus* has a split C/D distribution, but the reasons for this are different from those of other species; for Mediterranean Gull it represents the colonisation of coastal waters of the temperate Atlantic from the mid-latitude range of largely inland waters. *Ichthyaetus* gulls are all migratory.

Chroicocephalus

Three species (27.27%) from this globally distributed group breed in the Palearctic. The Black-headed Gull *C. ridibundus* is a Holarctic generalist with a continuous pan-palearctic range. Most populations are highly migratory except for those of the Western Palearctic, which are largely sedentary or short-distance migrants. The Slender-billed Gull *C. genei* is a migratory bioclimatic specialist of the mid-latitude belt (Latitude Category D) with a discontinuous distribution across the Western and Central Palearctic. The Grey-headed Gull *C. cirrocephalus*, a bioclimatic semi-specialist resident, breeds on both sides of the Atlantic, in Africa and the Americas; it barely reaches the Western Palearctic, at its southern edge, where it has a subtropical (Latitude Category E) range.

Saundersilarus

Saunders's Gull *S. saundersi* is in a single-species genus. It is a migratory specialist of the mid-latitude belt (Latitude Category D) in the Eastern Palearctic.

Hydrocoloeus

The two species in this genus are Palearctic. Little Gull *H. minutus* is a bioclimatic semi-specialist of the Boreal belt (Latitude Category B) and is highly migratory. The Little Gull has a pan-palearctic range which is continuous in the Central Palearctic. Ross's Gull *H. rosea* is a semi-specialist of the Arctic belt – it occupies a fragmented range across the Central and Eastern Palearctic and is also migratory.

Pagophila

This is a Holarctic single-species genus. The Ivory Gull *P. eburnea* is a semi-specialist that breeds in the Arctic (Latitude Category A) belt of the Central Palearctic and is migratory.

Xema

This is another Holarctic single-species genus. Sabine's Gull *X. sabini* is closely related to the Ivory Gull. It is a bioclimatic semi-specialist of the Arctic (Latitude Category A) belt of the Eastern Palearctic, and is highly migratory.

Rissa

The two kittiwakes both breed in the Palearctic. The Black-legged Kittiwake *R. tridactyla* is a bioclimatic moderate at 40°N (Latitude Category B) and has a continuous pan-palearctic range. It also breeds in north-west North America. The Red-legged Kittiwake *R. brevirostris* is a specialist of the Arctic (Latitude Category A) belt in the Eastern Palearctic and Nearctic. The kittiwakes are partially migratory.

STERNIDAE

The branch that was to become the tern lineage separated from the gulls very early in the Tertiary, in the Late Palaeocene (around 60 mya; Baker *et al.*, 2007). The radiation of the terns may have started during the Eocene and Oligocene but it seems to have accelerated during the Middle Miocene (around 17 mya) and continued into the Pliocene (Bridge *et al.*, 2005). An early branching leading to the genera *Gygis* and *Anous* left no representatives in the Palearctic. Subsequent branching in the Late Miocene led to 'grey' terns of the genus *Onychoprion* (represented in the Palearctic by the Aleutian Tern *O. aleutica*) and, later, *Sternula* (represented in the Palearctic by the Little Tern *S. albifrons*). The Late Miocene and Early Pliocene saw a major radiation of terns leading to the modern genera *Gelochelidon, Hydroprogne, Larosterna, Chlidonias, Thalasseus* and *Sterna*. Radiations within these genera took place during the Pliocene (Bridge *et al.*, 2005).

Onychoprion

This is a widespread genus of three tropical and one Arctic species. The Aleutian Tern breeds along the Arctic coasts of north-eastern Siberia and north-western America. It is a migratory bioclimatic specialist of the Arctic belt (Latitude Category A) and, in the Palearctic, confined to the extreme north-east.

Sternula

This is another small genus of four warm, largely tropical, terns. One species – Little Tern *S. albifrons* – is the most bioclimatically tolerant (semi-generalist) in the genus; it breeds in the Western and Central Palearctic and also in the south-east of the Eastern Palearctic. It is highly migratory and represents the northernmost limit of the genus's range. Southern, tropical populations of Little Tern are sedentary. The success of this genus away from the tropics has depended on bioclimatic tolerance and migration.

Gelochelidon

This is a single-species genus. The Gull-billed Tern *G. nilotica* breeds in the Palearctic and Nearctic, including the Caribbean. It radiated from other terns in the Late Miocene and is closest to the genus *Hydroprogne.* Like the Mediterranean Gull it has a split Temperate/Warm (C/D) range, which probably reflects occupation of mid-latitude and temperate wetlands across the Palearctic and colonisation of mild oceanic coastal areas in the west. This seems to be a pattern among species in this group of birds. The Gull-billed Tern therefore has a highly fragmented pan-palearctic range, reflecting former connections along the mid-latitude and temperate belts at times when wetlands occupied greater surface areas (due to large ice-dammed lakes in Siberia, for example). Human destruction of wetlands has added significant stresses to birds such as the Gull-billed Tern.

Hydroprogne

The Caspian Tern *H. caspia* is the only species in its genus and has an even wider range than the Gull-billed Tern, breeding in the Americas, Africa, Asia and Australasia. It is a bioclimatic semi-generalist with a discontinuous split Boreal/mid-latitude belt (B/D) range in the Palearctic. It is therefore similar to the Gull-billed Tern, although it reaches further north in the west. Its range in the Central and Eastern Palearctic borders on the northern edge of the mid-latitude belt, close to the Temperate belt. The Caspian Tern is highly migratory.

Chlidonias

Three of four species in this genus breed in the Palearctic (only the Black-fronted Tern *C. albostriatus* of New Zealand is absent). This is another globally widespread genus. The Whiskered Tern *C. hybridus* has populations across the Palearctic, Africa, Asia and Australia; the Black Tern *C. niger* has a Holarctic range, and the White-winged Tern *C. leucopterus* has a Palearctic distribution. The three Palearctic species are bioclimatic moderates: Whiskered has a discontinuous pan-palearctic range across the mid-latitude belt (Latitude Category D). Black and White-winged Terns have ranges to the north of Whiskered, centred on the Temperate (Latitude Category C) belt. The Black Tern's range is continuous in the Central Palearctic and adjacent regions of the Western Palearctic; further west the range is discontinuous and it is absent from the Eastern Palearctic. The White-winged Tern's range is continuous in the Central Palearctic and in the Eastern but the populations are disconnected from each other; its range is marginal in the Western Palearctic. All three species are highly migratory.

Thalasseus

This is a genus of six globally widespread, largely tropical terns. The Sandwich Tern *T. sandvicensis* resembles the Gull-billed Tern in having breeding populations in the Americas, including the Caribbean, and also in the Palearctic. The Palearctic thus represents the northern edge of the species (and genus) range. It is a bioclimatic generalist with a split Temperate/Warm (C/D) range that is similar to the Gull-billed Tern's, except that it is missing from the Eastern Palearctic. Its distribution also recalls that of the Mediterranean Gull. It is a migratory tern, though some south-western birds may be sedentary. The other two species with Palearctic ranges – Lesser Crested Tern *T. bengalensis* and Royal Tern *T. maximus* – are marginal semi-specialists along the southern fringe of the Western Palearctic, along the subtropical belt (Latitude Category E).

Sterna

Three (18.75%) of sixteen species of *Sterna* terns breed in the Palearctic. This is another globally widespread genus with most species in tropical and southern waters. The Common Tern *S. hirundo* is the only generalist (most species are specialists or semi-specialists) in the genus, which seems to follow the 'single generalist per genus' pattern. Common Tern has an American–Caribbean and Palearctic range, recalling other terns such as Gull-billed, Sandwich and Black. It has a pan-palearctic, multi-latitude (Latitude Category F) range, which is discontinuous in the west. Like most species in this genus (bar some tropical species), the Common

Tern is highly migratory. Roseate Tern *S. dougallii* is one of two bioclimatic moderates in the genus, and has a fragmented Atlantic–Indian Ocean distribution. In the Palearctic it is exclusive to the temperate (Latitude Category C) west; it is also highly migratory. The most migratory tern (and probably the most migratory bird) is the Arctic Tern *S. paradisea,* a Holarctic semi-specialist of the Arctic (Latitude Category A) belt. It has a continuous, pan-palearctic range.

STERCORARIIDAE

Most classifications place the skuas as an early divergent group from the gulls, terns and auks (Paton *et al.*, 2003, 2006; Thomas *et al.*, 2004 a, b; Baker *et al.*, 2007). They are closest to the auks but the separation of the two lineages dates back to the Palaeocene, over 60 mya. The skuas therefore represent a separate cluster of species within the Lari.

Stercorarius

The three *Stercorarius* skuas breed in the Palearctic, and they all have circumpolar Holarctic ranges. Arctic *S. parasiticus* and Long-tailed *S. longicaudus* are bioclimatic semi-specialists of the Arctic (Latitude Category A) belt. The Pomarine Skua *S. pomarinus* is a bioclimatic specialist, also of the Arctic belt. Arctic and Long-tailed Skuas (marginal in the west) have pan-palearctic ranges, while Pomarine Skua only breeds in the Central and Eastern Palearctic. All three are highly migratory and are among the birds with the most dramatic seasonal changes of lifestyle, from tundra predators of small mammals during the breeding season to pelagic kleptoparasites on passage and in the winter months.

Catharacta

This genus comprises five species, one of which – Great Skua *C. skua* – breeds in the Palearctic. The other species are all birds of the southern hemisphere, which was colonised relatively recently from the northern hemisphere during a mild climatic period; this was followed by a period of diversification between 210,000 and 150,000 years ago when Antarctic birds were forced away from the continent to establish populations on the subantarctic islands, Tristan da Cunha, Patagonia and the Falklands. Speciation has been incomplete as neighbouring populations of different species still exchange genes (Ritz *et al.*, 2008). The Great Skua is a bioclimatic specialist that breeds in the Boreal (Latitude Category B) belt of the Western Palearctic, where it is a partial migrant.

DROMADIDAE

Dromas

The Crab Plover *Dromas ardeola* is a unique and unusual species that falls between the auks and skuas (Thomas *et al.*, 2004a) and must have diverged in the Late Palaeocene. It is a partially migratory specialist of the subtropical (Latitude Category E) belt of the Western Palearctic.

ALCIDAE

The auks are an ancient lineage that has its origins in the Palaeocene (around 60 mya) and radiated in stages from the Early Eocene (around 55 mya) to the end of the Pliocene (Pereira and Baker, 2008). From an origin in the Pacific the family is characterised by multiple trans-oceanic dispersals. An early branch separated into two main lineages around 55 mya: (a) a lineage that would lead to the Pacific auklets (*Aethia, Ptycoramphus, Cyclorrhynchus*), the Pacific and Atlantic puffins (*Fratercula*) and the Rhinoceros Auklet *Cerorhinca monocerata* of the Pacific; and (b) the remaining auks – *Alle, Alca, Uria, Pinguinus*, murrelets *Synthliboramphus, Brachyramphus* and pigeon guillemots *Cepphus.*

The *Brachyramphus* murrelets radiated from the rest of its group in the Pacific low Arctic during the Early Eocene, around 48 mya. The next split, leading to the pigeon guillemots around 45 mya, also happened in the Pacific low Arctic; the Black Guillemot *Cepphus grylle* line then split off around 20 mya, with Pigeon *C. columba* and Spectacled *C. carbo* splitting in the Pacific much later, in the Pliocene.

The branch leading to the large auks (including the Palearctic species) branched from the *Synthliborhamphus* lineage of murrelets either in the Atlantic or Pacific low Arctic in the Late Eocene (around 40 mya). The murrelet radiation then proceeded in the Pacific subtropical region in the Miocene and Pliocene. The *Uria* guillemots branched off in the Atlantic low Arctic during the Oligocene (around 30 mya). The Common *U. aalge* and Brunnich's *U. lomvia* Guillemot branches then split in the Miocene (around 15 mya). In the other branch, the ancestor of the Little Auk *Alle alle* diverged shortly after the Oligocene split of the guillemots, leaving a lineage that would diverge into Razorbill *Alca torda* and Great Auk *Pinguinus impennis* at the Oligocene–Miocene boundary (around 25 mya). The closest living relative of the Great Auk is therefore the Razorbill (Moum *et al.*, 2002).

The other large auk branch separated from the previous one in the Pacific low Arctic in the Early Eocene (around 55 mya). It was followed by a separation of auklet (*Aethia, Ptychoramphus, Cyclorrhynchus*) and puffin (*Fratercula, Cerorhinca*) lineages in the Pacific low Arctic around 50 mya. The murrelet radiation then proceeded, somewhat later, in the Pacific low Arctic during the Miocene and Pliocene. On the other side of this branch the Rhinoceros Auklet diverged from the puffins in the Oligocene (around 30 mya) in the Pacific low Arctic; this region also saw the *Fratercula* puffins subsequently radiate in the Late Miocene, with the Atlantic Puffin *F. arctica* finally splitting away during the Pliocene in the Atlantic low Arctic.

Alca

This is a single-species genus with a Holarctic (Atlantic) distribution. The Razorbill is a bioclimatically moderate Western Palearctic species of the Boreal (Latitude Category B) belt. Its northernmost populations are highly migratory but southern ones disperse only locally.

Pinguinus

The extinct Great Auk had a similar distribution to the Razorbill. Based on our knowledge of this species I have provisionally placed it as a bioclimatic moderate of the Boreal (Latitude Category B) belt and, within the Palearctic, exclusive to the Western.

Alle

This is another single-species genus. The Little Auk is a circumpolar Holarctic semi-specialist species. It breeds in the Arctic belt (Latitude Category A) of the Central Palearctic and is highly migratory.

Uria

The two Holarctic circumpolar guillemot species breed in the Palearctic. The Common Guillemot is a bioclimatic moderate that breeds in the Western and Eastern Palearctic, but not the Central, along the Boreal (Latitude Category B) belt. Its northern populations are migratory and southern ones tend to disperse within local waters in the winter. Brunnich's Guillemot is a semi-specialist of the Arctic (Latitude Category A). It also breeds in the Western and Eastern Palearctic and, breeding further north than the Common Guillemot, is fully migratory.

Synthliboramphus

Two of the four murrelets in this Holarctic genus breed along the Pacific mid-latitude coasts (Latitude Category D) of the Eastern Palearctic. The Ancient Murrelet *S. antiquus* is a bioclimatic semi-specialist, and the Japanese Murrelet *S. wumizusume* is a specialist. Both species are partially migratory.

Cepphus

All three pigeon guillemots (another Holarctic genus) breed in the Palearctic. The Black Guillemot is a bioclimatic moderate and the only Western Palearctic species. It breeds along the Arctic (Latitude Category A) belt of the entire Palearctic (fragmented in the centre) and is partially migratory. The Pigeon Guillemot is also a moderate and breeds in the Eastern Palearctic around 40°N (Latitude Category B); it is also partially migratory. The third species, Spectacled Guillemot, is a partially migratory semi-specialist of the Temperate belt (Latitude Category C) of the Eastern Palearctic.

Brachyramphus

The three species of this Holarctic murrelet genus breed in the Eastern Palearctic. The Marbled Murrelet *B. marmoratus* is a bioclimatic moderate of the Temperate (Latitude Category C) belt, while Long-billed *B.*

perdix and Kittlitz's *B. brevirostris* Murrelets are Boreal (Latitude Category B) semi-specialists and specialists respectively. All three murrelets are partially migratory.

Aethia

The three members of this Holarctic genus breed in the Palearctic. All three are partially migratory bioclimatic specialists of the Boreal (Latitude Category B) belt of the Eastern Palearctic – Crested *A. cristalleta*, Whiskered *A. pygmaea* and Least *A. pusilla* Auklets.

Cyclorrhynchus

The Parakeet Auklet *C. psittacula* is the only species of this Holarctic genus. It is a partially migratory bioclimatic specialist of the Boreal (Latitude Category B) belt of the Eastern Palearctic.

Cerorhinca

The Holarctic Rhinoceros Auklet is a mid-latitude (Latitude Category D) specialist that breeds in the Eastern Palearctic and is partially migratory.

Fratercula

The three species of puffins breed in the Palearctic. The Atlantic Puffin *F. arctica* is the only Western Palearctic species with a Holarctic Atlantic range. It is a bioclimatic moderate of the Boreal Belt (Latitude Category B) and is partially migratory. The Tufted Puffin *F. cirrhata* is a migratory bioclimatic moderate with a Holarctic Pacific range; it breeds in the Boreal belt (Latitude Category B) of the Eastern Palearctic. The Horned Puffin *F. corniculata* is a bioclimatic specialist of the Boreal belt (Latitude Category B) with a Holarctic Pacific range. It breeds in the Eastern Palearctic.

GLAREOLIDAE

The lineage of the pratincoles and coursers dates from the Late Cretaceous (and thus predates the K/T event). The courser and pratincole lineages split soon after, during the Early Palaeocene, around 60-65 mya. The Australian Pratincole *Stiltia isabella* separated from *Glareola* during the Oligocene (Baker *et al.*, 2007). So the pratincoles and coursers represent ancient lineages that have remained largely unchanged since the beginning of the Tertiary.

Cursorius

The coursers are inhabitants of dry, open country in Africa and southern Asia. One of four species, the Cream-coloured Courser *C. cursor,* occupies the northern part of the genus's range, across the arid belt of North Africa and Arabia and into Central Asia. It is a bioclimatic specialist of the subtropical (Latitude Category E) belt of the Western and Central Palearctic; northern populations, close to the mid-latitude belt, are migratory.

Glareola

Pratincoles are birds of warm climates, ranging from dry to wet, of Africa and southern Asia. Those that breed in the Palearctic (3 species, 42.86%) are, like the Cream-coloured Courser, at the northern end of the genus's range. Collared Pratincole *G. pratincola* in the Western and Central Palearctic and Oriental Pratincole *G. maldivarum* in the Eastern Palearctic are bioclimatic moderates of the mid-latitude belt (Latitude Category D). The Black-winged Pratincole *G. nordmanni* occupies a higher latitude belt (Temperate, Latitude Category C) than the other two species; it is a Central Palearctic species with a marginal distribution in the Western Palearctic. The three Palearctic pratincoles are the only migratory species of the genus.

TURNICIDAE

The buttonquails form a unique group at the base of the Lari. They are a family of birds that emerged in the Late Cretaceous and survived the K/T event.

Turnix

Two (12.5%) of 16 species of this African, southern Asian, New Guinean and Pacific genus breed in the Palearctic (and are therefore at the northern end of the genus's range). The Small Buttonquail (sometimes called the Andalucian Hemipode) *T. sylvatica* is a bioclimatic semi-specialist that breeds in a small area of the south-western Palearctic (southern Iberia, where it is probably extinct, and Morocco). The species is also found in Africa and southern Asia. The Yellow-legged Buttonquail *T. tanki* is a southern Asian bioclimatic moderate that breeds in the Eastern Palearctic, and whose Palearctic populations are migratory. This species is unusual in a genus of residents.

Taxa in the Palearctic: Suborder Scolopaci

Two major clusters form this suborder, and they appear to represent ancient lineages that split in the Late Cretaceous and survived the K/T event (Baker *et al.*, 2007). The first line includes jacanas (Jacanidae) and painted snipes (Rostratulidae) and a sister group comprising the Plains Wanderer (Pedionomidae) and the seed snipes (Thinocoridae). The second line is represented by the Scolopacidae (sandpipers, snipes and related species). The first line is largely absent from the Palearctic. Only the Painted Snipe *Rostratula benghalensis*, with a tropical African and Asian distribution, penetrates the south-eastern Palearctic (Latitude Category D). It is a bioclimatic specialist of warm and wet climates and depends on the migratory strategy in its limited Palearctic range, being sedentary elsewhere.

SCOLOPACIDAE

Genera are listed in the approximate order of branching off from the main line, with the oldest first.

Numenius

The first split in the Scolopacidae represents a line that led to the genera *Bartramia* (Upland Sandpiper of the Nearctic) and *Numenius*. This is an ancient group that separated in the Early Palaeocene (60–65 mya), with the two genera splitting early in the Eocene (*c.* 50 mya). The eight species in this Holarctic genus are highly migratory and five of them (62.5%) breed in the Palearctic. Three species appear to have split from the rest of the genus first and independently of each other (Thomas *et al.*, 2004a): the Eskimo Curlew *N. borealis* from the Nearctic, the Little Curlew *N. minutus* and the Slender-billed Curlew *N. tenuirostris*. The three species appear to have had localised Holarctic distributions but it is difficult to know exactly the ancestral position, given that they have been seriously affected by human action. The Little Curlew is a bioclimatic specialist of the Arctic (Latitude Category A) belt and is restricted to a small area of the Eastern Palearctic. Even more restricted is the rare Slender-billed Curlew which, on present distribution, appears to be a specialist of the Temperate (Latitude Category C) of the Central Palearctic.

The remaining cluster then splits into two, one group containing the Nearctic Bristle-thighed *N. tahitiensis* and the Palearctic Far-eastern *N. madagascariensis* Curlews. Of the three remaining species, the Whimbrel *N. phaeopus* branches off first, leaving the closely related Eurasian *N. arquata* and Long-billed *N. americanus* Curlews. The Far-eastern Curlew is a bioclimatic semi-specialist of the Boreal (Latitude Category B) belt of the Eastern Palearctic. The Whimbrel is a bioclimatic moderate of the Boreal (Latitude Category B) belt with a pan-palearctic range that is discontinuous in the centre and east. The Eurasian Curlew is also a bioclimatic moderate of the Boreal belt but it is larger than the Whimbrel and is separated from it by habitat. This species also has a pan-palearctic range, discontinuous in the west and east. The westernmost oceanic populations of Curlew include sedentary or short-distance migratory populations; this is atypical of the genus.

Limosa

The next split in the Scolopacidae involves the *Limosa* lineage, which branched off in the Late Palaeocene (*c.* 56 mya, Baker *et al.*, 2007). The four species in this genus form a tight cluster with the two Nearctic species (Marbled *L. fedoa* and then Hudsonian *L. haemastica*) breaking off first, leaving the two Palearctic sister-species. Bar-tailed Godwit *L. lapponica* is a migratory bioclimatic specialist of the Arctic (Latitude Category A) belt and has a pan-palearctic range which is discontinuous in the west. Black-tailed Godwit *L. limosa* is a migratory bioclimatic moderate of the Temperate (Latitude Category C) belt. It also has a pan-palearctic range, which is discontinuous in the west and the east.

Limnodromus

This Holarctic genus branched off during the Early Eocene (*c.* 50 mya) (along with *Lymnocryptes, Coenocorypha, Gallinago* and *Scolopax*), and this genus and *Lymocryptes* separated from the others shortly afterwards. The split between *Limnodromus* and *Lymnocryptes* followed in the Middle Eocene (*c.* 45 mya). Two of the three dowitcher species breed in the Palearctic. They are all highly migratory. The Asian Dowitcher *L. semipalmatus* is a bioclimatic specialist of the Temperate (Latitude Category C) belt of the Central and Eastern Palearctic. The Holarctic Long-billed Dowitcher *L. scolopaceus* is a bioclimatic specialist of the Arctic (Latitude Category A) belt which breeds in the Eastern Palearctic. This genus is not represented in the Western Palearctic.

Lymnocryptes

The Jack Snipe *L. minimus* is the only species of this Palearctic genus. It is a migratory bioclimatic semi-specialist of the Boreal (Latitude Category B) belt, with a pan-palearctic distribution that is discontinuous in the west.

Scolopax

This is a predominantly South-east Asian–New Guinean genus, with a Palearctic (Eurasian Woodcock *S. rusticola*) and a Neartic (American Woodcock *S. minor*) species. It is another ancient lineage, which split off from the snipes in the Middle Eocene (45 mya). These are, atypically, birds of forest habitats and, even though we lack details of their history (all species seemingly branching off at a similar time, Thomas *et al.*, 2004a), it would seem that they follow a similar pattern to a number of the passerine and other forest groups. In this case it seems that a species broke away from the South-east Asian forests, presumably during a warm period (which given the ancient history of these birds was probably earlier than similar expansions of forest passerines) and colonised the Palearctic. From there it would have made it into the Nearctic with subsequent, probably multiple, isolations during cold glacials. The Pleistocene history of the American Woodcock would seem to bear this out (Rhymer *et al.*, 2005). The Eurasian Woodcock is among the few wader species that has become a bioclimatic generalist (the others are the Common Snipe *Gallinago gallinago* and the Little Ringed Plover *Charadrius dubius*). It has a pan-palearctic, Temperate (Latitude Category C) belt range that is discontinuous in the west. It is highly migratory, the only migrant other than the American Woodcock in a genus of sedentary birds.

The Amami Woodcock *Scolopax mima* is a resident in the extreme southeast of the Palearctic. It is a bioclimatic specialist of the subtropical (Latitude Category E) belt.

Gallinago

This is a globally distributed genus of 16 species, 6 of them (37.5%) breeding in the Palearctic. The Common Snipe is a bioclimatic generalist and the only one in a genus dominated by specialists and semi-specialists; *Gallinago* follows the 'single generalist per genus' rule. It is a species of the Boreal (Latitude Category B) belt with a pan-palearctic range that is discontinuous in the west. It is highly migratory with sedentary or short-distance migrants in the extreme west of the range; other, non-Palearctic populations are also sedentary. The Solitary Snipe *G. solitaria* is a sedentary, multi-latitude (Latitude Category F) bioclimatic moderate of the Central and Eastern Palearctic. Three snipes are bioclimatic semi-specialists: Great Snipe *G. media,* a Boreal (Latitude Category B) belt migrant of the Western and Central Palearctic; Pintail Snipe *G. stenura,* a Boreal (Latitude Category B) migrant of the Central and Eastern Palearctic which is marginal in the west; and Latham's Snipe *G. hardwickii,* a migrant of the mid-latitude (Latitude Category D) belt of the Eastern Palearctic. The remaining species, Swinhoe's Snipe *G. megala,* is a migratory bioclimatic specialist of the Temperate (Latitude Category C) belt of the Central and Eastern Palearctic.

Phalaropus

The next branch to split off, in the Middle Eocene (45 mya) includes this genus along with *Steganopus, Prosobonia, Xenus, Actitis, Catoptrophorus, Tringa* and *Heteroscelus* (Baker *et al.*, 2007). *Phalaropus* was the first lineage to split away from the group, still during the Middle Eocene. Two species of phalarope in this Holarctic genus – Red-necked *P. lobatus* and Grey (or Red) *P. fulicaria* – breed in the Palearctic. Both are highly migratory bioclimatic semi-specialists with Arctic (Latitude Category A) pan-palearctic ranges, continuous in Red-necked and discontinuous in the west and centre in the case of the Grey Phalarope.

Xenus

This genus split off next, during the late Middle Eocene (Baker *et al.*, 2007). It is a single-species Palearctic genus. The Terek Sandpiper *X. cinereus* is a highly migratory Boreal (Latitude Category B) pan-palearctic species that is peripheral in the west.

Actitis

This Holarctic genus branched off next within this group, towards the end of the Eocene (36 mya, Baker *et al.*, 2007). One of two species breeds in the Palearctic. The Common Sandpiper *A. hypoleucos* is a migratory bioclimatic semi-generalist with a multi-latitude (Latitude Category F) range. Its range spans the Palearctic and is discontinuous in the west.

Tringa

This was the last genus with Palearctic species to branch off, in the Early Miocene (20 mya, Baker *et al.*, 2007). Seven (58.33%) species breed in the Palearctic, and the remaining five are Nearctic. All are highly migratory. The Common Redshank *T. totanus* is a bioclimatic semi-generalist with a multi-latitude (Latitude Category F) pan-palearctic range that is discontinuous in the west and the east. It is the only bioclimatically tolerant species in the genus, suggesting a version of the 'single generalist per genus' model. Likewise there is only one bioclimatic moderate – Common Greenshank *T. nebularia*. It is a species of the Boreal (Latitude Category B) belt with a continuous pan-palearctic range. Spotted Redshank *T. erythropus* and Marsh Sandpiper *T. stagnatilis*, along with the two American yellowlegs, form a cluster of species within the genus, with the Common Redshank and the Common Greenshank. Spotted Redshank is a specialist of the Arctic (Latitude Category A) with a pan-palearctic range, discontinuous in the west; Marsh Sandpiper is a specialist with a Temperate (Latitude Category C) Western and Central Palearctic range.

A second cluster groups the two tattlers (sometimes classified in a separate genus, *Heteroscelus*) and the Wood Sandpiper *T. glareola*. This species is a semi-specialist of the Boreal (Latitude Category B) belt with a pan-palearctic range that is peripheral in the west. The Grey-tailed Tattler *T. brevipes* is a specialist of the Boreal belt with a discontinuous range in the Central and Eastern Palearctic. The Green Sandpiper *T. ochropus* was the first species to branch off within the genus. It is a semi-specialist of the Boreal belt with a pan-palearctic range (discontinuous in the west).

Arenaria

The second branch that split from the previous cluster of genera in the Middle Eocene is represented by the Holarctic genera *Arenaria, Tryngites, Micropalama, Philomachus, Limicola, Eurynorhynchus, Calidris* and *Aprhiza* (Baker *et al.*, 2007). The first split within this group is *Arenaria*. This lineage separated early on, in the Middle Eocene some time between the *Phalaropus* and *Xenus* splits in the previous group. One of two Holarctic species – Ruddy Turnstone *A. interpres* – breeds in the Palearctic. It is a highly migratory bioclimatic specialist of the Arctic (Latitude Category A) belt and has a continuous pan-palearctic range.

Tryngites

The next to split was the separation of the *Tryngites–Micropalama, Philomachus–Limicola* and *Eurynorhynchus–Calidris–Aphriza* lineages, during the Middle Oligocene (*c.* 28 mya); the *Tryngites–Micropalama* lineage then split further into the Nearctic Stilt Sandpiper *Micropalama himantopus* and the Holarctic Buff-breasted Sandpiper *Tryngites subruficollis* lines at the end of the Oligocene (*c.* 24 mya). The Buff-breasted Sandpiper is another migratory bioclimatic specialist with an Arctic (Latitude Category A) Eastern Palearctic range.

Philomachus

Philomachus separated from *Limicola* at about the time of the *Tryngites–Micropalama* split. It is another Holarctic single-species genus. The Ruff *P. pugnax* is a bioclimatic semi-specialist of the Boreal (Latitude Category B) belt. It has a pan-palearctic range that is discontinuous in the west and the east. Like other Arctic-Boreal waders, it is highly migratory.

Limicola
This is a single-species Palearctic genus. The Broad-billed Sandpiper *L. falcinellus* is a migratory bioclimatic specialist with an Arctic (Latitude Category A) range. This range is discontinuous but pan-palearctic.

Eurynorhynchus
This Palearctic genus split off from the *Calidris–Aphriza* lineage about the time of the *Philomachus–Limicola* separation. The Spoon-billed Sandpiper *E. pygmeus* represents another single-species lineage. Like the Buff-breasted Sandpiper, this species is a migratory specialist of the Arctic (Latitude Category A) belt of the Eastern Palearctic.

Calidris
This genus branched from *Aphriza* (Surfbird) in the early Middle Miocene (*c.* 15 mya). Fifteen (83.33%) species of this Holarctic genus breed in the Palearctic. It is the genus with most species (18) in the Scolopacidae (only approached by *Tringa* with 12 species), and it is well-represented in the Palearctic. All *Calidris* are all highly migratory bioclimatic specialists and semi-specialists. The Long-toed Stint *C. subminuta* is a semi-specialist of the Boreal (Latitude Category B) belt of the Central and Eastern Palearctic, while Dunlin *C. alpina* and Purple Sandpiper *C. maritima* are semi-specialists of the Arctic (Latitude Category A) belt. The Dunlin has a continuous pan-palearctic range but the Purple Sandpiper is missing from the Eastern Palearctic. The remaining 12 species are all specialists of the Arctic (Latitude Category A) belt. Of these only Temminck's Stint *C. temminckii* has a continuous pan-palearctic range. The remaining species are regionally localised and do not breed in the Western Palearctic: Little Stint *C. minuta* (Central); Pectoral Sandpiper *C. melanotos,* Curlew Sandpiper *C. ferruginea,* Red Knot *C. canutus* and Sanderling *C. alba* (Central and Eastern); Red-necked Stint *C. ruficollis,* Sharp-tailed Sandpiper *C. acuminata,* Great Knot *C. tenuirostris,* Western Sandpiper *C. mauri,* Baird's Sandpiper *C. bairdii* and Rock Sandpiper *C. ptilocnemis* (Eastern).

Taxa in the Palearctic: Suborder Charadrii

The Charadrii branched from the Lari and Scolopaci in the Late Cretaceous, pre-dating the K/T event (Baker *et al.*, 2007). Several splits within the suborder occurred as early as the Late Cretaceous: (a) the first branched off the lineages leading to *Burhinus, Esacus* and the *Chionis* sheathbills, and *Pluvianellus* (Magellanic Plover); (b) the separation of the *Pluvianus* (Egyptian Plover) lineage from the remaining group; (c) the separation of two lineages – one leading to *Cladorhynchus* (Banded Stilt), *Himantopus, Recurvirostra, Haematopus, Ibidorhyncha* and *Pluvialis*, and a second leading to *Elseyornis, Thinornis, Charadrius, Eudromias, Phegornis, Oreopholus, Anarhynchus, Peltohyas, Erythrogonys* and *Vanellus*; and (d) still within the Late Cretaceous, the *Pluvialis* lineage branched off. There was therefore a greater subdivision of lineages in the Late Cretaceous in this suborder than in the Scolopaci or Lari.

BURHINIDAE

This is an ancient family with its origins in the Late Cretaceous. The numerically dominant genus, *Burhinus,* separated from *Esacus* (a genus that today occupies parts of southern Asia and Australia) in the Late Eocene.

Burhinus
Like many ancient lineages that pre-date the final separation of the continents, the thick-knees are globally distributed. The two Palearctic species – Eurasian Stone Curlew *B. oedicnemus* and the Senegal Thick-knee *B. senegalensis* – are sister species that together occupy a large area of Africa, southern Asia and the southern Palearctic. The Stone Curlew is at the northern end of this large range. Its southernmost populations, including those of the south-western Palearctic and southern Asia, are sedentary, but those of the northern and continental parts of the Palearctic range are migratory. The Eurasian Stone Curlew is a species of the mid-latitude (Latitude Category D) belt of the Palearctic; it has a discontinuous range, typical of many mid-latitude belt species, in the Western and Central Palearctic and is missing in the east. The Senegal Thick-knee is marginal in our region, occuring along the southern fringe of the Western Palearctic. It is the

geographical counterpart of the Eurasian Stone Curlew in sub-Saharan Africa, and in the Palearctic occupies a small area of the subtropical (Latitude Category E) belt.

RECURVIROSTRIDAE

The two main branches of this family, leading on the one hand to the stilts (*Cladorhynchus, Himantopus*) and on the other to the avocets (*Recurvirostra*) branched off in the Late Eocene (Baker *et al.*, 2007). Both ancient genera have global distributions.

Himantopus

One of five species in the genus breeds in the Palearctic. The Black-winged Stilt *H. himantopus* is a species with a large African-southern Asian-Palearctic range. As happens in a number of these ancient lineages, a single species appears to have a range which in more modern genera would be taken up by several species. Palearctic populations of the Black-winged Stilt are at the northern end of the species range and are highly migratory, the species elsewhere being sedentary. It is a bioclimatic moderate (the most tolerant of the genus) of the mid-latitude belt (Latitude Category D). Like other mid-latitude belt species (e.g. Eurasian Stone Curlew) it has a discontinuous distribution in the Western and Central Palearctic and is absent in the east.

Recurvirostra

The Pied Avocet *R. avosetta* is a migratory Palearctic species, and one of four globally distributed (albeit fragmented) species. It is surprisingly absent as a breeding species from much of Africa and southern Asia. The Pied Avocet is a bioclimatic moderate of the mid-latitude (Latitude Category D) belt; the range is fragmented in the Western and Central Palearctic and the species barely reaches the Eastern Palearctic.

HAEMATOPODIDAE

Haematopus

The oystercatcher lineage separated from the stilts and avocets in the early Middle Eocene (*c.* 48 mya). Like the stilts and avocets, the oystercatchers constitute a globally distributed genus, probably reflecting an ancient range that pre-dates the final separation of the continents. Two of eleven (18.2%) species breed in the Palearctic. The Eurasian Oystercatcher *H. ostralegus,* whose closest relative is rather surprisingly the Variable Oystercatcher *H. unicolor* of New Zealand (Thomas *et al.*, 2004a), is a bioclimatic semi-generalist with a multi-latitude (Latitude Category F) discontinuous pan-palearctic range. It is highly migratory except in the mild, oceanic west. The recently extinct Canary Islands Oystercatcher *H. meadewaldoi* was a sedentary endemic of the south-western extreme of the Palearctic. Curiously, this was a species that branched off at an early stage of the oystercatcher radiation, and it was not closely related to either the Eurasian or the African Oystercatcher *H. moquini* (Thomas *et al.*, 2004a). It was a bird of the subtropical (Latitude Category E) belt.

PLUVIALIDAE

We have seen that the genus *Pluvialis* is an ancient one with little connection to other plovers. It has to be removed from the Charadriidae, unless we were to subsume the Haematopodidae and Recurvirostridae into the Charadriidae. Given the deeply rooted separation of these lineages, I prefer to allocate *Pluvialis* into its own family.

Pluvialis

This is a Holarctic genus of four species that has its origins near the K/T boundary. Three species breed in the Palearctic. The Eurasian Golden Plover *P. apricaria* is a bioclimatic moderate of the Boreal (Latitude Category B) belt. It is a highly migratory species that breeds in the Western and Central Palearctic. The Pacific Golden Plover *P. fulva* and the Grey Plover *P. squatarola* are semi-specialist and specialist migrants, respectively, of the Arctic (Latitude Category A) belt of the Central and Eastern Palearctic.

CHARADRIIDAE

The three genera, *Charadrius* and *Eudromias* on the one hand and *Vanellus* on the other, include Palearctic representatives of two lineages that split in the Palaeocene and which should perhaps best be considered as different families (Charadriidae and Vanellidae). I am keeping to convention, however, as most authors place them in a single family.

Vanellus

Six lapwing species (25%) of this ancient, globally distributed genus breed in the Palearctic. Most species are sedentary bioclimatic specialists and semi-specialists. The Northern Lapwing *V. vanellus* is the most widespread in the Palearctic; it is a bioclimatic moderate with a pan-palearctic, multi-latitude (Latitude Category F) range. It is a migratory species except in the mild, oceanic west and south-west, where populations range from short-distance migrants to residents. The Grey-headed Lapwing *V. cinereus* is also a partially migratory bioclimatic moderate (Warm, Latitude Category D) but it is restricted to the Eastern Palearctic.

Three species occupy the subtropical (Latitude Category E), mid-latitude (D) and Temperate (C) belts of the Western and Central Palearctic – Spur-winged Lapwing *V. spinosus* (migratory, semi-specialist), White-tailed Lapwing *V. leucurus* (partially migratory, semi-specialist) and Sociable Lapwing *V. gregarius* (migratory, specialist). The Red-wattled Lapwing *V. indicus* is a fringe resident of the subtropical belt (Latitude Category E) of the Central Palearctic.

Charadrius

This is the wader genus with the most species (31), nine of which (29.03%) breed in the Palearctic. Among them is the only bioclimatic generalist of the entire suborder Charadrii – Little Ringed Plover *C. dubius,* a pan-palearctic, multi-latitude (Latitude Category F) migratory species. Populations in southern Asia are sedentary. This genus appears to follow the 'single generalist species rule', and also includes, typical in such genera, a single semi-specialist – Kentish Plover *C. alexandrinus*; this Holarctic species has a discontinuous pan-palearctic range, centred on the mid-latitude belt (Latitude Category D). It is highly migratory although the southernmost Palearctic populations, as well as those from south-east Asia, are sedentary.

Three species are bioclimatic moderates. The Common Ringed Plover *C. hiaticula* has an Arctic (Latitude Category A) pan-palearctic range and is highly migratory; the Long-billed Plover *C. placidus* is an Eastern Palearctic partially migratory species of the mid-latitude belt (Latitude Category D); and the Lesser Sand Plover *C. mongolus* is a migratory species with a multi-latitude (Latitude Category F) range in the Eastern Palearctic. The remaining four species are bioclimatic specialists – Kittlitz's *C. pecuarius* is an African species at the northern edge of its range in the Western Palearctic; it is a sedentary species of the sub-tropical (Latitude Category E) belt. Greater Sand *C. leschenaultii* and Caspian *C. asiaticus* Plovers are mid-latitude (Latitude Category D) migrants of the Central Palearctic, with marginal presences in the Western Palearctic. Oriental Plover *C. veredus* is a Temperate (Latitude Category C) migrant of the Eastern Palearctic.

Eudromias

This is a single-species genus. The Eurasian Dotterel *E. morinellus* has a pan-palearctic (split latitude A/D) range. It is a bioclimatic moderate (C) and is fully migratory.

CONCLUSIONS

(a) **Bioclimatic generalisation and migration**. Bioclimatic generalists and semi-generalists make up a very low proportion of the Palearctic shorebirds – 12 species, just 8.33%. Most species, as in other groups, have pan-palearctic ranges. 11 of 12 fall into this category, with seven having multi-latitude distributions – Black-headed Gull, Common Tern, Little Tern, Common Sandpiper, Common Redshank, Eurasian Oystercatcher and Little Ringed Plover. The other four are Common Snipe (Latitude Category B), Eurasian Woodcock (Latitude Category C), Kentish Plover (Latitude Category D) and Caspian Tern (Split Category B/D). The twelfth species does not breed in the Eastern Palearctic – Sandwich Tern

(Latitude Category C/D). The lack of bioclimatic tolerance is compensated by migratory behaviour: 94 species (66.67%) are highly migratory and a further 42 (29.79%) are partially migratory. In contrast only five (3.54%) are sedentary.

(b) High latitude representation. Shorebirds are characterised by the large number of species with ranges centred in the high latitudes, with 41 (29.08%) species centred on the Arctic belt (Latitude Category A) and 36 on the Boreal belt (Latitude Category B). Together they represent over half of all the Palearctic species. It seems that bioclimatic generalisation has been sacrificed for specialisation in high-latitude environments, which has only been possible by becoming highly migratory. By contrast there are few (10, 7.09%) multi-latitude species (Latitude Category F). Mid-latitude belt species are not uncommon (25, 17.73%) and constitute a second group of shorebirds. The high-latitude species have, as we would expect, many more pan-palearctic species (27, 35.07%) than mid-latitude species (4, 16%). There are also many Holarctic species in this category, showing the continuity of conditions between Palearctic and Nearctic at these high latitudes.

(c) Nature of the genera. The lineages of shorebirds are ancient and many pre-date the K/T event. Some appear to have remained conservative (e.g. Turnicidae, Dromadidae) with few species, although it is possible that this could represent 'pruning' after an initial radiation. A number of genera (e.g. *Himantopus, Recurvirostra, Haematopus*) appear to have speciated after continental separation and are global, sometimes with ranges fragmented and missing from large areas, and their phylogenetic relationships seem strange (e.g. geographically distant species being more closely related than close ones, as in the oystercatchers). Given their age, these genera may have gone through several phases of expansion, fragmentation and extinction, leading to superimposed phylogeographic patterns that are difficult to unravel. The complex geographical and historical distribution patterns of the Herring Gull group (once considered a ring species but now shown not to be so) or the southern *Catharacta* skuas, relatively recent evolutionary phenomena compared to some of the older lineages, illustrate the difficulty of interpretion. Other genera have radiated and generated many species (e.g. *Charadrius, Calidris, Tringa*).

(d) Representation in the Western Palearctic. The shorebirds are best represented in the Eastern Palearctic (100 species) against 82 in the Central and 84 in the Eastern Palearctic. This is largely due to the large number of auks breeding around the Arctic and Boreal Pacific rim, and also to the extension of the range of a number of Nearctic waders into the Western Palearctic. Only 17 (12.06%) species are exclusively found in the Western Palearctic. Of these, seven are marginal species with subtropical ranges – Kittlitz's Plover, Canary Islands Oystercatcher, Crab Plover, Senegal Thick-knee, Grey-headed Gull, Lesser Crested Tern and Royal Tern. Six others are gulls – Pallas's, Mediterranean, Yellow-legged, Iceland, Audouin's and Armenian. Four species have Boreal ranges along the Atlantic coast – Atlantic Puffin, Razorbill and the extinct Great Auk, plus Greater Black-backed Gull. The remaining species, Roseate Tern, has a Temperate Atlantic distribution. These findings show that there has been a far greater level of speciation and diversification among Pacific shorebirds than among those of the Atlantic.

CHAPTER 14

Divers, tubenoses and waterbirds

Group 3

This group of species, many not previously considered to be related, constitutes Hackett *et al.'s* (2008) 'waterbird' clade (Node H in their definition). Ancestral to this group are two groups of terrestrial and arboreal birds that will be discussed in the next chapter. It includes the penguins (Sphenisciformes, not discussed due to lack of Palearctic representation). In Hackett *et al.'s* (2008) classification the divers (Gaviidae) form the basal group which, using Pratt *et al.'s* (2009) time calibration (their phylogeny resembles Hackett *et al.'s*, 2008, with some modifications – in this chapter I use Hackett *et al.'s* phylogeny and apply Pratt *et al.'s* time estimates), would have split just after the K/T event, estimated at 64.89 mya. This means that this group of waterbirds has its origins *after* that of the shorebirds of the previous chapter.

The next families to branch off were the shearwaters (Procellariidae) and storm-petrels (Hydrobatidae) along with the penguins, around the same time (Pratt et *al.* actually place it slightly before the K/T event at around 68 mya). The next lineage to separate was the storks (Ciconiidae), still in the Palaeocene at 61 mya. In a separate estimate, Fain *et al.* (2007) place the separation of the storks from the herons (Ardeidae) much earlier, at 80.7 mya. At the end of the Palaeocene (*c.* 57 mya) the line leading to the anhingas (Anhingidae), frigatebirds (Fregatidae), cormorants (Phalacrocoracidae) and gannets (Sulidae) split off. Shortly afterwards (probably in the Early Eocene) the lineage that would lead to ibises (Threskiornithidae) and herons (Ardeidae) branched from the pelicans (Pelecanidae). The separation of ibises and herons was somewhat later.

Climate

As in Chapter 13, here I have sorted the species into ecological groups: a) waterbirds (herons, ibises, storks and pelicans); b) shearwaters and storm petrels; and c) smaller groups (cormorants, gannets and divers). Globally, waterbirds are bioclimatic specialists, but the Palearctic species include the most bioclimatically tolerant, and all the generalists (Figure 14.1i). In this respect, they resemble the larids and shorebirds. The shearwaters and storm-petrels are different, however, and resemble the auks. Globally, these seabirds are bioclimatically intolerant, with Palearctic species replicating the wider pattern (Figure 14.1ii).

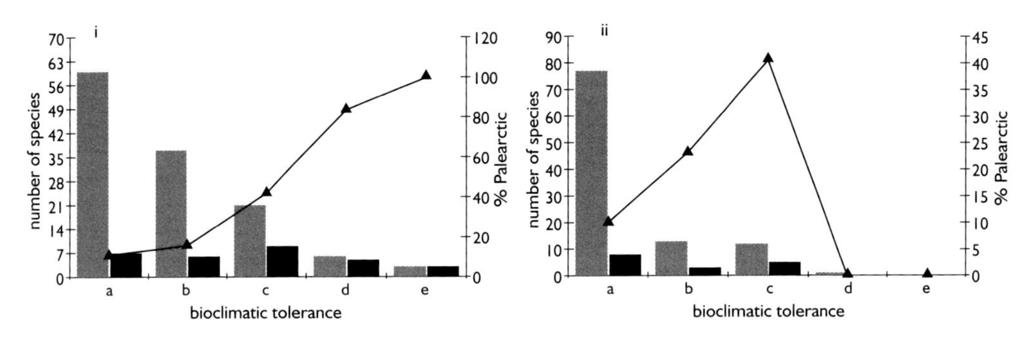

Figure 14.1 *Bioclimatic tolerance of waterbirds (i) and storm-petrels and shearwaters (ii). Definitions as on Figure 4.1 (see p. 38). Key as in Figure 13.1.*

The waterbirds are species of warm and humid climates although they are spread out across the humidity gradient (Figure 14.2). Palearctic species tend away from the warmest and wettest climates but are not frequent at the other end of the climatic spectra. The shearwaters and storm-petrels are also birds of warm and wet climates but Palearctic species are clearly birds of cold climates, tending also away from the wettest (Figure 14.3). Cormorants are spread across the temperature and humidity range: Temperature Categories A (4), B (3), C (15), D (2), E (14); Humidity Categories A (10), B (8), C (5), D (2), E (13). Palearctic species are spread across this range (Appendix 1). Gannets are birds of warm (7 of 9 are Category E) and humid (7 of 9 are categories E and D) climates, with the Northern Gannet *Morus bassanus* occupying the cold (unusual) and wet bioclimate. This is a species that has shifted its position along the temperature gradient, keeping itself within the typical end of the humidity gradient. Finally, the divers are birds of cold and dry climates (Appendix 1). In bioclimatic terms they resemble some of the specialist Arctic shorebirds.

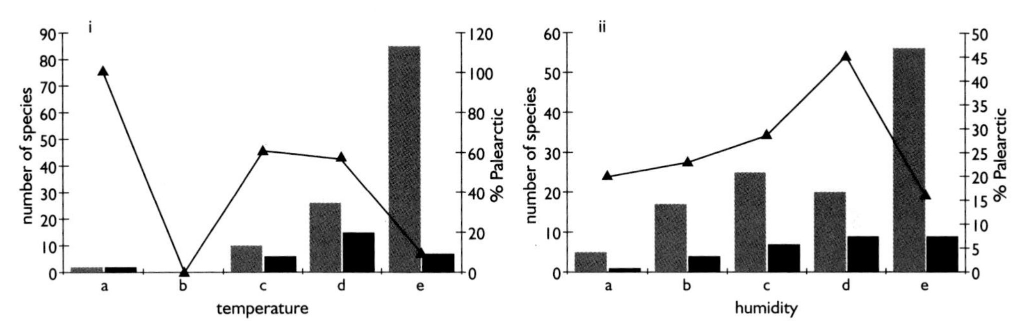

Figure 14.2 *Temperature (i) and humidity (ii) tolerance of waterbirds. Definitions as in Figure 4.2(see p. 38). Key as in Figure 13.1.*

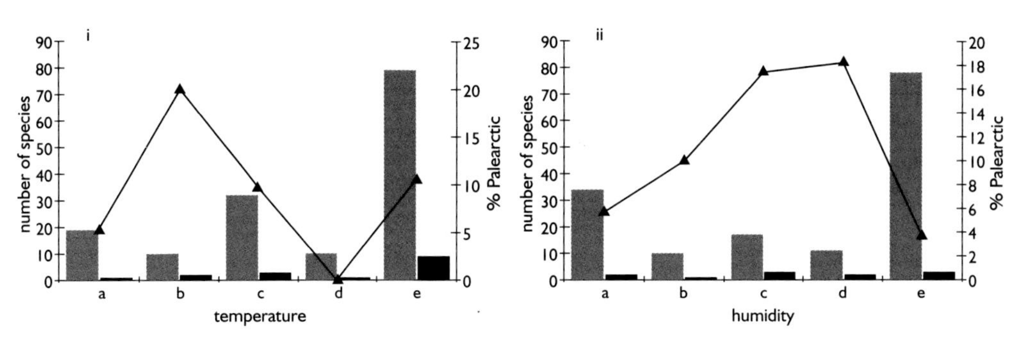

Figure 14.3 *Temperature (i) and humidity (ii) tolerance of storm-petrels and shearwaters. Definitions as in Figure 4.2 (see p. 38). Key as in Figure 13.1.*

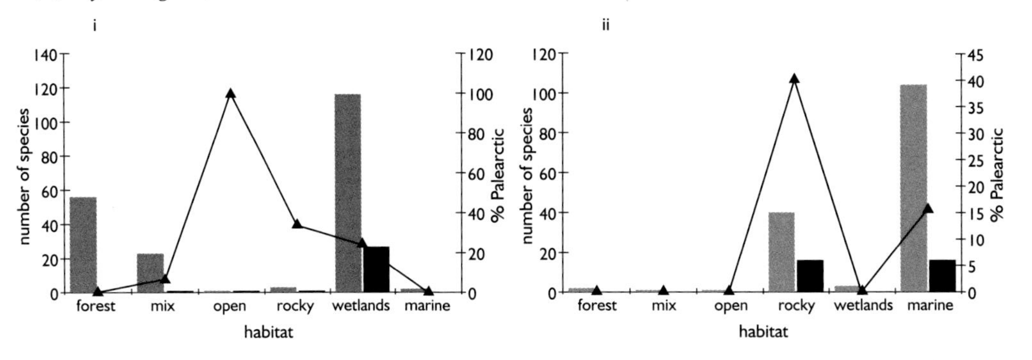

Figure 14.4 *Habitat occupation by waterbirds (i) and storm-petrels and shearwaters (ii). Definitions as in Figure 4.3(see p. 39). Key as in Figure 13.1.*

Waterbirds	Global	Palearctic	Global %	Palearctic %	Palearctic of global %
Migratory	62	27	48.82	90	43.55
Sedentary	116	7	91.34	10	6.03
Total Species	127	30			

Shearwaters	Global	Palearctic	Global %	Palearctic %	Palearctic of global %
Migratory	104	16	99.05	100	15.39
Sedentary	1	0	0.95	0	0
Total Species	105	16			

Table 14.1 *Migratory behaviour in global and Palearctic Procelariiformes and allies.*

Habitat

The Palearctic waterbirds show little departure from the typical wetland habitats that all species exploit except that a few occur in the mixed or forested habitats alongside (Figure 14.4i). The shearwaters resemble the auks in their concentration on rocky and marine habitats (Figure 14.4ii). The cormorants are birds of rocky (22 species), wetland (35) and marine (35) habitats, with the Palearctic species replicating the global picture (Appendix 1). The gannets are similar, while the divers are birds of wetlands.

Migratory behaviour

The waterbirds are mainly sedentary, but they include a significant number of migratory species or species with some migratory populations. Palearctic species are overwhelmingly migratory (Table 14.1). This behaviour has permitted the exploitation of seasonal water habitats and, along with broad bioclimatic tolerance, has marked the success of these birds in the Palearctic. The shearwaters and storm-petrels derive their success from the ability to wander over huge distances of ocean, and the Palearctic species reflect this nature (Table 14.1). This wide-ranging behaviour and the ability to exploit rocky surfaces for nesting have enabled the success of these bioclimatic specialists in the Palearctic, in a similar manner to the auks. The cormorants include migratory (14) and sedentary (35) populations globally and also in the Palearctic (Appendix 1). Their success seems to include features of the other groups – use of rocky habitats, migratory behaviour, and some level of bioclimatic tolerance. The gannets include migratory (9) and sedentary (6) populations; the Northern Gannet is partially migratory, depending on age (Finlayson, 1992), and also exploits rocky habitats for breeding. Finally, the divers are highly migratory. Their success in Arctic waters has depended on this behaviour.

Fossil species

Some species are present in the fossil record of the Late Pliocene, with most others present from the Early Pleistocene (Appendix 2). Little can be gleaned from species that provide little in the way of fossils, except for the southern presence of Red-throated Diver down to the Strait of Gibraltar, where today it is a rare winter visitor.

Taxa in the Palearctic

GAVIIDAE

Gavia

All five species in the genus breed in the Palearctic. The divers are a Holarctic genus of migratory bioclimatic specialists and semi-specialists. The Black-throated Diver *G. arctica* is the most bioclimatically tolerant species (moderate) and has a continuous, Boreal (Latitude Category B), pan-palearctic range. Red-throated *G. stellata,* Great Northern *G. immer* and Pacific *G. pacifica* Divers are Arctic (Latitude Category A) semi-specialists; Red-throated has a continuous pan-palearctic range, while Great Northern Diver is confined to the Western Palearctic, Pacific Diver to the Eastern Palearctic. The fifth species, White-billed Diver *G. adamsii,* is an Arctic belt specialist that breeds in the Central and Western Palearctic.

PROCELLARIIDAE

Fulmarus

This is the basal genus of the Palearctic tubenoses. It is closely related to the southern hemisphere genera *Macronectes, Daption, Thalassoica* and *Pagodroma* (Kennedy and Page, 2002). The Northern Fulmar *F. glacialis* represents a dispersal into the North Atlantic from the southern hemisphere, where the only other species in the genus, the Southern Fulmar *F. glacialoides,* breeds. Northern Fulmar is a migratory bioclimatic semi-specialist of the Boreal (Latitude Category B) belt, breeding in the Western and Central Palearctic.

Calonectris

All three species in this genus – Cory's Shearwater *C. diomedea,* Streaked Shearwater *C. leucomelas* and Cape Verde Shearwater *C. edwardsii* – breed in the Palearctic. Cory's is a migratory bioclimatic semi-specialist of the mid-latitude belt (Latitude Category D) and is exclusively Western Palearctic within the region. Streaked and Cape Verde Shearwaters are specialists of subtropical (Latitude Category E) warm and wet climates in the Eastern and Western Palearctic respectively. The ancestral species seems to have been the Streaked or its ancestor, and the divergence of Atlantic from Pacific populations appears to have been linked to the closing of the Panama isthmus around 3 mya. Subsequent range contractions during glaciations separated the Atlantic (*C. d. borealis*), Mediterranean (*C. d. diomedea*) and Cape Verde populations (Gómez-Díaz *et al.*, 2006).

Puffinus

These shearwaters are derived from two ancestral groups that separated soon after the split of the genus from the other shearwaters. A northern hemisphere-derived population established itself in the southern hemisphere, from which four groups of species emerged. The northern hemisphere group, which includes the four species (20% of the world total) that breed in the Palearctic, was formed when water bodies around the Atlantic and Central Europe fragmented in the Late Tertiary. It was followed by secondary dispersals into the Indian and Pacific Oceans (Austin, 1996). The Little Shearwater *P. assimilis baroli* of the North Atlantic is apparently not closely related to its southern hemisphere counterpart but is instead close to Audubon's Shearwater *P. lherminieri*; together they form the basal group of the Palearctic species. The other three (Manx *P. puffinus,* Balearcic *P. mauretanicus* and Yelkouan *P. yelkouan* Shearwaters) are closely related, and the latter two are sister-species (Heidrich *et al.*, 1998; Kennedy and Page, 2002).

Palearctic *Puffinus* are all confined to the west, in the Atlantic and Mediterranean. Two are bioclimatic moderates – Manx Shearwater centred on the Temperate belt (Latitude Category C) and Little Shearwater on the mid-latitude belt (Latitude Category D). The other two, Balearic and Yelkouan, are mid-latitude belt bioclimatic specialists.

Pterodroma

This is a large genus of largely warm and wet bioclimatic specialists of the southern hemisphere, two of which (6.9% of the species) breed on the Atlantic islands of the south-western Palearctic – Zino's Petrel *P. madeira* and Fea's Petrel *P. feae.* Like all species of the genus, they are migratory. They are bioclimatic specialists of the subtropical (Latitude Category E) belt.

Bulweria

Bulwer's Petrel *B. bulwerii* is one of two species in the genus. Like the Palearctic *Pterodroma* it is a specialist of warm and wet bioclimates of the subtropical belt (Latitude Category E), breeding on the islands of the south-western Palearctic.

HYDROBATIDAE

Oceanodroma

Three species (23.08%) of this largely Pacific genus breed in the Palearctic. Leach's Storm-petrel *O. leucorhoa* is a migratory bioclimatic semi-generalist that breeds in the Boreal (Latitude Category B) belt of the Western and Eastern Palearctic. Swinhoe's Storm-petrel *O. monorhis* is a migratory bioclimatic moderate of the subtropical (Latitude Category E) belt of the Eastern Palearctic. Finally, the Band-rumped (or Madeiran) Storm-petrel *O. castro* is a bioclimatic specialist that breeds along the sub-tropical belt of the Western and Eastern Palearctic.

Hydrobates

This is an exclusively Western Palearctic single-species genus. The European Storm-petrel *H. pelagicus* is a migratory bioclimatic moderate with a multi-latitude (Latitude Category F) range.

Pelagodroma

The White-faced Storm-petrel *P. marina,* the only species of its genus, is a widespread migratory pelagic seabird of southern waters that reaches the Atlantic islands of the south-western Palearctic. It is a bioclimatic moderate which, in the Palearctic, occupies the mid-latitude (Latitude Category D) belt position in the extreme west.

CICONIIDAE

Ciconia

Three (42.86%) of the seven species in this genus breed in the Palearctic. The Black Stork *C. nigra* appears to represent an early split in the genus, which then forms two clusters – (a) a tropical African–southern Asian group consisting of Abdim's *C. abdimii,* Storm's *C. stormi* and Woolly-necked *C. episcopus* Storks; and (b) a cluster including the only American species – the Neotropical Magauri Stork *C. magauri* – and the Palearctic sister-species White *C. ciconia* and Oriental *C. boyciana* Storks (Slikas, 1997). The closeness of the Magauri Stork to Palearctic species rather than the Afro-Asian ones suggests an entry into the Americas from the Old World after the Palearctic–tropical species had split. Given the absence of storks of this genus in the Nearctic, the Palearctic storks represent the northern limit of the genus range.

The Palearctic storks are the only migratory species in the genus. Recent behavioural changes among south-western Palearctic White Storks, which attend refuse tips, have permitted these populations to remain sedentary. The Black Stork, the most divergent and ancient species, is a bioclimatic semi-generalist with a Temperate (Latitude Category C) belt range which spans the Palearctic, though it is discontinuous in the west. It is the most bioclimatically tolerant and widespread species. The White Stork is a moderate of the Western and Central Palearctic and has a multi-latitude (Latitude Category F) range. The bioclimatic specialist Oriental Stork replaces it in the Eastern Palearctic, but its range is centred on the Warm (Latitude Category D) belt.

FREGATIDAE

Fregata

The Magnificent Frigatebird *F. magnificens* is the only species of five in the genus that reaches the Palearctic, on the islands of the south-west. It is a bioclimatic specialist of the warm and wet subtropical (Latitude Category E) belt.

PHALACROCORACIDAE

Phalacrocorax

Seven of 38 species (18.42%) breed in the Palearctic. Five species form part of a global cluster of species that include large and medium-sized cormorants, as well as cliff shags (Kennedy *et al.*, 2000). At its base are two Palearctic sister-species – Red-faced Cormorant *P. urile* and Pelagic Cormorant *P. pelagicus.* They are both Eastern Palearctic species. The first is a bioclimatic semi-specialist of the Warm (Latitude Category D) belt and is partially migratory; the second is a semi-generalist of the Boreal (Latitude Category B) belt and is largely sedentary. These two species cluster with the American species, Brandt's Cormorant *P. penicillatus*. The European Shag *P. aristotelis* branches next on its own. It is a sedentary bioclimatic multi-latitude (Latitude Category F) semi-generalist of the Western Palearctic. The most recent separation in this cluster separates the sister-species Great *P. carbo* and Japanese *P. capillatus* Cormorants. The Great Cormorant is a semi-generalist with a split B/D range. The mid-latitude (Latitude Category D) Great Cormorants (subspecies *sinensis*) have a discontinuous pan-palearctic range and are highly migratory. Boreal (subspecies *carbo*) Great Cormorants are partially migratory and are restricted to the Western Palearctic (Atlantic coasts).

Two other species of cormorants lie outside this cluster. The Long-tailed Cormorant *P. africanus* is an African species with the northern edge of its range along the southern limits of the Western Palearctic. It is a sedentary semi-specialist with a subtropical (Latitude Category E) Palearctic range. The Pygmy Cormorant *P. pygmaeus* is a partially migratory bioclimatic specialist of the mid-latitude (Latitude Category D) belt of the Western and Central Palearctic.

SULIDAE

Morus

The gannets (*Morus*) and boobies (*Sula*) are distinct lineages within the Sulidae. Of the three gannets, the Northern Gannet separated out first from the other two species, which have southern-hemisphere distributions (Friesen and Anderson, 1997). The Northern Gannet is a migratory bioclimatic specialist of the Boreal (Latitude Category B) belt of the Western Palearctic.

THRESKIORNITHIDAE

Platalea

Two main divisions of this family are represented by the subfamilies Plateinae (spoonbills) and Threskiornithinae (ibises). The spoonbills of the genus *Platalea* have a wide Old World distribution that includes tropical Africa and Asia, New Guinea and Australasia. The Palearctic is therefore the northern part of the genus range. Two of the five species breed in the Palearctic – Eurasian Spoonbill *P. leucorodia* and Black-faced Spoonbill *P. minor*. The Eurasian Spoonbill is a bioclimatic moderate with a mid-latitude (Latitude Category D) pan-palearctic range, which is discontinuous in the west and centre. Palearctic Eurasian Spoonbills are highly migratory, except in the extreme south-west. Outside the Palearctic, the southern Asian populations are sedentary. The success of the spoonbills in the Palearctic has been heavily dependent on the ability to migrate. The Black-faced Spoonbill *P. minor* is a migratory bioclimatic semi-specialist of the Boreal (Latitude Category B) belt of the Eastern Palearctic.

Plegadis

This genus is globally distributed and the sole Palearctic species – Glossy Ibis *P. falcinellus* – also breeds in North and South America, tropical Africa, and southern Asia to Australasia. Palearctic Glossy Ibises are at the northern edge of the genus and species range. These birds are migratory bioclimatic moderates of the mid-latitude (Latitude Category D) belt, breeding in the Western and Central Palearctic where they are discontinuously distributed. Elsewhere they are sedentary.

Geronticus

In contrast to *Plegadis*, *Geronticus* has a very limited range, which is due in part to human disturbance. It must have once had a wide distribution across the mid-latitude belt and the southern extension through East Africa. Today the Southern Bald Ibis *G. calvus* is restricted to southern Africa, and the Northern Bald Ibis or Waldrapp *G. eremita*, a migratory bioclimatic semi-specialist, to very limited areas of the mid-latitude belt of the Western Palearctic. *Geronticus* is therefore part of the mid-latitude belt rocky habitat avifauna.

Threskiornis

The five species in this genus have a wide Old World distribution that includes tropical Africa, southern Asia and Australasia. The Black-headed Ibis *T. melanocephalus* is a bioclimatic specialist of tropical southern Asia with an isolated migratory Eastern Palearctic (Temperate belt, Latitude Category C) population, which is at the northernmost edge of the genus range.

Nipponia

The Crested Ibis *N. nippon* belongs to a localised, single-species genus. It is a sedentary bioclimatic specialist of the Warm (Latitude Category D) belt of the Eastern Palearctic, which had a wider range in the past but has suffered severely from human activity.

ARDEIDAE

Botaurus

This genus is closely related to *Ixobrychus,* and basal to the rest of the genera of Palearctic herons (Sheldon *et al.*, 2000). It is a global genus that is missing from tropical southern Asia and New Guinea, but which re-appears in Australasia. The Great Bittern *B. stellaris* is the only Palearctic representative. It is also found in the northern Afrotropical region, and is at the northern end of the genus and species range. Great Bittern is a bioclimatic generalist of the Temperate (Latitude Category C) belt and has a pan-palearctic range that is discontinuous in the west. Palearctic populations are highly migratory.

Ixobrychus

This is another globally distributed genus with representatives across the Old World to Australasia and in the New World. Four of nine (44.44%) species breed in the Palearctic. The Little Bittern *I. minutus* is a bioclimatic semi-generalist and the most bioclimatically tolerant of the genus, which seems to follow the 'single bioclimatically tolerant species' rule (in this case a semi-generalist and not a generalist). The Little Bittern's Palearctic range follows the mid-latitude belt (Latitude Category D) in the Western (discontinuous) and Central Palearctic; typically for birds along this range that are not rocky-habitat species, they are missing from the east. Palearctic Little Bitterns are highly migratory, contrasting with African and southern Asian populations of the species, which are sedentary. Thus, it is the birds at the northern end of the range that rely on migration for survival. The absence of the Little Bittern across large areas of the Sahara, for example, illustrates the degree to which aridity has fragmented a range that must have once connected Palearctic and African Little Bitterns.

The remaining three *Ixobrychus* species are all Eastern Palearctic – Schrenck's Bittern *I. eurhythmus* (migratory, bioclimatic moderate, Latitude Category D), Yellow Bittern *I. sinensis* (migratory, bioclimatic semi-specialist, Latitude Category E) and Cinnamon Bittern (migratory, bioclimatic specialist, Latitude Category E).

Gorsachius

This is a genus of five tropical south-east Asian and American night herons that are closer to the bitterns and little bitterns than they are to the true night herons (McCracken and Sheldon, 1998). One species – Japanese Night Heron *G. goisagi* – breeds in the Eastern Palearctic. It is a migratory bioclimatic specialist of the subtropical (Latitude Category E) belt. This genus does not appear to have reached (or survived in) the Central and Western Palearctic, even though it is present in Africa and has also reached the Americas.

Nycticorax

The remaining herons form three clusters (Sheldon *et al.*, 2000). The first includes the *Nycticorax* night herons and the non-Palearctic *Nyctanassa.* One of two *Nycticorax* – the Black-crowned Night Heron *N. nycticorax* – breeds in the Palearctic. It has a global range that includes the New World, Palearctic, Africa and southern Asia and some Pacific islands. It is replaced in parts of south-east Asia, New Guinea and Australasia by the Rufous Night Heron *N. caledonicus.* The Black-crowned Night Heron is sedentary in tropical regions but in the Palearctic, the northern part of the species and genus range, it is highly migratory. It is a bioclimatic generalist with a discontinuous pan-palearctic range centred on the mid-latitude belt (Latitude Category D). Unusually for such a species it has occupied the east, although this may reflect a separate colonisation from the south by this widely distributed bird.

Butorides

The second of the three clusters includes *Butorides, Ardeola, Bubulcus, Casmerodius* and *Ardea. Butorides*, which is basal to the cluster, has three species, two American and a globally distributed third – Striated, or Green-backed, Heron *B. striatus*. This is a bioclimatic moderate that is sedentary across large areas of tropical Africa and southern Asia, from where the Eastern Palearctic is reached by a population that is highly migratory. It occupies the mid-latitude belt (Latitude Category D) and its distribution is continuous with the tropical south. It is another genus, like *Gorsachius*, that is missing from the Western and Central Palearctic. The connectivity of the Eastern Palearctic with the tropical south, unlike the rest of the Palearctic which is separated from the tropical world by mountains, seas and deserts, must have been a critical factor in the species' Palearctic distribution.

Ardeola

Another early separation from the second cluster is the genus *Ardeola* (McCracken and Sheldon, 1998). This is an Old World genus of six species that are mainly tropical African–southern Asian; they are absent from the New World, New Guinea and Australasia. The Palearctic represents the northern end of the genus range. Two species breed in the Palearctic. The Squacco Heron *A. ralloides* breeds in tropical Africa, where it is sedentary, and across the Western and Central Palearctic, where it has a discontinuous range. It is the most bioclimatically tolerant species of its genus (moderate) and occupies the mid-latitude belt (Latitude Category D), which may account for its absence in the Eastern Palearctic; there it is replaced by the Chinese Pond Heron *A. bacchus*, a bioclimatic specialist (Latitude Category D). This species also has sedentary populations to the south, but within the Palearctic it, like the Squacco, is highly migratory. The range of the Chinese Pond Heron is much more continuous than the heavily fragmented range of the Squacco Heron to the west. This probably reflects the continuity of Palearctic with tropical populations in the east and not in the west. This would mean that the fragmented populations of Western and Central Palearctic Squacco Herons are a stage behind *Gorsachius* and *Butorides*, which are missing altogether from these regions. In this case the Squacco Heron's relatively good bioclimatic tolerance may be permitting it to retain a foothold in the Western and Central Palearctic.

Bubulcus

This single-species genus forms the nucleus of the second cluster with *Casmerodius* and *Ardea*. It provides an example of a rapid (including trans-oceanic) colonisation of regions of the world; the spread of Cattle Egret *B. ibis* may have been favoured by its unheronlike habit of associating with ungulates, so its spread has in some measure to do with cattle farming. This expansion has been spectacular, involving the colonisation of the Americas (Crosby, 1972) and South Africa (Kopij, 2008) during the 20th Century. This gives us a glimpse of the powers of dispersal of the herons, which may account for the global ranges of many species. The Cattle Egret is a bioclimatic semi-generalist with a very fragmented mid-latitude (Latitude Category D) pan-palearctic range. Tropical populations and those of the extreme south-western Palearctic (Iberian Peninsula) are largely sedentary but those of the continental interior of the Palearctic are migratory. A combination of bioclimatic tolerance and migratory potential (as well as dispersal abilities) represents the key to the species' recent success.

Casmerodius

This is another global single-species genus that occurs in the New World, the Palearctic and much of tropical Africa and southern Asia. The Great White Egret *C. albus* is a bioclimatic generalist with a mid-latitude (Latitude Category D) pan-palearctic range, discontinuous in the west and centre. Tropical populations are sedentary but Palearctic birds, with the exception of those that have recently established themselves in south-western Iberia, are highly migratory.

Ardea

Two of eleven (18.2%) species of this globally distributed genus breed in the Palearctic. Goliath Heron *A. goliath* could perhaps be considered a third species on account of a small population that is borderline with the southern edge of the Western Palearctic. The Grey Heron *A. cinerea* is a widespread sedentary species of the tropical Asian region and Madagascar but is absent from much of Africa. It has a multi-latitude (Latitude Category F) pan-palearctic range that is discontinuous in the west. Grey Herons are bioclimatic semi-generalists and are highly migratory in the Palearctic except in the oceanic west, where most are sedentary or short-distance migrants. In contrast, the slightly smaller Purple Heron *A. purpurea* is fully migratory in the Palearctic, but tropical African and Asian populations are sedentary. The Purple Heron is also a semi-generalist, providing one of the few exceptional cases of more than one bioclimatically tolerant species in a genus (notably neither are full generalists). Purple Heron has a fragmented pan-palearctic range that is centred on the mid-latitude belt (Latitude Category D).

Egretta
This globally distributed genus forms the third heron cluster, along with *Syrigma*, which is basal to the cluster. Five of 13 (38.5%) species breed in the Palearctic. The Little Egret *E. garzetta* has a broad distribution across Africa and southern Asia through to Australasia; the Palearctic Little Egrets are at the northern end of the species and genus range and, in contrast with sedentary tropical populations, are highly migratory. This species is a bioclimatic moderate with a discontinuous, mid-latitude belt (Latitude Category D), range in the Western and Central Palearctic. The Pacific Reef Egret *E. sacra* is also a bioclimatic moderate but its range is limited to the subtropical belt (Latitude Category E) of the Eastern Palearctic, where it has remained sedentary. A comparison of Little and Pacific Reef Egrets, both bioclimatic moderates, would strongly suggest that it is the migratory habit that has permitted the wider Palearctic range of the Little Egret. This conclusion seems strengthened by the marginal range of the semi-specialist sedentary Western Reef Egret *E. gularis* on the subtropical belt of the Western Palearctic.

Two other species – Intermediate *E. intermedia* and Chinese *E. eulophotes* Egrets, semi-specialist and specialist respectively – occupy the Eastern Palearctic. The Intermediate Egret occupies the subtropical (Latitude Category E) and the Chinese the mid-latitude (Latitude Category D) belts. Both are migratory but have sedentary populations to the south.

PELECANIDAE

Pelecanus
All eight pelicans are included in this globally distributed genus, which includes two Palearctic species on the northern edge of the genus range. The pelicans make up a group of bioclimatic specialists of which the Great White *P. onocrotalus* of the Palearctic is the most bioclimatically tolerant (moderate). It is sedentary across tropical Africa and Asia but highly migratory in the Palearctic. Its range is centred on the mid-latitude (Latitude Category D) belt, where is has a discontinuous distribution, like many other waterbirds, along the Western and Central Palearctic. The semi-specialist Dalmatian Pelican *P. crispus* is exclusive to the mid-latitude belt of the Western and Central Palearctic, where it is migratory with the exception of those in the extreme south-west of the range.

CONCLUSIONS

(a) **Bioclimatic generalisation and migration**. Twelve species (19.67%) are bioclimatically tolerant (Appendix 1), a figure comparable to the raptors and much higher than the shorebirds. Eight of the twelve have pan-palearctic ranges but only one – Grey Heron – has a multi-latitude range. The others are mostly mid-latitude – Great White Egret, Black-crowned Night Heron, Purple Heron and Cattle Egret – although these ranges are often fragmentary. To these we should add the Great Cormorant; though it has a split B/D distribution, its pan-palearctic range is only along the mid-latitude belt. Great Bittern and Black Stork belong to the Temperate (Latitude Category C) belt. The other generalists have restricted ranges – European Shag (multi-latitude, Latitude Category F), Pelagic Cormorant and Leach's Storm-petrel (Boreal, Latitude Category B). Migratory behaviour is an added factor favouring survival in the Palearctic. It is even more highly developed in this group than in the shorebirds (fully migratory 45 species, 73.77% against 66.67% in the shorebirds). The higher proportion of partial migrants among shorebirds means, however, that overall this group has more sedentary species (7, 11.48%) than the shorebirds (5, 3.54%). A combination of bioclimatic generalisation and migratory behaviour has marked the success of the Palearctic species in this group.

(b) **Mid- and low-latitude representation.** The birds of this group contrast strongly with the shorebirds in another way. A very low proportion are Arctic (Latitude Category A, 4 species, 6.56%) or Boreal (Latitude Category B, 6 species, 9.84%). In contrast, the group is dominated by species of the mid-latitude belt (Latitude Category D, 27 species, 44.26%) with a good representation of subtropical belt species (Latitude Category E, 15 species, 24.59%). Because many are highly migratory, they have succeeded in colonising and remaining in the arid regions of the mid-latitude belt, particularly by exploiting seasonal water bodies.

(c) **Nature of the genera.** A striking feature of this group is the large number of genera and species with global distributions for which the Palearctic is the northern part of the global range – *Ciconia, Platalea, Plegadis, Threskiornis, Botaurus, Ixobrychus, Gorsachius, Nycticorax, Butorides, Ardeola, Casmerodius, Ardea, Egretta* and *Pelecanus*. Connectivity between Eastern Palearctic and the tropical south has permitted the survival of species and genera that are missing to the west (e.g. *Gorsachius, Butorides*) or have marginal or fragmented ranges in the Central and Western Palearctic (e.g. *Ardeola*). One genus, now heavily fragmented, seems to follow the pattern of rocky-dwelling mid-latitude belt and southern extension (to South Africa) species (as seen in e.g. *Gypaetus, Oenanthe* or *Monticola*)- it is the ibis genus *Geronticus*.

A number of marine taxa appear as marginal in the Palearctic, either in terms of species or genera, being mainly birds of the southern hemisphere. *Pterodroma, Fulmarus, Fregata* and *Morus* fall under this category. They contrast with the auks, a northern-hemisphere group. Another genus – *Oceanodroma* – is largely Pacific in distribution, with some species reaching the Palearctic. Two marine genera – *Calonectris* and *Puffinus* – have species and subspecies that reflect the break-up of ancient oceans and seas.

(d) **Representation in the Western Palearctic.** The species in this group are best represented in the Western Palearctic (41 species) against 33 in the Eastern Palearctic and only 22 in the Central Palearctic. The presence of seabirds in the Western and Eastern Palearctic but not in the Central is largely responsible for this disparity.

CHAPTER 15

Cranes, rails, bustards and cuckoos

Group 4

This group is ancestral to the waterbirds of the previous chapter (Hackett *et al.*, 2008). Cranes, rails, bustards and cuckoos all have their origins in the Cretaceous and are recognisable as separate lineages before the K/T event (van Tuinen *et al.*, 2006; Brown *et al.*, 2008).

Climate

All four groups are characterised by low bioclimatic tolerance, and the Palearctic species are in all cases at the most tolerant end of the spectrum for each family (Table 15.1). With the exception of the cranes, which occupy a range of temperatures, these families are typical of the warmest climates, with Palearctic species shifting in all cases away towards cooler positions on the gradient. Cuckoos and Rails are at the humid end of the gradient with the Palearctic species drifting away from the most humid climates but not occupying the driest either. The bustards and cranes are at the dry end of the spectrum, which is where the Palearctic species are situated.

Habitat

Cuckoos are forest birds and the Palearctic birds are not exceptional in this respect (Table 15.2). Cranes and rails are predominantly wetland birds and Palearctic species are, once again, not exceptional in this respect. The bustards occupy open country preferentially and this includes the Palearctic species. Overall, the Palearctic birds do not deviate from the global pattern.

		A	B	C	D	E
Tolerance	Cuckoos	0/51	1/21	3/5	1/1	1/1
	Cranes	5/9	0/3	2/3	0/0	0/0
	Rails	1/105	3/21	5/12	3/3	0/1
	Bustards	1/13	2/12	1/1	0/0	0/0
Temperature	Cuckoos	0/0	1/1	4/5	0/1	1/72
	Cranes	2/2	0/0	4/5	0/0	0/6
	Rails	1/2	2/3	6/20	4/4	2/106
	Bustards	0/0	0/0	0/0	3/3	1/22
Humidity	Cuckoos	0/3	1/6	1/8	1/4	3/58
	Cranes	4/6	2/4	0/1	0/1	1/3
	Rails	0/5	0/10	5/14	5/9	3/97
	Bustards	3/11	1/6	0/5	0/3	0/3

Table 15.1 *Bioclimatic characteristics of cranes, cuckoos and allies (Palearctic species/global species).*

	Forest	Mixed	Open	Rocky	Wetland
Cuckoos	5/67	1/64	0/2	0/0	0/15
Cranes	0/1	0/5	1/11	0/0	6/13
Bustards	0/0	0/19	4/23	0/3	0/0
Rails	0/38	0/44	1/46	0/1	11/116

Table 15.2 *Habitats occupied by cranes, cuckoos and allies (Palearctic species/global species).*

Migratory behaviour

Cuckoos, bustards and rails are predominantly sedentary species, with the Palearctic species exceptional in being largely migratory (Table 15.3). The cranes are balanced between sedentary and migratory species but Palearctic ones are all migratory. Migratory behaviour seems the key to the success of the few species that occupy the Palearctic, these species being stereotypic with regards to habitat and showing a limited degree of bioclimatic tolerance.

Cuckoos	Global	Palearctic	Global %	Palearctic %	Palearctic of global %
Migratory	18	6	22.78	100	33.33
Sedentary	72	0	91.14	0	0
Total Species	79	6			

Cranes	Global	Palearctic	Global %	Palearctic %	Palearctic of global %
Migratory	9	7	60	100	77.78
Sedentary	8	0	53.33	9	0
Total Species	15	7			

Bustards	Global	Palearctic	Global %	Palearctic %	Palearctic of global %
Migratory	3	3	12	75	100
Sedentary	25	3	100	75	12
Total Species	25	4			

Rails	Global	Palearctic	Global %	Palearctic %	Palearctic of global %
Migratory	19	10	13.38	83.33	52.63
Sedentary	137	4	96.48	33.33	2.92
Total Species	142	12			

Table 15.3 *Migratory behaviour in Palearctic and global cranes, cuckoos and allies.*

Fossil species

A number of species first appear in the fossil record in the Pliocene: Common Crane *Grus grus*, Little *Tetrax tetrax* and Great *Otis tarda* Bustards, Spotted Crake *Porzana porzana*, Water Rail *Rallus aquaticus* and Common Moorhen *Gallinula chloropus* (Mlíkovský, 2002). The rest are present by the Early Pleistocene. A feature of interest is the high frequency of sites with birds of open and grassland country in the Late Pleistocene, including Corncrake *Crex crex* and the bustards (Appendix 2). There is also a high frequency of wetland species, especially rails and Common Crane. These results may be indicative of the westward spread of steppe and of significant areas taken up by wetlands.

Taxa in the Palearctic

OTIDIDAE

The present day species shared a common ancestor at the end of the Oligocene or beginning of the Miocene (22–26 mya, Pitra *et al.*, 2002), arguing against a pre K/T event origin for the bustard lineage. But van Tuinen *et al.*'s (2006) more recent analysis places the origin of the lineage in the Cretaceous. I interpret these results to show a radiation of the lineage during the Tertiary from a line with an ancient origin. The radiation, if Pitra *et al.*'s estimates are correct, would have occurred at a time of increasing aridity and loss of forests, which would have favoured these birds of open habitats.

The initial radiation seems to have been rapid, and its origin has been placed in tropical Africa (Pitra *et al.*, 2002). The result is dependent on the way in which the various regions were defined, and I would suggest that a more parsimonious conclusion would be that the origin and radiation of modern bustards took place along the drying belts, where tropical forests were giving way to open habitats, along what is today northern, eastern and southern Africa and Arabia, with a subsequent spread along the opening corridors of habitat across the mid-latitude belt, including the southern Palearctic. Two main clusters of bustard genera emerged. One is represented today mainly in tropical Africa but also South-east Asia and Australasia, while the other has African, southern Asian and Palearctic species.

Tetrax

The bustards are all specialists or semi-specialists of warm, usually dry, climates, with the exception of the Little Bustard, which is a bioclimatic moderate. This is the only species of its genus and clusters in a subgroup of species that do not include the other Palearctic bustards, branching off early and seemingly standing alone. It is exclusively Palearctic and has a fragmented range across the mid-latitude belt (Latitude Category D) of the Western and Central Palearctic. It therefore follows the pattern observed among many other species of the mid-latitude belt (fragmented range, absence in the east). It is highly migratory over much of its range except the south-west, where it is largely sedentary in Iberia.

Otis

This genus is sister to the other Palearctic genus, *Chlamydotis*. They stand alone, with the closest genus being the South African *Afrotis*. The Great Bustard is the only species of this exclusively Palearctic genus. It is a bioclimatic semi-specialist of the mid-latitude belt (Latitude Category D), with a discontinuous pan-palearctic range. Central and Eastern Palearctic populations are fully migratory but those in the west are largely sedentary.

Chlamydotis

The two species in this genus are geographical counterparts and sister-species. The pattern of distribution recalls other species pairs (Tawny and Steppe Eagles, African and Asian Desert Warblers). These species pairs may be the product of a latitude step between the Western and Central Palearctic, the deserts occupying a more northerly position in the Central than in the Western. In the bustards, MacQueen's Bustard *C. macqueenii* is a migratory semi-specialist of the mid-latitude belt (Latitude Category D) of the Central Palearctic, and Houbara Bustard *C. undulata* is a sedentary specialist of the subtropical belt (Latitude Category E) of the Western Palearctic.

GRUIDAE

The cranes are closely related to the rails (Rallidae). The two lineages split before the K/T event around 73.2 mya (Fain *et al.*, 2007). The cranes then separated from the Aramidae (Limpkin *Aramus guarauna*) in the Middle Eocene (48.5 mya). The radiation of the cranes started with the split of the Gruinae and Balearicinae (crowned cranes) in the Early Oligocene (*c.* 31.4 mya), which was followed by a rapid radiation of the cranes leading to the genera *Grus, Anthropoides* and *Bugeranus.* These cranes are mostly bioclimatic specialists and semi-specialists, with three species (including two Palearctic cranes – Sandhill *G. canadensis,* a Holarctic species, and Common Crane *G. grus*) and the Nearctic Whooping Crane *G. americana* being bioclimatic moderates. A feature of all Palearctic cranes is that they are highly migratory.

Leucogeranus

The Siberian Crane *L. leucogeranus* is the basal species of the cranes, separating and remaining on its own in the phylogenetic tree (Fain *et al.*, 2007; Krajewski *et al.*, 2010). It is a bioclimatic specialist of the Arctic (Latitude Category A) belt and has a fragmented Central and Eastern Palearctic range.

Grus

The White-naped Crane *G. vipio* is part of a basal cluster of species in this genus with the Sarus Crane *G. antigone* of southern Asia and New Guinea and the Brolga Crane *G. rubicunda* of Australasia. Together they form an eastern cluster of which White-naped is at the northern end of the joint range. White-naped is a specialist of the Temperate (Latitude Category C) belt of the Eastern Palearctic. The Sandhill Crane is the next to branch off. It is a bioclimatic moderate of the Arctic (Latitude Category A) belt of the Eastern Palearctic and is therefore a more northerly counterpart of the White-naped Crane.

A second cluster of cranes includes three Palearctic species – Common Crane, a moderate with a split Boreal/mid-latitude (Latitude Category B/D) range in the Western and Central Palearctic; and Hooded *G. monacha* and Red-crowned *G. japonensis* Cranes, specialists of the Temperate (Latitude Category C) belt of the Eastern Palearctic.

Anthropoides

Another group branches off from the first of the two *Grus* clusters and leads to the genera *Bugeranus* and *Anthropoides.* One of two *Anthropoides* cranes breeds in the Palearctic. The Demoiselle Crane *A. virgo* is a Central Palearctic specialist of the steppe with a marginal range in the Western Palearctic. The other species, Blue Crane *A. paradisea,* is a southern African specialist and seems to indicate a former wider distribution of this genus across savannas and steppes from South Africa north and across the southern Palearctic. The Demoiselle Crane, at the northern end of the range of the genus, is highly migratory while the Blue Crane is sedentary.

RALLIDAE

The Rallidae split from the Heliornithidae (the Neotropical Sungrebe *Heliornis fulica*) in the Late Eocene (*c.* 42.8 mya) and this was followed much later, in the Early Miocene (*c.* 21.8 mya) by the radiation of the Rallidae. The sister genera *Rallus* and *Gallirallus* are basal to the family, forming the earliest part of the radiation. This was followed by a split into two groups that encompassed the remaining genera: (a) *Porzana, Fulica, Gallinula,* and (b) *Laterallus, Porphyrio.* Within the first group, *Porzana* is basal to the sister genera *Fulica* and *Gallinula.* In the absence of more detailed information I place the Corncrake *Crex crex* on its own but closer to *Rallus* than to *Porzana*, following Sibley and Monroe (1990), with *Coturnicops* in a similar position.

Coturnicops

This is a Holarctic genus of three species with only one species in the Palearctic. Swinhoe's Rail *C. exquisitus* is a migratory bioclimatic semi-specialist of the Temperate (Latitude Category C) belt of the Eastern Palearctic. It is probably derived from a lineage that entered the Palearctic from North America during a warm period.

Rallus

This genus has a fragmented global distribution, being present in the Neotropics and the Nearctic, parts of Africa and Madagascar but largely missing from large parts of Africa and southern Asia. One species (11.11%) breeds in the Palearctic. The Water Rail *R. aquaticus* is exclusively Palearctic and it marks the northern part of the genus range. It is a bioclimatic moderate with a Temperate (Latitude Category C) pan-palearctic range that is fragmented in the Western Palearctic. It is highly migratory (the only species in the genus to migrate) except in the oceanic west and south-west, where it is sedentary or a short-distance migrant.

Crex

The Corncrake represents a single-species genus of exclusively Palearctic rails. It is a highly migratory bioclimatic moderate of the Temperate (Latitude Category C) belt of the Western (discontinuous) and Central Palearctic. It seems likely that it is a species that evolved *in situ* within the Palearctic, as a terrestrial version of a previously aquatic rail, and benefited from the expansion of grasslands and related habitats as these spread in the Miocene and, especially, the Pliocene.

Porzana

This is a globally distributed genus of bioclimatic specialists and semi-specialists with a few more tolerant species, which includes those that breed in the Palearctic. Five species (31.25%) breed in the Palearctic. Baillon's Crake *P. pusilla* is a semi-generalist, exceptional for a *Porzana*. It has a wide tropical distribution, in Africa, southern Asia, Australasia and New Zealand. It is sedentary across much of this range, which contrasts with the northernmost populations, those of the Palearctic, which are highly migratory. It has a pan-palearctic range discontinuous in the west and centred on the mid-latitude belt (Latitude Category D).

Little *P. parva* and Spotted *P. porzana* Crakes are similar in that they are both bioclimatic moderates with discontinuous ranges in the Western and Central Palearctic. They are highly migratory and their ranges are centred on the Temperate (Latitude Category C) belt. The two remaining species are migrants that are exclusive to the mid-latitude belt of the Eastern Palearctic – Band-bellied Rail *P. paykullii* is a semi-specialist and Ruddy-breasted Rail *P. fusca* is a specialist.

Fulica

The coots constitute another global genus, although in this case most species are Neotropical. Two species (18.2%) breed in the Palearctic. The Common Coot *F. atra* has a wide distribution across tropical Asia and Australasia. Palearctic populations mark the northern limit of the species and genus range. The Common Coot is a semi-generalist (only the generalist American Coot *F. americana* is more tolerant) with a multi-latitude (Latitude Category F) pan-palearctic range. Western Palearctic populations, like the tropical ones, are sedentary but those of the continental interior of the Palearctic across to the east are migratory. The Red-knobbed Coot *F. cristata* is a tropical African semi-specialist which has fringe populations at the extreme northern edge of the range, in the south-western Palearctic (Morocco and southern Iberia). This is therefore a sedentary species of the subtropical (Latitude Category E) belt of the Western Palearctic.

Gallinula

This is another globally distributed genus with a single Palearctic (11.11%) species. The Common Moorhen is widespread in the Americas (including the Caribbean), the Palearctic, southern Asia and the Pacific. Throughout much of its tropical range it is sedentary, as it is also across much of the Western Palearctic, but it is highly migratory in continental parts of the rest of the Palearctic. The Common Moorhen resembles the Common Coot in having a pan-palearctic, multi-latitude (Latitude Category F) range.

Porphyrio

This is another globally distributed genus. One species, Purple Swamphen *P. porphyrio*, breeds across tropical Africa, southern Asia, New Guinea, Australasia, the Pacific and New Zealand. It has a limited range in the south-western Palearctic, which is the northern limit of the species and genus range. It is a sedentary bioclimatic moderate of the mid-latitude (Latitude Category D) belt. Presumably its sedentary behaviour has prevented its spread further into the Palearctic.

CUCULIDAE

The cuckoos form the other large branch of this ancient lineage. Two genera of this largely tropical family – *Cuculus* and *Clamator* – are represented in the Palearctic.

Cuculus

This is a largely tropical South-east Asian genus, with some African representation, of forest insectivores. The distribution of species, with several on the Himalayan slopes, recalls that of forest passerines and would seem to make this a candidate genus for the 'launch-pad model' (see p. 41): one species is pan-palearctic, another is Central and Eastern, and three others are Eastern. It also conforms with the 'single generalist species per genus', also having a single semi-generalist. In all these aspects *Cuculus* resembles a number of forest passerine genera. The Common Cuckoo *C. canorus* is a migratory bioclimatic generalist with a pan-palearctic, multi-latitude (Latitude Category F) range.

The Oriental Cuckoo *C. saturatus* is a migratory semi-generalist with a broad multi-latitude (Latitude Category F) range across the Central and Eastern Palearctic. The other three Palearctic cuckoos – Hodgson's Hawk *C. fugax*, Indian *C. micropterus* and Lesser *C. poliocephalus* – are migratory bioclimatic moderates of the mid-latitude belt (Latitude Category D) of the Eastern Palearctic. The success of this genus in the Palearctic, despite a specialised insectivorous diet, has depended on bioclimatic tolerance and migration.

Clamator

This is a tropical African and Asian genus of two bioclimatically intolerant species. The Great Spotted Cuckoo *C. glandarius* is a bioclimatic semi-specialist with a tropical African range that reaches the southern Palearctic. Palearctic populations are highly migratory, contrasting with those to the south. In the Palearctic, the Great Spotted Cuckoo occupies the mid-latitude belt (Latitude Category D) of the Western and Central Palearctic.

CONCLUSIONS

(a) **Bioclimatic generalisation and migration**. Five species (17.2%, Appendix 1) are bioclimatically tolerant, a figure comparable to the waterbirds of the previous group. There are no bioclimatically tolerant bustards or cranes, so it is from the rails and cuckoos that these species are derived. Four of the five – Baillon's Crake, Common Coot, Common Moorhen and Common Cuckoo have pan-palearctic ranges; these are all multi-latitude (Latitude Category F), except Baillon's Crake, which has a mid-latitude (Latitude Category D) belt range. The fifth species, Oriental Cuckoo, has a multi-latitude range but is absent from the Western Palearctic. As in the previous waterbird group, migratory behaviour has been an added factor in the success of species in the Palearctic. Twenty-one species (72.42%) are fully migratory, which is almost the same proportion as in the previous group and higher than in the shorebirds. The proportion of residents (3 species, 10.34%) is also comparable.

(b) **Mid- and temperate-latitude representation**. The majority of birds in this group occupy ranges that are to the north of the waterbirds (dominated by belts D and E) and to the south of the shorebirds (dominated by belts A and B). Most – 11 species (37.93%) – belong to the mid-latitude belt (Latitude Category D), and another nine (31.03%) to the Temperate (Latitude Category C) belt. Like the waterbirds, a number of migratory species have managed to colonise and survive in the arid regions of the mid-latitude belt, particularly by exploiting seasonal water bodies. In some exceptional cases (e.g. Corncrake) species have turned terrestrial to exploit grassland and similar open habitats. In other cases (bustards and cranes) expansion and diversification has been permitted by the opening of areas, during the Miocene and Pliocene, that had previously been woodland and forest.

(c) **Nature of the genera.** As in the previous group, a number of genera and species have global distributions for which the Palearctic represents the northern part of the global range – Spotted Crake, Common Coot, Red-knobbed Coot, Common Moorhen, Common Cuckoo, Purple Swamphen and Great Spotted

Cuckoo. These are almost exclusively migratory populations of otherwise sedentary taxa. There is also a high representation of genera and species that are exclusively Palearctic – *Tetrax, Otis, Chlamydotis, Crex,* Water Rail, Spotted Crake and Little Crake. Two crane genera – *Leucogeranus* and *Grus* – are Holarctic.

(d) Representation in the Western Palearctic. The species in this group are evenly represented across the Palearctic, with a slightly higher number in the Eastern (18) than in the Western or Central Palearctic (16 each).

CHAPTER 16

Nightjars and swifts

Group 5

The nightjars, swifts and hummingbirds have deep roots in the Late Cretaceous. The swifts (Apodiformes) and hummingbirds (Trochiliformes) are closer to each other than each is to the nightjars (Caprimulgiformes) and diverged during the Eocene–Oligocene, but their lineage diverged from the nightjars before the K/T event (van Tuinen *et al.*, 2006; Brown *et al.*, 2008). Pratt *et al.* (2009) place the hummingbird-swift separation at 47.97 mya, in the Middle Eocene.

Climate

Swifts and nightjars are very similar bioclimatically, being intolerant and living largely in warm and humid climates (Table 16.1). Palearctic species are the most tolerant in their respective groups and show the greatest tendency to depart from the climatic optimum.

Habitat

Swifts and nightjars are aerial feeders, and the Palearctic swifts hunt over most available habitats (Table 16.2). The nightjars occupy a variety of habitats, mainly mixed, with the Palearctic species birds of forest, mixed and open habitats.

	A	B	C	D	E
Swifts (Tolerance)	2/77	2/15	3/6	1/1	0/0
Nightjars (Tolerance)	2/52	1/12	2/4	0/0	0/0
Swifts (Temperature)	0/0	1/1	2/4	1/2	4/86
Nightjars (Temperature)	0/0	0/0	2/3	1/3	2/57
Swifts (Humidity)	1/4	1/6	3/4	2/2	1/77
Nightjars (Humidity)	3/9	1/4	0/8	1/1	0/41

Table 16.1 *Bioclimatic characteristics of swifts and nightjars (Palearctic species/global species).*

	Forest	Mixed	Open	Rocky	Wetland
Swifts	8/67	8/87	8/17	8/53	8/11
Nightjars	1/25	2/57	2/17	0/5	0/10

Table 16.2 *Habitats occupied by swifts and nightjars (Palearctic/global). Both groups feed on aerial insects so the habitat given is the substrate below which they feed and where they breed.*

Swifts	Global	Palearctic	Global %	Palearctic %	Palearctic of global %
Migratory	14	8	14.14	100	57.14
Sedentary	95	1	95.96	12.5	1.05
Total Species	99	8			

Nightjars	Global	Palearctic	Global %	Palearctic %	Palearctic of global %
Migratory	13	5	19.12	100	38.46
Sedentary	65	0	95.59	0	0
Total Species	68	5			

Table 16.3 *Migratory behaviour in Palaearctic and global swifts and nightjars.*

Migratory behaviour

Swifts and nightjars are predominantly sedentary groups, though Palearctic species stand out in direct contrast as they are highly migratory (Table 16.3). A combination of limited improvement in tolerance and a strong dependence on migration are the features that have permitted a few species to survive in the Palearctic.

Fossil species

The presence of swifts and nightjars in the Pleistocene would seem to indicate seasonally favourable climates, at least at times, for these migratory species (Appendix 2).

Taxa in the Palearctic

CAPRIMULGIDAE

The Caprimulgidae may have radiated during the Eocene, followed by independent evolution on separate continents, with many basal morphological features having been retained (Larsen *et al.,* 2007). The New and Old World nightjars are clearly separated on genetic lines (Barrowclough *et al.*, 2006; Han *et al.*, 2010). In the most recent taxonomic revision (Han *et al.*, 2010) the genus *Caprimulgus* is retained only for the Old World nightjars, which constitute almost one half of the world's species (Barrowclough *et al.*, 2006).

Caprimulgus

Three species (9.1% of the species in the genus) breed in the Palearctic. It has been suggested that the Eurasian Nightjar *C. europaeus* has an African origin and that the current distribution is post-glacial, the species having survived the last glacial in a southern European or African refugium (Larsen *et al.*, 2007). This interpretation is flawed for the simple reason that Africa does not form part of the Eurasian Nightjar's breeding range. So to postulate a gradual post-glacial northward expansion begs the question of why none were left behind in Africa? In my interpretation, the Eurasian Nightjar is one of a genus of tropical African and Asian species that included the Palearctic in their ancestral range in the Tertiary, when the world was warmer. As the climate cooled, the northern populations that survived did so by a combination of bioclimatic tolerance and migration. That subsequently, during the glaciations, Eurasian Nightjar must have had a reduced range in southern Palearctic refugia, like so many other species, is undeniable, but there is no evidence of breeding to the south, in Africa, which would have been separated from the current range by the Sahara Desert. There were other species of nightjar (Nubian *C. nubicus,* Egyptian *C. aegypticus* and even Red-necked *C.*

ruficollis) adapted to arid and semi-arid climates that survived in Saharan refugia to the south of the Eurasian Nightjar and still remain there today. Part of the problem with interpretations of 'southern ancestral homes' is that they rely on applying political divisions (Africa, Asia, etc.) when these are largely irrelevant to our understanding of biogeography. The large land mass that included the Palearctic, the rest of Asia and Africa needs to be seen as a single, changing, evolving unit, with regional idiosyncrasies (Finlayson, 2009). Only then can we begin to understand the history of the Palearctic's birds.

The Eurasian Nightjar is a bioclimatic moderate, one of only two in a genus of specialists and semi-specialists. It has a multi-latitude (Latitude Category F), discontinuous, pan-palearctic range. It is highly migratory. The Grey Nightjar *C. indicus* of the Eastern Palearctic is the other bioclimatic moderate and, like the Eurasian Nightjar, has a multi-latitude range. These two species have managed to colonise the temperate and southern fringes of the Boreal latitude bands. The remaining species include a semi-specialist – Egyptian Nightjar (mid-latitude belt, Latitude Category D, Western and Central Palearctic) - and two specialists – Red-necked (mid-latitude belt, Latitude Category D, Western Palearctic) and Nubian (subtropical belt, Latitude Category E, Western Palearctic).

APODIDAE

The basal genus to this family is the Neotropical *Streptoprocne.* The next genus to branch off from the main cluster of Apodidae after this is the Asian *Hirundapus* which includes a Palearctic species. The other Palearctic genera – *Apus* and *Tachymarptis* – are sisters, with *Collocalia* their closest genus.

Hirundapus

This is a tropical Asian genus with one species that reaches the Eastern Palearctic. The White-throated Needletail *H. caudacutus* is a migratory bioclimatic moderate with a multi-latitude (Latitude Category F) range. Tropical populations to the south are resident so the migratory habit has, as in *Cuculus,* the Hirundinidae and some flycatchers, permitted a broad latitudinal Palearctic range that approaches the Boreal belt.

Tachymarptis

This is a tropical African and Asian genus of two species, with one occupying the southern Palearctic. The Alpine Swift *T. melba* is a migratory bioclimatic moderate of the mid-latitude (Latitude Category D) belt of the Western and Central Palearctic.

Apus

Six species (35.29%) breed in the Palearctic. The Palearctic species include the most bioclimatically tolerant in a genus of specialists and semi-specialists, and six of seven migratory swifts. The Fork-tailed Swift *A. pacificus* is an Eastern Palearctic semi-generalist that has achieved, like the White-throated Needletail, a multi-latitude (Latitude Category F) distribution. The genus seems to conform to the 'single generalist per genus' rule and also has a single moderate – the Common Swift *A. apus* – which also has a multi-latitude range. Common Swift is pan-palearctic although discontinuous in the Central and Western Palearctic.

The remaining species include two semi-specialists and two specialists. Little *A. affinis* and White-rumped *A. caffer* Swifts are semi-specialists of the sub-tropical belt (Latitude Category D) of the Western and Central (only the Western for the White-rumped) Palearctic. The White-rumped Swift is migratory as are the Central Palearctic populations of Little, but those to the west are sedentary. The White-rumped Swift has undergone a rapid colonisation of south-western Iberia, starting during the 1960s (Allen and Brudenell-Bruce, 1967; Benson *et al.*, 1968). Its nearest breeding areas were south of the Sahara, so this colonisation represented a significant jump in range from south to north. Little Swifts have also colonised south-western Europe in recent decades but they have always bred on the southern shore of the Strait of Gibraltar, so the colonisation does not compare with that of the White-rumped but is instead part of a randomly fluctuating range edge.

The Pallid Swift *A. pallidus* is a migratory bioclimatic specialist of the mid-latitude belt (Latitude Category D) of the Western Palearctic while the Plain Swift *A. unicolor* is a migratory specialist of the sub-tropical belt of the Western Palearctic, breeding on the Atlantic islands of the south-west.

CONCLUSIONS

(a) **Bioclimatic generalisation and migration.** There is only one bioclimatically tolerant species (7.69%) and no generalists among them. This low figure is comparable to the shorebirds. This intolerance is compensated by a very strong migratory tendency. All species are migratory and only the Little Swift is partially so within the Palearctic. Thus we can conclude that the success of these specialised aerial insectivores has depended on the ability to migrate.

(b) **Low- and multi-latitude representation.** The most striking feature of the distribution of these birds is the capacity of many to occupy wide latitudinal ranges, made possible by the ability to migrate. Five species (38.46%) are in this category – Eurasian Nightjar, Grey Nightjar, Common Swift, Fork-tailed Swift and White-throated Needletail. The remaining species are restricted to low latitudes and are bioclimatically intolerant – Egyptian Nightjar, Red-necked Nightjar, Alpine and Pallid Swifts (Latitude Category D) and Nubian Nightjar, Little, White-rumped and Plain Swifts (Latitude Category E).

(c) **Representation in the Western Palearctic.** The Western Palearctic dominates, with 10 species compared with only five in the Central and Eastern Palearctic. It reflects the 'arid-belt' nature of the nightjars and the dominance of the genus *Apus* in the west.

CHAPTER 17

Pigeons, sandgrouse, tropicbirds, flamingos and grebes

Group 6

The pigeons and doves, sandgrouse, tropicbirds, flamingos and grebes are independent lineages that have their origins in the Late Cretaceous (Pereira *et al.*, 2007; Brown *et al.*, 2008). Within this ancient group, the Flamingo (Phoenicopteridae)–Grebe (Podicipedidae) lineage is basal and branches off first. It is followed by the tropicbirds (Phaethontidae), then sandgrouse (Pteroclididae), leaving the pigeons (Columbiformes). These divisions pre-date the K/T event.

Climate

These birds are bioclimatically intolerant, with the Palearctic species being the least confined. This is especially noticeable among the pigeons, five of the six most tolerant species being Palearctic species (Table 17.1). Flamingos and tropicbirds are also intolerant. The pigeons and sandgrouse are warm-climate species, but they differ from each other in that the pigeons are birds of humid climates and sandgrouse of arid ones. Grebes occupy a wide spectrum of climates. All Palearctic species in these families conform with the global picture.

Habitat

Pigeons are predominantly birds of forest and mixed habitats. The presence of two species in rocky habitats in the Palearctic is noteworthy (Table 17.2). The other families are equally conservative of habitat, grebes and flamingos of wetlands, tropicbirds of marine environments and sandgrouse of open habitats.

	A	B	C	D	E
Pigeons (Tolerance)	6/260	2/37	2/6	4/5	1/1
Grebes (Tolerance)	0/8	0/3	4/8	1/1	0/1
Sandgrouse (Tolerance)	3/9	4/7	0/0	0/0	0/0
Pigeons (Tolerance)	0/0	2/3	4/7	1/2	8/281
Grebes (Tolerance)	0/2	2/2	0/3	3/7	0/4
Sandgrouse (Tolerance)	0/0	0/0	0/0	3/3	4/13
Pigeons (Tolerance)	0/16	1/7	4/25	4/5	6/239
Grebes (Tolerance)	0/2	3/6	0/3	2/3	0/4
Sandgrouse (Tolerance)	6/10	0/1	1/3	0/0	0/2

Table 17.1 *Bioclimatic characteristics of pigeons, sandgrouse and grebes (number of Palearctic species/number of global species).*

	Forest	Mixed	Open	Rocky	Wetland
Pigeons	6/251	5/153	2/26	2/9	0/30
Grebes	0/0	0/0	0/0	0/0	5/21
Sandgrouse	0/0	0/7	7/15	0/4	0/0

Table 17.2 *Habitats occupied by pigeons, grebes and sandgrouse (Palearctic species/global species).*

Pigeons	Global	Palearctic	Global %	Palearctic %	Palearctic of global %
Migratory	16	7	5.18	46.67	43.75
Sedentary	306	12	99.03	80	3.92
Total Species	309	15			

Grebes	Global	Palearctic	Global %	Palearctic %	Palearctic of global %
Migratory	8	5	38.09	100	62.50
Sedentary	19	3	90.48	60	15.79
Total Species	21	5			

Sandgrouse	Global	Palearctic	Global %	Palearctic %	Palearctic of global %
Migratory	4	2	23.53	28.57	50
Sedentary	17	7	100	100	41.18
Total Species	17	7			

Table 17.3 *Migratory behaviour in global and Palearctic pigeons, grebes and sandgrouse.*

Migratory behaviour

Tropicbirds, being pelagic, are migratory and are unusual within this group. Grebes are next most migratory, in a group dominated by sedentary species (Table 17.3). The Palearctic species show an increase in migratory tendency but this is not as pronounced as in other groups. Overall, a few pigeons seem to have succeeded in the Palearctic by becoming tolerant and combining this feature with migration. Grebes have succeeded by a combination of tolerance and migration within the context of occupation of wetlands. Flamingos have been more restricted than grebes in this respect. The sandgrouse have become specialists of the driest parts of the world.

Fossil species

The most remarkable feature of the fossil history of this group is the high frequency of Rock Dove sites (Appendix 2), something that is not surprising given the species' predilection for rocky habitats, where fossils often preserve best. Their numbers are comparable to those of the choughs at these sites.

Taxa in the Palearctic

PHOENICOPTERIDAE

The flamingos are a sister group of the grebes (van Tuinen et al., 2001; Mayr 2004). Brown *et al.* (2008) place the separation of the two groups in the Late Cretaceous, although Pratt *et al.* (2009) place it after the K/T event, at the start of the Eocene (*c.* 55-56 mya). Either way these are two related lineages that diverged at an early stage.

Phoenicopterus

This is the only genus of flamingos, and it contains five species. They have a fragmented global distribution that includes the New World, the Palearctic and Africa. Flamingos do not breed in southern Asia or Australasia. Three species are Andean and one is African – the Lesser Flamingo *P. minor* – which occasionally strays north into the Palearctic. The only Palearctic species is the Greater Flamingo *P. ruber* which has a large range that encompasses the Americas and tropical Africa as well. The Palearctic is the northern limit of the genus and species range. Greater Flamingo is a bioclimatic moderate, which makes it the most bioclimatically tolerant flamingo, with a fragmented Western and Central Palearctic range centred on the mid-latitude belt (Latitude Category D). It is therefore another species of wetland bird with these geographical characteristics (see for example some herons and pelicans). It is nomadic, moving large distances to breed in wetlands with appropriate water levels.

PODICIPEDIDAE

Podiceps

Four of nine species (44.44%) of this globally distributed genus breed in the Palearctic. They are the most bioclimatically tolerant in an otherwise specialist or semi-specialist group. The Great Crested Grebe *P. cristatus* has a wide tropical range across Africa, southern Asia and Australasia. The Palearctic is the northern part of this huge range. It is sedentary across much of the tropics but is highly migratory in the Palearctic, except in the oceanic south-west. The Great Crested Grebe is a bioclimatic semi-generalist with a pan-palearctic, multi-latitude (Latitude Category F) range. *Podiceps* therefore fits into the 'single bioclimatically tolerant species per genus' model.

The other three species are bioclimatic moderates. The Black-necked Grebe *P. nigricollis* breeds in the Americas and in Africa so the Palearctic is the northern part of its Old World range. It is sedentary in tropical areas and also in the south-western Palearctic but it is highly migratory in the continental parts of the Palearctic. Its Temperate (Latitude Category C) range is continuous in the Central Palearctic but discontinuous in the east and the west. The Red-necked Grebe *P. grisegena* contrasts with the Black-necked Grebe in being Holarctic. It is a species that has survived in the northern part of the genus's range. Like the Black-necked Grebe, it occupies the Temperate belt and has a highly fragmented pan-palearctic range. The Slavonian Grebe *P. auritus* is another Holarctic species and its pan-palearctic range is continuous. This is not surprising as the range is centred to the north of the other grebes, in the Boreal (Latitude Category B) belt. The Slavonian Grebe marks the extreme limit of *Podiceps* survival. Both Slavonian and Red-necked Grebes are highly migratory.

Tachybaptus

This is a globally widespread genus of four species that is found in the Americas, Africa, southern Asia, Australasia and the Pacific islands as well as the Palearctic. The bioclimatically moderate Little Grebe *T. ruficollis* is the only Palearctic representative of the genus. It breeds across the tropical Old World, from Africa through southern Asia to New Guinea, where it is largely resident. In the Palearctic it is curiously missing from the arid centre. It is highly migratory in the east and the continental part of the Western Palearctic, but it is sedentary in the oceanic west.

PHAETHONTIDAE

Tropicbirds were once widely distributed across ancient European seas from the Palaeocene to the Middle Eocene, and again in the Middle to Late Miocene (Bourdon *et al.*, 2008; Mlíkovský, 2009). They are an excellent example of the range contraction and loss of diversity with the changes that affected the Palearctic from the Late Miocene. Tropicbirds are now relicts in the Palearctic, confined to tropical waters.

Phaethon

One of the three species of tropicbirds reaches the south-western islands of the Palearctic. The Red-billed Tropicbird *P. aethereus* is a migratory bioclimatic semi-specialist of the subtropical (Latitude Category E) belt of the Western Palearctic. In the east, the Red-tailed Tropicbird *P. rubicauda*, in a similar situation to the Red-billed in the west, barely touches the southern edge of the Palearctic and is not included here.

PTEROCLIDIDAE

The sandgrouse – *Syrrhaptes* and *Pterocles* – are a family of birds that have primarily evolved within the arid belt that stretches across North Africa, Arabia, India and Central Asia. In this respect they resemble the bustards. Sandgrouse are bioclimatic specialist and semi-specialist sedentary herbivores (seed-eaters) that have benefitted from the aridity that engulfed these huge areas from the Late Miocene. The most striking of sandgrouse adaptations is the male's ability to hold water in its belly feathers and transport it long distances to the young (Maclean, 1983).

Syrrhaptes

The two species in this genus include a Palearctic species and a Tibetan one – Tibetan Sandgrouse *S. tibetanus* – which is extralimital. Pallas's Sandgrouse *S. paradoxus* is a bioclimatic specialist of the mid-latitude belt (Latitude Category D) of the Central and (marginally) Eastern Palearctic. Its northern fringe populations are, unusually among sandgrouse, migratory.

Pterocles

Six of 14 species (42.86%) breed in the Palearctic. Two are species of the mid-latitude belt (Latitude Category D) – Pin-tailed *P. alchata* and Black-bellied *P. orientalis* Sandgrouse are semi-specialists with Central and Western (discontinuous) Palearctic ranges. They are, along with Pallas's, partly migratory, those from continental and northerly areas moving south in the autumn. The other four *Pterocles* in the region are birds of the subtropical (Latitude Category E) belt – Chestnut-bellied *P. exustus* and Crowned *P. coronatus* in the Western and Central Palearctic, and Spotted *C. senegallus* and Liechtenstein's *P. lichtensteinii* only in the west. The sandgrouse are another example of species occupying mid-latitude (centre) and subtropical (west) parts of the arid Palearctic.

COLUMBIDAE

Treron

One species (4.5%) of this large genus of tropical, mainly southern Asian arboreal fruit-eating pigeons breeds in the Palearctic. The White-bellied Green Pigeon *T. sieboldii* is a partially migratory bioclimatic specialist of the Warm (Latitude Category D) belt of the Eastern Palearctic. It is the only species of this genus of bioclimatically intolerant pigeons that exhibits migratory behaviour, which is limited to the northernmost populations. This genus is part of a cluster that separated at an early stage from a large group of mainly tropical fruit-eating pigeons, including large tropical genera such as *Ptilinopus* and *Ducula* (Gibb and Penny, 2010).

Oenas

Another early branch of this cluster includes this single-species genus. The terrestrial seed-eating Namaqua Dove *Oena capensis* is a sedentary African bioclimatic semi-specialist that reaches the southern edge (Latitude Category E) of the Western Palearctic.

Columba

A second major cluster incorporates the Old World (*Columba*) and New World (*Patagoenas*) pigeons, as well as the *Streptopelia* doves. Here I include the Pink Pigeon in *Streptopelia* and exclude the New World pigeons from *Columba* (Johnson *et al.*, 2001; Gibb and Penny, 2010). Eight species (21.62%) breed in the Palearctic. *Columba* is a genus of sedentary bioclimatic specialists and semi-specialists with a wide Old World distribution that includes Africa, tropical Asia, New Guinea, Australasia and the Pacific. The Palearctic species include the most bioclimatically tolerant and migratory of the genus.

The Wood Pigeon *C. palumbus* is a bioclimatic semi-generalist of the Western and Central (discontinuous) Palearctic with a multi-latitude (Latitude Category F) range. It is exclusively Palearctic and its continental and northern populations are highly migratory, while those of the oceanic south-west are sedentary. The Rock Dove *C. livia* has similar characteristics to the Wood Pigeon except that it reaches the Eastern Palearctic and is sedentary; its present multi-latitude range may reflect expansion associated with human dwellings, which would mean that its original latitudinal categorisation would be mid-latitude (Latitude Category D). This species is also native to southern Asia.

The Stock Dove *C. oenas* and the Hill Pigeon *C. rupestris* are bioclimatic moderates. The Stock Dove is exclusively Palearctic; it has a discontinuous, multi-latitude (Latitude Category F) range in the west and centre and, like the Wood Pigeon, has highly migratory continental populations. The Hill Pigeon is a largely migratory (some southern populations being sedentary) pigeon of the warm belt (Latitude Category D) of the Eastern Palearctic.

The other Palearctic pigeons are bioclimatic specialists and island endemics (or almost so). Three belong to the Atlantic islands of the south-western Palearctic and thus occupy the subtropical belt, where they are sedentary – Trocaz *C. trocaz*, Bolle's *C. bollii* and Laurel *C. junoniae* Pigeons. The Japanese Wood Pigeon *C. janthina* is a resident of the subtropical belt (Latitude Category E) of the Japanese islands and adjacent mainland.

Streptopelia

Five species (31.25%) breed in the Palearctic. This is an Old World genus of species that are widely distributed across tropical areas of Africa and southern Asia as well as the Palearctic. Most are sedentary bioclimatic specialists and semi-specialists but the Palearctic species are exceptional, including bioclimatically tolerant and migratory doves. This genus appears to follow the 'single generalist per genus' rule, and also has two semi-generalists. The Oriental Turtle Dove *S. orientalis* is a bioclimatic generalist with a large tropical Asian and Eastern and Central Palearctic range that reaches the Boreal belt. It has a multi-latitude (Latitude Category F) range within the Palearctic. The Eurasian Collared Dove *S. decaocto* is a sedentary bioclimatic semi-generalist with a well-documented history of geographical expansion across Europe in the 20th Century (Burton, 1995). It is a mainly Western Palearctic bird with isolated populations in the Central and Eastern Palearctic, and has a multi-latitude (Latitude Category F) range. The Eurasian Turtle Dove *S. turtur* is another semi-generalist with a multi-latitude Western and Central (discontinuous) Palearctic range. It differs from Eurasian Collared Dove in being highly migratory. While the Eurasian Collared Dove's range includes southern Asia, the Turtle Dove's is exclusively Palearctic.

The two remaining Palearctic species are bioclimatically intolerant. The Laughing Dove *S. senegalensis* is a tropical African and Asian sedentary semi-specialist that reaches southern (Latitude Category E) parts of the Western and Central Palearctic. The Red Turtle Dove *S. tranquebarica* is a migratory specialist of the subtropical (Latitude Category E) belt of the Eastern Palearctic. Its range includes tropical southern Asia where it is sedentary.

CONCLUSIONS

(a) **Bioclimatic generalisation and sedentary behaviour.** Six species (20.69%) are bioclimatically tolerant, a figure comparable to the water bird group and significantly higher than the shorebirds and the swifts-nightjars. All six species (including five pigeons) have multi-latitude (Latitude Category F) ranges – Great

Crested Grebe, Wood Pigeon, Rock Dove, Eurasian Collared Dove, Oriental Turtle Dove and Eurasian Turtle Dove. The big difference with previous groups is the high representation of sedentary species (13, 44.83%), which partly explains the limitation in high-latitude ranges. These birds have instead relied on bioclimatic tolerance. A comparison of the Eurasian Collared Dove (semi-generalist), which managed to colonise much of the Western Palearctic while its congener the Red Turtle Dove (a migratory specialist) is restricted to the southern edge of the Palearctic, provides a useful contrast.

(b) **Low- and multi-latitude representation**. This group follows the pattern of the swifts and nightjars, of a good representation of low-latitude and multi-latitude species with few exclusively high-latitude ones. Eleven species occur in the subtropical belt – Chestnut-bellied, Spotted, Liechtenstein's and Crowned Sandgrouse, Laughing and Red Turtle Dove, Trocaz, Bolle's, Laurel and Japanese Wood Pigeon, and Red-billed Tropicbird. Another six occur in the mid-latitude belt (Latitude Category D) – Greater Flamingo, Pin-tailed, Black-bellied and Pallas's Sandgrouse, Hill Pigeon and White-bellied Green Pigeon. No fewer than eight have multi-latitude ranges – Great Crested and Little Grebe, Oriental Turtle, Eurasian Turtle and Eurasian Collared Doves, Rock and Stock Dove, and Wood Pigeon. Three grebes depart from the theme – Black-necked and Red-necked are Temperate (Latitude Category C) and Slavonian is Boreal (Latitude Category B).

(c) **Representation in the Western Palearctic**. The Western Palearctic dominates the group, with 22 species compared with 17 in the Central Palearctic and only 13 in the east. This reflects the dominance of arid-habitat families (Pteroclididae) and species (doves), which are largely missing in the Eastern Palearctic.

CHAPTER 18

Geese, swans, ducks and gamebirds

Group 7

Waterfowl (Anseriformes) and gamebirds (Galliformes) form the last of the avian groups (other than the ratites, which we do not consider, see p. 34), and both have their origins in the Cretaceous (van Tuinen *et al.*, 2006; Brown *et al.*, 2008). The diversification of the anseriform lineages seems to have commenced shortly after the K/T event with the split between the *Dendrocygna* (whistling ducks) lineage and the other Anseriformes very early in the Palaeocene (*c.* 65–60 mya; Donne-Goussé *et al.*, 2002). The branching of the geese and swans (Anserinae) from the ducks (Anatinae) probably took place in the Early Eocene (*c.* 55 mya). The major goose genera *Branta* and *Anser* separated in the Early Miocene (*c.* 23 mya). Using this as a reference point, we can infer that the swans separated from the geese slightly earlier (probably in the Late Oligocene) as did the stifftails (*Oxyura*, Hackett *et al.*, 2008), that the tribes Tadornini, Mergini and Cairinini radiated at about the time of the *Branta-Anser* separation in the early Miocene, and that the branching of the ducks into the major tribes (Tadornini, Mergini, Cairinini, Aythyini and Anatini) probably took place in the Late Eocene-Early Oligocene.

The origins of the galliforms lie in the Cretaceous, as do the early splits into the megapodes (Megapodiidae), cracids (Cracidae) and guineafowl (Numididae). A major radiation of lineages seems to date to the Early Eocene (*c.* 56 mya) leading to the pavonines (*Meleagris, Pavo, Afropavo, Argusianus, Rheinardia*), junglefowl (*Gallus, Bambusicola*), pheasants (Phasianidae), grouse (Tetraonidae), and the line leading to the francolins (*Francolinus*), Old World quail and partridges (Phasianidae: *Alectoris, Perdix, Coturnix*, etc.) and New World quails (Odontophoridae), which branched on several occasions afterwards in a rapid and complex manner that is not fully resolved (Crowe *et al.*, 2006; Shen *et al.*, 2010).

Climate

Ducks and geese and, particularly, gamebirds are bioclimatically intolerant species (Figure 18.1). In the case of the Anseriformes, the Palearctic species include the most bioclimatically tolerant including the only generalist – Mallard *Anas platyrhynchos*.

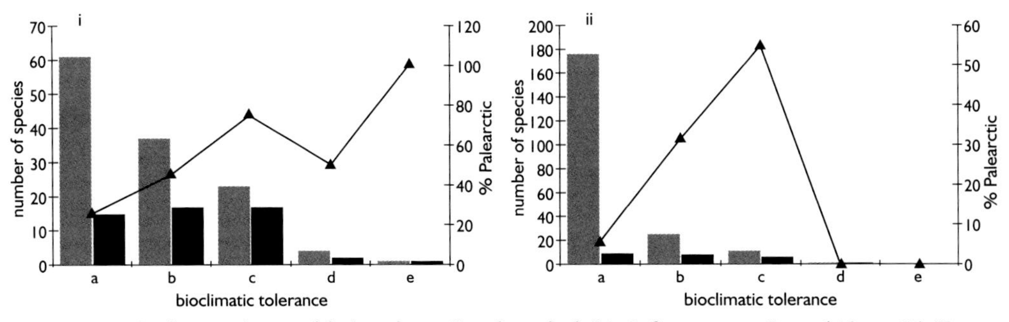

Figure 18.1 *Bioclimatic tolerance of ducks and geese (i) and gamebirds (ii). Definitions as on Figure 4.1(see p. 38). Key as in Figure 13.1.*

The picture is similar for the Galliformes, except that they do not reach the higher levels of bioclimatic tolerance. The Anseriformes show a bimodal pattern on the temperature gradient with most species in the warmest part but a substantial number also at the cold end, which is where most Palearctic species are situated (Figure 18.2i). The Galliformes, on the other hand, are birds of warm climates and the subset of Palearctic species consists of the small number that has made it to survive in the coldest climates (Figure 18.3i). Anseriformes are broadly distributed across the humidity gradient. Palearctic species appear distributed mainly along the arid portions of this gradient (Fig. 18.2ii). The Galliformes are birds of humid habitats with a small peak of birds that have adapted to the driest. Palearctic birds tend towards the dry half of the humidity range (Figure 18.3ii).

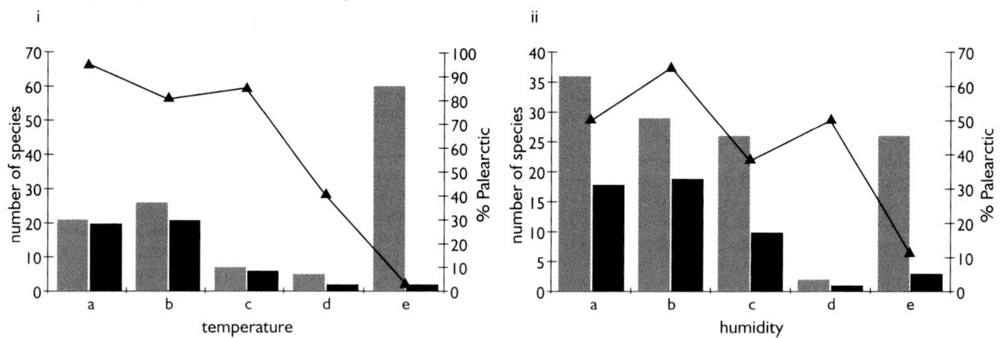

Figure 18.2 *Temperature (i) and humidity (ii) tolerance of ducks and geese. Definitions as in Figure 4.2 (see p. 38). Key as in Figure 13.1.*

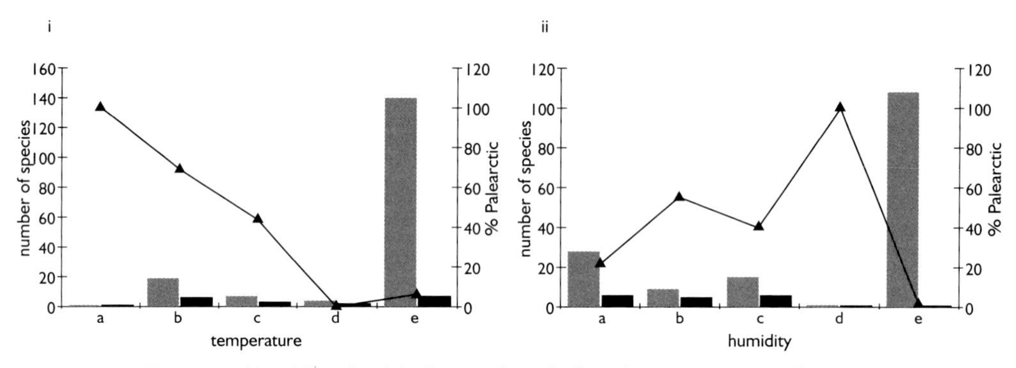

Figure 18.3 *Temperature (i) and humidity (ii) tolerance of gamebirds. Definitions as in Figure 4.2 (see p. 38). Key as in Figure 13.1.*

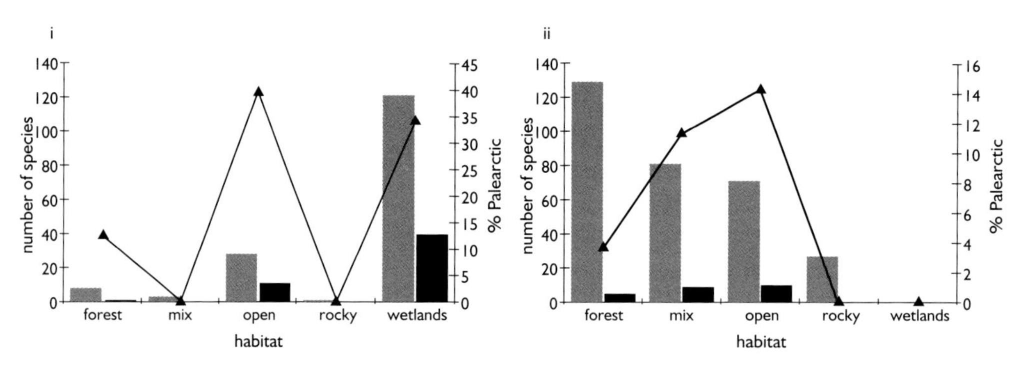

Figure 18.4 *Habitat occupation by ducks and geese (i) and gamebirds (ii). Definitions as in Figure 4.3 (see p. 39). Key as in Figure 13.1.*

Habitat

The Anseriformes are birds of open country and wetlands and the Palearctic birds reflect this, but that they are proportionately more birds of open country than of wetlands (Figure 18.4i), is a reflection of the contribution made by geese. Galliformes avoid water habitats and occupy a range of terrestrial habitats, from forest to open. Palearctic species tend towards mixed and, especially, open habitats (Figure 18.4ii), thus resembling the Anseriformes in this respect.

Migratory behaviour

There is a strong contrast between Anseriformes and Galliformes. The latter are strict residents with only the quails being migratory. Two of the three qualis breed in the Palearctic, which therefore holds a large section of all migratory gamebirds (Table 18.1). Globally, there are more Anseriformes with sedentary than with migratory populations, which is in complete contrast with the Palearctic subset; this is dominated by migratory species (Table 18.1). The success of the Palearctic Anseriformes has been due to increased bioclimatic tolerance, on the one hand, and specialisation towards cold and dry climates coupled with strong migratory behaviour on the other. The Galliformes have been more limited than the Anseriformes with regard to tolerance and migratory behaviour but, like the Anseriformes, the occupation of open habitats seems to have been the key to the success of those that survived in the Palearctic.

Fossil species

The anseriform fossil record appears older than for most other groups, with a number of species identifiable from the Late Miocene – Eurasian Teal *Anas crecca* and Shoveler *A. clypeata* – with the rest appearing in the Pliocene and Early Pleistocene. Most gamebirds appear during the Pliocene and Early Pleistocene (Mlíkovský, 2002). Several noteworthy aspects of this fossil record include the high frequency of birds of open ground and tundra, especially geese and grouse, in the Late Pleistocene of Europe. To this we may add a good representation of other open-ground species, particularly partridges, as well as ducks, concordant with the presence of significant wetlands in the Late Pleistocene. Several species – Long-tailed Duck *Clangula hyemalis* and Velvet Scoter *Melanitta fusca* in particular – indicate important southward range shifts in the Late Pleistocene, reaching the latitude of Gibraltar, where they are extremely rare today (Finlayson, 1992).

Anseriformes	Global	Palearctic	Global %	Palearctic %	Palearctic of global %
Migratory	76	50	60.32	96.15	65.79
Sedentary	100	20	79.37	38.46	20
Total Species	126	52			

Galliformes	Global	Palearctic	Global %	Palearctic %	Palearctic of global %
Migratory	3	2	1.41	8.33	66.67
Sedentary	211	12	99.06	50	5.69
Total Species	213	24			

Table 18.1 *Migratory behaviour in Anseriformes and Galliformes.*

Taxa in the Palearctic

ANATIDAE

Oxyura

This is an ancient lineage with a fragmented global distribution that includes the Palearctic, Nearctic, Neotropical, Africa and Australasia but is missing from southern Asia. One of seven (14.29%) species breeds in the Palearctic. The White-headed Duck *O. leucocephala* is a bioclimatic semi-specialist (the Nearctic Ruddy Duck *O. jamaicensis*, a moderate, is bioclimatically more tolerant) of the mid-latitude (Latitude Category D) belt of the Western and Central Palearctic. Its range is fragmented, in part due to human activity. This species is highly migratory except for the southern and south-western populations, which are sedentary.

Cygnus

Three of the six swans of this genus breed in the Palearctic. The genus has a global, but strongly fragmented, distribution, with species in the Neotropics and Australasia and the rest in the Holarctic. This distribution must represent significant ancient regional extinction. The three Palearctic species sort themselves out by latitudinal belts. The Mute Swan *C. olor* is a bioclimatic semi-specialist of the Temperate (Latitude Category C) belt and has a discontinuous pan-palearctic range. It is highly migratory except in the oceanic west. The Whooper Swan *C. cygnus* is also a semi-generalist but its range is centred on the Boreal (Latitude Category B) belt and it has a continuous pan-palearctic range. It is fully migratory. The third species – Bewick's Swan *C. columbianus* – occupies the Arctic (Latitude Category A) belt and is absent from the Western Palearctic. Its range continues eastwards into North America.

Anser

The geese of this genus and *Branta* are the only true avian grazing herbivores of the Palearctic. Nine of the 10 species in this highly migratory Holarctic genus breed in the Palearctic. The Greylag Goose *A. anser* is the most bioclimatically tolerant of this genus of specialists and semi-specialists of cold climates. It is a moderate with a Temperate (Latitude Category C) pan-palearctic range, the most southerly of ranges except for the specialist high-altitude Bar-headed Goose *A. indicus* of the mid-latitude belt (Latitude Category D). These are the only two species that are partially migratory, with south-western and southern populations respectively being sedentary. The Swan Goose *A. cygnoides*, a semi-specialist of the Eastern Palearctic, also occupies a Temperate range, but it is migratory.

The Bean Goose *A. fabalis* occupies a latitude intermediate between these species and those of the High Arctic. It is a semi-specialist of the Boreal Belt (Latitude Category B) of the Central and Eastern Palearctic. The remaining species are all highly migratory specialists of the Arctic (latitude Category A) belt – Pink-footed Goose *A. brachyrhynchus* (Western), Lesser White-fronted Goose *A. erythropus* (Central), White-fronted Goose *A. albifrons* (Central and Eastern), Emperor Goose *A. canagicus* and Snow Goose *A. caerulescens* (Eastern).

Branta

This is a Holarctic genus with an exceptional species – Nene *B. sandvicensis* – on Hawaii. They are more restricted to high latitudes than the *Anser* geese. Three of the five species breed in the Palearctic and are highly migratory bioclimatic specialists of the Arctic (latitude Category A) belt – Barnacle Goose *B. leucopsis* and Red-breasted Goose *B. ruficollis* in the Central Palearctic, and Brent Goose *B. bernicla* in the Central and Eastern Palearctic.

Alopochen

The Egyptian Goose *A. aegyptiacus* is in a single-species genus, closely related to the shelducks (*Tadorna*). Along with *Chloephaga* and *Tadorna* it constitutes the tribe Tadornini (Donne-Goussé *et al.*, 2002). It is a sedentary tropical African semi-specialist that reaches the southern edge of the Western Palearctic.

Tadorna

This genus has an unusual distribution, which must reflect a larger former range. It is today represented in the Palearctic, Australasia and New Zealand. Three of the seven species breed in the Palearctic. The Common

Shelduck *T. tadorna* is the most bioclimatically tolerant (along with Ruddy Shelduck *T. ferruginea*) of a genus of specialists and semi-specialists. It is a highly migratory Palearctic species with sedentary populations in the oceanic west. Its range is split Boreal/mid-latitude (latitude Category B/D) Western and Central Palearctic, fragmented along the mid-latitude belt and the west of the Boreal belt. The Ruddy Shelduck *T. ferruginea* is sedentary in southern Asia and in the south-western Palearctic; it is migratory across much of its pan-palearctic range, across the mid-latitude belt (Latitude Category D). Its range is very fragmented in the west. The probably extinct Crested Shelduck *T. cristata* has (or had) a highly localised, mid-latitude (Latitude Category D), range in the Eastern Palearctic. It is believed to have been migratory.

Clangula

This is a single-species Holarctic genus that is most closely related to the eiders. Together with the scoters, mergansers, goldeneyes and the South American Ringed Teal *Callonetta leucophrys*, it constitutes the tribe Mergini. The Long-tailed Duck *C. hyemalis* is a migratory bioclimatic semi-specialist of the Arctic (Latitude Category A) belt with a continuous pan-palearctic range.

Somateria

All three species of this Horactic genus breed in the Palearctic. They are migratory bioclimatic specialists and semi-specialists, only the westernmost and easternmost populations of Common Eider *S. mollissima* being sedentary. All three are birds of the Arctic (Latitude Category A) belt. The Common Eider is a bioclimatic semi-specialist that breeds in the Western and Eastern Palearctic; King Eider *S. spectabilis* is a semi-specialist of the Central and Eastern Palearctic; and Spectacled Eider *S. fischeri* is a specialist of the Eastern Palearctic.

Polysticta

This is a single-species Holarctic genus. Steller's Eider *P. stelleri* is a migratory bioclimatic specialist of the Arctic belt that breeds in the Eastern Palearctic.

Histrionicus

This single-species genus appears to belong to the Mergini (Livezey, 1995). This bird is a bioclimatic moderate of the Boreal belt (Latitude Category B) of the Western and Eastern Palearctic, where it is largely sedentary.

Melanitta

This is a Holarctic genus of three migratory bioclimatic semi-specialists, two of which breed in the Palearctic. The Common Scoter *M. nigra* occupies the Arctic (Latitude Category A) belt across the Palearctic, and the Velvet Scoter *M. fusca* overlaps with the Common but has a range centred on the Boreal Belt (Latitude Category B).

Mergus

This is a genus with representatives in the Palearctic, Nearctic, Neotropics and a historically extinct species (Auckland Islands Merganser *M. australis*) in New Zealand. Three of the five species breed in the Palearctic – Goosander *M. merganser* is a semi-generalist and the Red-breasted Merganser *M. serrator* a bioclimatic moderate of the Boreal (Latitude Category B) belt, with a continuous pan-palearctic range; the two species are highly migratory except for the southernmost and the most oceanic populations. The Scaly-sided (or Chinese) Merganser *M. squamatus* is a migratory semi-specialist of the Temperate (Latitude Category C) belt of the Eastern Palearctic.

Mergellus

This is a single-species, exclusively Palearctic, genus. The Smew *M. albellus* is a highly migratory semi-specialist of the Boreal (Latitude Category B) belt of the Central and Eastern Palearctic, being marginal in the west.

Bucephala

This is the sister genus of *Mergus* and *Mergellus*. Its range is Holarctic. Two of the three species breed in the Palearctic – the Common Goldeneye *B. clangula* is a highly migratory bioclimatic moderate of the Boreal (Latitude Category B) belt with a continuous pan-palearctic range. Barrow's Goldeneye *B. islandica* is similar in characteristics to the Common Goldeneye but is restricted to a region of the Arctic (Latitude Category A) belt of the Western Palearctic.

Aix

This genus clusters with its distant relative the Muscovy Duck (*Cairina moschata*). It is a Holarctic genus of two species, one of which – the Mandarin Duck *A. galericulata* – breeds in the Warm (Latitude Category D) belt Eastern Palearctic, where it is partially migratory.

Marmaronetta

The Marbled Duck *M. angustirostris* is a migratory bioclimatic specialist that is unique to the arid and semi-arid regions of the mid-latitude belt (Latitude Category D) of the Western and Central Palearctic, where its range is fragmented, as in other waterbirds of this belt. This genus stands alone and is basal to *Netta, Aythya* and *Anas.*

Netta

This genus is a sister to *Aythya* and together they form the tribe Aythyini. It is a genus of three species with representatives in the Neotropics (including the Andean region), Africa and the Palearctic. A single species – Red-crested Pochard *N. rufina* – breeds in the Palearctic. It is a bioclimatic moderate with a range that is centred on the mid-latitude belt of the Western (discontinuous) and Central Palearctic. It is highly migratory except in the south-western part of the range.

Aythya

This genus has a fragmented global distribution with representatives in the Nearctic, Palearctic, Madagascar, Australasia and New Zealand. Five species (41.67%) breed in the Palearctic. Four species are bioclimatic moderates. Ferruginous Duck *A. nyroca* has the most southerly range, centred on the mid-latitude belt (Latitude Category D), although some populations reach north into the Temperate belt; the range is pan-palearctic but fragmented in the east and the west and the continental populations are highly migratory while those to the south and south-west are sedentary. The Common Pochard *A. ferina* has a range centred on the Temperate (Latitude Category C) belt; this range is also pan-palearctic and fragmented in the west and marginal in the east. Continental populations are highly migratory while those of the oceanic west are sedentary or short-distance migrants. The Tufted Duck *A. fuligula* and the Greater Scaup *A. marila* occupy the Boreal (Latitude Category B) belt and have continuous pan-palearctic ranges. The Greater Scaup's range is latitudinally narrower than the Tufted Duck's. The scaup is fully migratory but western populations of Tufted Duck are sedentary or short-distance migrants. Together, the four species provide an excellent example of pan-palearctic range fragmentation with latitude.

The fifth Palearctic species, Baer's Pochard *A. baeri,* is a migratory semi-specialist of the Temperate (Latitude Category C) belt of the Eastern Palearctic.

Anas

This is a large, globally distributed genus of 42 species, 10 of which (23.81%) breed in the Palearctic. The Mallard is a Holarctic bioclimatic generalist and among the most successful of all Palearctic birds. Its range is continuous across the Palearctic and it occupies all latitude belts (Latitude Category F). It is a migratory duck except in the oceanic south-west of the Western Palearctic. It is the only generalist in the genus (which has two semi-generalists), which conforms to the 'single generalist per genus' rule.

The Shoveler is one of the semi-generalists in the genus (the other is the Blue-winged Teal *A. discors* of the Americas). Like the Mallard it has a multi-latitude (Latitude Category F) pan-palearctic range but it is discontinuous in the west. The Shoveler is strongly migratory across most of this range but some populations along the extreme west of the Western Palearctic are sedentary.

Six species are bioclimatic moderates. Gadwall *A. strepera* occupies a pan-palearctic range centred on the Temperate (Latitude Category C) belt but this range is fragmented in the west and marginal in the east. It is mostly migratory except in the extreme south-west. Garganey *A. querquedula* is fully migratory with a pan-palearctic range, discontinuous in the west, also centred on the Temperate belt. Common Teal *A. crecca,* Pintail *A. acuta* and Wigeon *A. penelope* have continuous pan-palearctic ranges across the Boreal (Latitude

Category B) belt and further illustrate the latitudinal aspect of range fragmentation that can be seen in *Aythya.* Wigeon and Pintail are fully migratory and Common Teal only has sedentary populations in the extreme south-west. The fourth bioclimatic moderate is a sedentary duck of the Eastern Palearctic (Warm belt, Latitude Category D), the Spot-billed Duck *A. poecilorhyncha.*

Two *Anas* ducks are semi-generalists – Falcated Duck *A. falcata* is an Eastern Palearctic migratory species of the Temperate (Latitude Category C) belt, and Baikal Teal *A. formosa* is a migratory semi-specialist of the Boreal belt (Latitude Category B) of the Eastern Palearctic.

NUMIDIDAE

Numida

The Helmeted Guinea Fowl *N. meleagris* is the only species in its genus and the only Palearctic representative of this galliform family. It is a sedentary tropical African bioclimatic semi-specialist that just reaches the southern edge of the Western Palearctic.

TETRAONIDAE

All birds in this family are sedentary.

Dendragapus

One of three species of this Holarctic genus breeds in the Palearctic. The Siberian Grouse *D. falcipennis* is a bioclimatic specialist of the Temperate (Latitude Category C) montane belt of the Eastern Palearctic.

Lagopus

The ptarmigans constitute another Holarctic genus of three species of open habitats, two of which breed in the Palearctic. The Willow Grouse *L. lagopus* is a bioclimatic moderate with a continuous Boreal (latitude Category B) pan-palearctic range. The closely related Ptarmigan *L. mutus* is also a bioclimatic moderate but has a split Boreal/mid-latitude belt (B/D) pan-palearctic range. This range is fragmented in the west. It is a species, like other split-category species, that had a wider range during glacials, and retains isolated populations on the mountains of the mid-latitude belt.

Tetrao

This is an exclusively Palearctic genus of four forest species. They complement the alpine-tundra ptarmigans in the forests of the Palearctic. The Western Capercaillie *T. urogallus* is a bioclimatic moderate with a split Boreal/mid-latitude belt (B/D) range. Its range spans the Western and Central Palearctic and is fragmented in the west. Its sister-species – Black-billed Capercaillie *T. parvirostris* – is a bioclimatic semi-specialist and replaces it across the Boreal belt of the Eastern Palearctic. The Black Grouse *T. tetrix* is another bioclimatic moderate, of the Temperate belt (Latitude Category C) of the Palearctic, having a discontinuous range in the west. The Caucasian Black Grouse *T. mlokosiewiczi* is a mid-latitude specialist (Latitude Category D) of the Caucasus Mountains. It has a fragmented range that is limited to the Western Palearctic.

Bonasa

One of three species in this genus breeds in the Palearctic. The Hazel Grouse *B. bonasia* is a bioclimatic moderate of the Boreal belt (Latitude Category B) and has a pan-palearctic range that is discontinuous in the west.

PHASIANIDAE

All the species of this family, except the *Coturnix* quails, are sedentary.

Tetraogallus

Three of five species of this genus of mountain species of the mid-latitude belt and adjacent ranges breed in the Palearctic. The other two are close by, in Tibet and the Himalayas. They are all bioclimatic specialists and effectively geographical counterparts in different parts of the belt. The Palearctic species are Caspian Snowcock *T. caspius* and Caucasian Snowcock *T. caucasicus* in the Western Palearctic (Latitude Category D) and the Altai Snowcock *T. altaicus* in the Central Palearctic (Latitude Category C).

Alectoris

This is another mid-latitude belt genus of seven species, four of which breed in the Palearctic with the other three in adjacent ranges of the mid-latitude belt. The genus is thought to have radiated in the Pliocene, between 6 and 2 mya (Randi, 1996). They are geographical counterparts. The Chukar *A. chukar* is a bioclimatic semi-specialist and has the widest range (Latitude Category D), which is pan-palearctic though marginal in the west. The Red-legged Partridge *A. rufa* is a bioclimatic moderate (Latitude Category D) that has managed to spread further north than other congeners in the mild oceanic west. The Rock Partridge *A. graeca* (semi-specialist, Latitude Category D) replaces Chukar and Red-legged Partridge in the mountains of central and south-eastern Europe, and the Barbary Partridge *A. barbara* is the North African counterpart across the Maghreb.

Ammoperdix

The two species of this genus are bioclimatic specialists of the subtropical (Latitude Category E) belt of the Western (Sand Partridge *A. heyi*) and Central (See-see Partridge *A. griseogularis*) Palearctic. They are to some extent arid climate southern counterparts of *Alectoris*.

Francolinus

This large African genus of 41 species has two representatives in the subtropical (Latitude Category E) belt of the Western Palearctic, which are at the very northern limits of the genus's range. The Double-spurred Francolin *F. bicalcaratus* is a bioclimatic semi-specialist, and the Black Francolin *F. francolinus* is a specialist of warm and dry climates.

Perdix

These are partridges of the steppe habitats of the Palearctic, with one species in the Tibetan-Himalayan uplands. The two Palearctic species are geographical counterparts – Grey Partridge *P. perdix* is a bioclimatic semi-specialist of the Temperate (Latitude Category C) belt of the Western and Central Palearctic, while the Daurian Partridge *P. dauricae* is a specialist of the mid-latitude (Latitude Category D) belt of the Eastern Palearctic.

Coturnix

Two of nine (22.22%) quails are Palearctic. The quails have a wide Old World distribution that includes tropical Africa, southern Asia, Australasia and New Zealand as well as the Palearctic. They are largely sedentary bioclimatic specialists and semi-specialists. The Common Quail *C. coturnix* is exceptional, being a semi-generalist, multi-latitude (Latitude Category F) migrant, although southern Asian populations are sedentary. Its range encompasses the Western and Central Palearctic. In the east it is replaced by the migratory semi-specialist Japanese Quail *C. japonica,* which also has a multi-latitude range.

CONCLUSIONS

(a) Bioclimatic generalisation, sedentary and migratory behaviour. One feature that characterises this group is the low level of bioclimatic tolerance (Appendix 1), which is limited to one generalist and three semi-generalists (5.41% of all species). The proportion alters little between the two large orders that make up this group. The significant difference between the two orders lies in migratory behaviour. The Anseriformes are largely migratory with only two species (4.08%) sedentary and 30 (61.22%) fully migratory. In contrast, only two Galliformes are migratory (8%) and the rest (92%) are sedentary. This difference has determined where these two lineages have been most successful.

(b) Low- and high-latitude representation. There is a clear split in latitudinal representation between the two orders. The sedentary Galliformes are restricted in the north. There are no Arctic (Latitude Category A) species and only three (12%) have occupied the Boreal (Latitude Category B) belt, though we may add two split B/D species to these. Most species are Temperate (4, 16%), mid-latitude (8, 32%) or subtropical (6, 24%), between them making up 72% of the species. In contrast, the dominant latitudinal belts in the Anseriformes are the Arctic (Latitude Category A), with 15 species (30.61%), and Boreal (Latitude Category

B), with 13 species (26.53%) belts. There is a secondary representation of Temperate (Latitude Category C, 8 species – 16.33%) and mid-latitude (latitude Category D, 9 species – 18.37%) belt species but subtropical (1) and multi-latitude (2) species are almost absent. There can be little doubt that among these two related (though distant) lineages of bioclimatically intolerant species, migratory ability has enabled access to higher latitudes.

(c) **Species of the mid-latitude belt.** We find two types of birds occupying the mid-latitude belt, of which we have seen examples from other lineages: (a) waterbirds that occupy the seasonal wetlands of the Western and Central Palearctic and have discontinuous ranges – *Oxyura, Tadorna, Marmaronetta, Netta* and Ferruginous Duck; and (b) birds of open and rocky (and some of forest) mountains and hills – Ptarmigan, *Tetrao, Tetraogallus* and *Alectoris* with *Perdix* and *Ammoperdix* on the northern and southern (respectively) fringes.

(d) **Representation in the Western Palearctic.** There is a clear difference between the two lineages in terms of Western Palearctic representation. It is poor among the Anseriformes (29 species compared to 31 in the Central and 39 in the Eastern Palearctic) and rich among the Galliformes (18 species compared with 10 each in the Central and Western Palearctic). The low representation among the first group is due to the large number of species of cold and dry climates, many of which are missing from the oceanic west. It is precisely the milder conditions of the west that may have facilitated the survival of more species of sedentary gamebirds than in the harsher Central and Eastern Palearctic.

CHAPTER 19
Climate and the history of the birds of the Palearctic

We have now reached the point when it is time to sum up the main conclusions of this book. The history of the birds of this large region is precisely that – a history. We have tried to establish trends and patterns that lineages have in common but ultimately it has been, and continues to be, a story in which chance events, contingency and serendipity have played a massive role. If we were to go back to the Cretaceous and replay the tape (as Gould, 1989, would have argued) we could be telling a very different story. There are parts of the story that are undoubtedly deterministic – for example, global cooling pushed many tropical bird groups out of the Palearctic – but many others depended on where and when lineages, families, genera, species, ultimately individuals, were when particular events happened. We can only imagine how the world of Palearctic birds might have panned out had the passerines, for example, reached it during the Eocene instead of the Miocene (Ericson *et al.*, 2001). So the best way to summarise this history is to briefly tell it as far our current state of knowledge, and giving it my personal interpretation, allows. The story starts in the Cretaceous some time before the asteroid impact and K/T–related events. References to fossil species below refer to Mlíkovský's (2002) summary.

Early origins – the Cretaceous (before 65 mya)

The Cretaceous was a world completely unlike ours. It was a warm world in which the great mountain chains had not started to build and in which the continents retained an unusual configuration. It was starting to shape up into a familiar pattern but it still had a long way to go. South America, India and Africa floated as islands that were disconnected from the major landmasses of North America and Eurasia, which occupied familiar positions by this time. This warm and flat world was occupied by the dinosaurs but all modern bird lineages that we have identified from Hackett *et al.'s* (2008) analysis, had already appeared on the scene. They probably looked nothing like their present-day counterparts but they soon started to split and diversify. The nightjar line branched off from the swift-hummingbird line and the geese and game birds also separated. Cranes, rails and cuckoos diverged. The buttonquails had, by then, embarked on their own particular adventure in evolution. The shorebirds had started to split along major lines – thick-knees, pratincoles-coursers, stilt-avocets, plovers and scolopacids. The divers separated from the penguins, shearwaters and storm-petrels, which themselves then started branching off in different directions. Six (if we add the ratites) of our eight main lines were branching before the asteriod impact.

A hot and humid world of waterbirds – the Palaeocene (65.0–55.8 mya)

The new world of the Palaeocene was a warm one which saw further re-arrangement along major lines. The waterbird lineage (our Group 3; see p. 166) seems to have been particularly active. The divers radiated at this very early stage, in a warm climate. Descended from a major lineage dominated by southern hemisphere birds (penguins, shearwaters and storm-petrels,) they seem to have appeared as a northern hemisphere family, presumably by immigration and isolation from the parent line. Today the living descendants are birds of cold climates that must have adapted to the progressively cooling world of the Tertiary. The genus *Gavia* first appeared, like many other modern genera, in the Early Miocene (*G. egeriana*). The Eocene genus *Colymboides* disappeared about this time. So the divers, descended from a southern hemisphere lineage, had ended up in the waters of the northern hemisphere and had gradually adapted to life there. The post-Miocene cooling confined the family, represented by a single genus, to the extreme north, where they adapted and became specialists of cool climates.

Other branches of this waterbird lineage (Group 3) started to emerge at this time. The storks radiated at this early stage and the gannet and cormorant lines split. The Ciconiidae itself emerged from the first of these lineages in the Early Miocene and the genus *Ciconia*, which includes the present-day Palearctic storks, in the Middle Miocene (*C. sarmatica*). The Sulidae first appeared in the Middle Eocene (*Eostega lebedinskyi*) and the genus *Morus* in the Middle Miocene. This genus seems to have been, along with *Sula,* the survivor of a greater Early Tertiary diversity that was subsequently pruned. The Phalacrocoracidae, and the genus *Phalacrocorax,* emerged in the Early Miocene. These observations show that the modern families and genera appeared long after the splits of their respective lineages. As seems to have been typical of most lineages, the emergence of many of these modern families and genera seems to have taken place in the Miocene.

The other lineage that was splitting in the Palaeocene was that of the shorebirds (of our Group 2; see p. 149). It seemed to be a continuation of the early diversification that had commenced in the Cretaceous. It included the split between skuas and auks, coursers and pratincoles, and the *Numenius* and *Limosa* lineages split from their respective parent stocks. The scolopacid lineage had separated in the Late Cretaceous, but the modern Scolopacidae emerged in the early Miocene. Among the early branches, the modern genera appear late – *Numenius* first appears in the Late Pliocene (*N. arquata*) and *Limosa* in the Early Pleistocene (*L. limosa*).

Overall, the Scolopaci are a group of northern hemisphere, mainly Arctic and Boreal, species that evolved along the long stretch of coastline from Scandinavia, across Eurasia to north-eastern America under warm conditions that pre-dated the polar ice caps. They emerged as a line of birds that occupied the land-water interface, to which they became superbly adapted. The intense climatic changes that they experienced throughout the Tertiary did not remove them from these northern coastlines. Instead, they adapted to the new climatic conditions without the need to make major changes in habitat use. Instead they exploited all land-water interfaces, including the tundra and inland wetlands. The cooling of the Miocene and Pliocene saw significant re-organisation at family and genus level, which explains the late arrival of modern types, and migration must have been part of the solution to the increased cooling. The Scolopaci are, as we have seen (see p. 192, Appendix 1), dominated by long-distance migrants. Unlike the terrestrial birds, which had to adapt to changing habitats, the Scolopaci stayed with the habitat and feeding method and adapted to the climate, or evolved ways of avoiding its worst elements.

Other lineages within the main shorebird line seemingly emerged to the south, probably along the southern coastlines of the Eurasian continent and land masses to the south (Africa, Australia, even South America). These lineages, which started to branch off in the Cretaceous and continued during the Palaeocene, are represented today by single-genus families with large, southern distributions. They include the stilts *Himantopus*, avocets *Recurvirostra*, and others that branched in the Eocene, like the oystercatchers *Haematopus.* These lineages, also exploiters of the land-water interface, did not experience the climatic upheavals of their northern cousins. It is not surprising to find a modern family like the Recurvirostridae and the avocet genus *Recurvirostra* already present in the Late Eocene (*R. sanctaneboulae*). The presence of essentially geographical replacements of the same body form (genus) and the absence of major radiation (few species) is to be expected among lineages that were not subjected to major climatic or ecological impacts. The Scolopaci, on the other hand, kept on changing and diversifying (producing species-rich genera), until recently.

Other shorebird lineages that had been splitting at an early stage included species that seemed well-capable of dealing with arid conditions, or at least adapted to these at an early stage. These lineages appear to have occupied middle and low latitude ranges and some strayed away from coasts and into arid environments. They included the thick-knees, pratincoles and coursers. In the case of the pratincoles, for which we have fossil evidence, we find the modern family Glaerolidae in the early Miocene (*Mioglareola dolnicensis)* and *Glareola* (*G. neogena*) in the Mid-Miocene. *Cursorius* is first encountered in the Late Miocene. The plovers (Charadrii) probably belong with this group. These shorebirds seemingly fall temporally between the early emerging and conservative 'southern' shorebirds and the rapidly adapting 'northern' Scolopaci. The radiation of the latter group continued in the Eocene, Oligocene and into the Miocene when the lineages leading to the large genera *Tringa* and *Calidris* took off. Speciation continued into the Early Pleistocene. There is a very clear parallel between these northern shorebirds and the divers, which occupied similar latitude belts and habitats, diverging very early with modern genera emerging in the Early Miocene. To these we could probably add the skuas.

Waterbird diversification in a slowly cooling world – the Eocene (55.8–33.9 mya)

The Eocene, which saw the progressive cooling of the climate, was dominated by branching and splitting of the shorebird lineages which had started in the previous periods. The following lineages (that led to these genera) split during this time: *Scolopax* from *Gallinago, Phalaropus, Xenus, Actitis, Arenaria, Burhinus, Recurvirostra-Himantopus, Haematopus, Bartramia-Numenius, Limnodromus* and *Lymnocryptes*. The *Scolopax* line is atypical in that it altered its habitats and moved into woodland. The genus first appeared in the Late Pliocene (*Scolopax carmesinae*). *Gallinago* appeared in the Late Miocene. The higher-latitude genera, once again, emerged more recently than the others, in the Early Pleistocene: *Phalaropus* (*P. fulicarius*) and *Lymnocryptes* (*L. minimus*).

The branches of waterbirds (Group 3) continued splitting early in the Eocene with the herons, ibises and pelicans diverging. These waterbird lineages, unlike the divers of the same group, occupied lower latitudes and are comparable to the 'southern' shorebirds in distribution. Like these, many are represented by single-species genera or single-genera families with large inter-continental ranges, which suggests that they too adapted at an early stage to water and water-land interfaces in situations that required little change compared to their northern counterparts. The Ardeidae appeared in the Late Oligocene-Early Miocene and *Ardea* in the Early Miocene (*A. aurelianensis*); Threskiornithidae in the Middle Eocene, *Plegadis* in the Early Miocene (*P. paganus*) and *Geronticus* in the Middle Miocene (*G. perplexus*); and Pelecanidae in the Early Miocene, with *Pelecanus* in the Late Miocene (*Pelecanus odessanus*). The waterbirds of our Group 3 and the Group 2 shorebirds thus represent a clear example of parallel evolution in relation to geography and climate change from the Eocene to the Early Pleistocene. Divers resemble sandpipers while ibises and herons resemble avocets and stilts. The modern genera of the first group tended to appear from the Miocene, but those of the latter were already around by then, and in some cases, long before.

The other important splits during the Eocene involved the cranes and rails of Group 4 (see p. 176), and between the swifts and hummingbirds and the radiation of the nightjars from Group 5 (see p. 183). The evolutionary activity in the cranes and rails, largely associated with water habitats, seems to indicate (alongside the evidence for shorebirds of Group 2 and waterbirds of Group 3) significant evolutionary pressure on species occupying these habitats at a very early stage.

The first big cooling – the Oligocene (33.9–23.0 mya)

The Oligocene saw the first significant period of sharp cooling and polar ice sheet formation. This is reflected very clearly within the shorebirds that had started to respond to cooling in the Eocene. Most importantly, it marks the start of the radiation of the auks and the gulls, much of the main lineage-sorting having taken place by the end of the Oligocene. The modern familiy Alcidae correspondingly appeared in the Late Oligocene (*Petralca austriaca*) and the Laridae in the Middle Oligocene (*Gaviota lipsiensis*). It is in the Oligocene that we pick up the first evolutionary activity among Group 1 (see p. 34), with the wryneck lineage splitting form the mainstream woodpeckers and the osprey lineage from the mainstream dirunal raptors. Early ospreys are present from the Oligocene.

The other lineage that continued diverging and radiating was Group 4 (see p. 176), with cranes, rails and bustards diversifying. The modern family Gruidae is identifiable in the fossil record from the Middle Eocene (*Palaeogrus princeps*) with *Grus* in the Late Miocene (*Grus moldavica*). The modern family Otididae, on the other hand, only emerges in the Middle Miocene and must represent a later radiation that coincides with the period of northern hemisphere aridification. *Otis* appears in the Middle Miocene (*O. affinis*), *Chlamydotis* in the Early Pliocene (*C. mesetaria*) and *Tetrax* in the Late Pliocene (*T. tetrax*). The Rallidae has its origins in the early Eocene with many genera emerging that are now extinct. *Crex* emerged in the Middle Miocene and *Porzana* and *Rallus* in the Late Pliocene (*P. porzana, R. aquaticus*), *Gallinula* in the Early Pliocene and *Fulica* in the early Pleistocene (*F. atra*).

The beginning of the modern world – the Miocene (23.0–5.3 mya)

The evolutionary activity of the ancient lineages seems largely over by the Miocene. We observe the *Caldris* and *Tringa* splits among the shorebirds and we see the final stages of the auk radiation, with the branching of the Common and Brünnich's Guillemot lineages. The terns radiated during the Miocene. However, it is Group 1 that comes to the fore and dominates the evolutionary activity of the Miocene. These are largely terrestrial groups that responded to the aridification of the Miocene, including the fragmentation of forests and the early expansion of deserts. These radiations involve, among others, the passerines, which had entered the Palearctic for the first time from South-east Asia during the Miocene (Ericson *et al.*, 2001).

Forest birds responded to these climatic changes with splitting and radiation of lineages. The owls (Group 1e; see p. 112) were especially active: *Strix, Bubo* and *Asio* split and radiate; New and Old World *Strix, Aegolius* and *Glaucidium* lineages split; and Old World *Otus* split from New World *Megascops*. The Striginae and Surninae lineages and the Barn and Bay Owls also diverged. We observe the colonisation and the diversification of woodpeckers in the new temperate forests and the isolation of the various intercontinental lineages. Other lineages responded to the newly opened habitats; the falcons *Falco* radiated and the modern bee-eaters emerged. It is among the newly arrived passerines that we see the greatest activity: shrikes, *Sylvia* warblers, rock thrushes, trumpeter finches, pipits and larks diversified into open habitats. Other genera also diversified in mixed and wetland habitats: *Locustella* and *Acrocephalus* warblers and the serins. Along with this burst of novelty and radiation came the extinction of many old genera. Waterbirds lost signifcant ground in this drying world, leaving ancient survivors (e.g. herons, ibises and stilts) stranded and new ones (e.g. sandpipers) appearing in a new world of cold, Arctic wetlands.

An arid world – the Pliocene (5.3–2.6 mya)

The Pliocene saw adaptation (of passerines in particular) to the opening up of the forested world and the expansion of the deserts. The passerine radiations included *Locustella, Sylvia,* the larks, tits, redstarts, thrushes, dippers, nuthatches, sparrows, snowfinches, petronias, shrikes and wagtails. The woodpecker colonisation of cold forests continued, with that of the tits in parallel. Forest (*Pernis*) and wetland (*Haliaeetus*) raptors were also involved probably alongside many other raptors (e.g. steppe and savanna eagles) that were adapting to this world.

The big freeze – the Pleistocene (2.60–0.01 mya)

The Pleistocene represents the tail-end of the Miocene-Pliocene explosion. Lineage splits (often incomplete) continued among closely related species, such as the spotted eagles, the White-tailed and Bald Eagles, the Eurasian *Buteo* buzzards, the *Catharacta* skuas, the great grey shrike complex, the Barn Swallow expansion, the *Fringilla* island colonisations and the separation of geographical populations of Lesser Kestrels and some further separation in the *Phylloscopus, Hippolais* and *Sylvia* lineages. However, these are relatively minor events in comparison with those that had preceded them.

Taxon						
Ciconia	*Threskiornis*	*Butorides*	*Pelecanus*	*Haematopus*	*Hieraaetus*	Aegypinae
Phalacrocorax	*Botaurus*	*Ardeola*	*Tachybaptus*	*Caprimulgus*	*Lophaetus*	Gypaetinae
Morus	*Ixobrychus*	*Casmerodius*	*Burhinus*	Pandionidae	Milvinae	Aquilinae
Platalea	*Gorsachius*	*Ardea*	*Himantopus*	Elaninae	Haliaeetinae	Large *Aquila* subgroup
Plegadis	*Nycticorax*	*Egretta*	*Recurvirostra*	Perninae	Circaetinae	

Table 19.1 *Genera and subfamilies of Palearctic birds with ancient distributions that transcend the Palearctic and which are often fragmented across large areas of the globe.*

The Holocene (0.01 mya to the present)

The last 10,000 years of the history of the Palearctic and its birds is that of the survivors, the species that had been shaped, first as lineages, then as orders and families, then as genera and finally species, over a protracted period from the Late Cretaceous to the Pleistocene. Those that made it to the Pleistocene were able to handle whatever was thrown at them – cold tundra, dry steppe and desert, temperate forests. They had independently evolved a range of tactics: bioclimatic tolerance, migration, diet diversity and the ability to occupy a wide range of habitats including open and treeless landscapes. Some were relatively recent products, like the passerines of the Miocene and Pliocene. Others were new versions of old 'production lines', like the Miocene and Pliocene sandpipers; yet others were from old lineages, like the herons and ibises, which had made it through the Miocene interchange.

Patterns of historical dispersion and distribution among Palearctic birds

We can observe repeating patterns among birds of different lineages, that indicate common themes that have recurred at various points during the history of Palearctic birds.

Ancient lineages

A number of Palearctic genera have wide global ranges, in some cases heavily fragmented, which reveal old distributions when the world was warmer and wetter. Invariably they include families and genera that predate the Miocene (Table 19.1). In the case of diurnal raptors they appear at the level of subfamilies, genera subsequently acting as geographically independent entities.

Auks and gulls – an Oligocene legacy

One consequence of the sharp cooling at the start of the Oligocene was a decrease in sea-water temperatures in boreal and arctic regions of the northern hemisphere. This resulted in a rearrangement of marine animal communities and an increase in productivity (Gladenkov and Sinel'nikova, 2009). It is not surprising to find that this was the time when the gulls and auks diversified into the present-day lineages. These became the dominant coastal and pelagic seabirds of the Palearctic, derived from a single lineage, the shorebirds (2). The other pelagic birds (shearwaters, storm-petrels) were immigrants from the southern hemisphere. Once established, these gull and auk lineages adapted, contracted or expanded ranges with changing climate. For example, the brief warming of the Pliocene oceans and the subsequent Pleistocene cooling (Filipelli and Flores, 2009) would have significantly affected the specialised auks.

The tundra

Tundra is a modern habitat that resulted from the cooling of the climate during the Tertiary, with its expansion taking place with the cooling of the Late Pliocene (Hopkins *et al.*, 1971). Several groups invaded this habitat and exploited it seasonally, depending on migration and habitat 'metamorphosis', which is shifting to a completely different kind of habitat (and food) outside the breeding season. The key exploiters of the

Species	Lineage	Species	Lineage
Shore Lark	1a (ii)	Red Knot	2
Arctic Redpoll	1a (v)	Sanderling	2
Lapland Longspur	1a (v)	Red-necked Stint	2
Snow Bunting	1a (v)	Sharp-tailed Sandpiper	2
Red-throated Pipit	1a (v)	Great Knot	2
Pechora Pipit	1a (v)	Western Sandpiper	2
Gyrfalcon	1c	Baird's Sandpiper	2
Snowy Owl	1e	Rock Sandpiper	2
Rough-legged Buzzard	1f	Common Ringed Plover	2
Ross's Gull	2	Pacific Diver	3
Sabine's Gull	2	Black-throated Diver	3
Arctic Skua	2	Great Northern Diver	3
Long-tailed Skua	2	White-billed Diver	3
Pomarine Skua	2	Pink-footed Goose	7b
Bar-tailed Godwit	2	Lesser White-fronted Goose	7b
Long-billed Dowitcher	2	White-fronted Goose	7b
Red-necked Phalarope	2	Emperor Goose	7b
Grey Phalarope	2	Snow Goose	7b
Spotted Redshank	2	Barnacle Goose	7b
Buff-breasted Sandpiper	2	Red-breasted Goose	7b
Spoonbill Sandpiper	2	Brent Goose	7b
Dunlin	2	Long-tailed Duck	7b
Purple Sandpiper	2	King Eider	7b
Little Stint	2	Spectacled Eider	7b
Pectoral Sandpiper	2	Steller's Eider	7b
Curlew Sandpiper	2	Common Scoter	7b

Table 19.2 *Species of Palearctic birds occupying tundra.*

tundra were sandpipers, skuas and geese, but other lineages (within Group 1) also spawned specialists of this habitat (Table 19.2).

The Himalayan launchpad

I have used this term throughout this book to describe a phenomenon that has repeatedly occurred, and which has led to the colonisation of the Palearctic, and in some cases the Nearctic afterwards, by forest birds that originated in tropical South-east Asia. From there they colonised altitude belts around the Himalayan-Tibetan massif and spread into the Palearctic when climatic conditions brought equivalent altitude and latitude belts together. The phenomenon is exclusive to this region, which is the only one with a direct physical connection between the tropics and the Palearctic. In a limited number of cases expansion into the Palearctic has not required altitude pre-adaptation, but in these cases a subsequent spread across the Palearctic has not occurred. These cases are rare in comparison and are indicated by species such as Fairy Pitta *Pitta nympha*, Black-winged Cuckoo Shrike *Coracina melaschistos*, Ashy Minivet *Pericrocotus divaricatus*, Asian Stubtail *Urosphena squameiceps*, the *Bradypterus* warblers, Père David's Laughing Thrush *Garrulax davidi*, Black Drongo *Dicrurus macrocercus*, Grey-faced Buzzard *Butastur indicus*, Mountain Hawk Eagle *Nisaetus nipalensis*, Painted Snipe *Rostratula benghalensis* and Brown Hawk Owl *Ninox scutulata*.

By contrast, the 'launchpad' has worked in many and diverse forest genera that have entered the temperate and boreal forests from here: *Perisoreus, Garrulus, Nucifraga, Oriolus, Phylloscopus, Aegithalos*, Paridae, *Luscinia, Tarsiger, Ficedula, Phoenicurus, Sitta, Certhia, Pyrrhula, Carpodacus, Loxia, Chloris, Picus, Dryocopus, Dendrocopos, Tyto, Scops, Strix, Pernis, Scolopax* and *Cuculus*. To these we may add, subject to confirmation, *Muscicapa, Zoothera, Turdus, Pinicola, Serinus* and *Accipiter*. It also seems to have involved non-forest genera: *Delichon, Cettia*, the ancestors of the Palearctic *Locustella*, and possibly *Acrocephalus*. The Himalayan launchpad has been one of the most powerful catalysts generating dispersals into the Palearctic. Those that arrived first usually became bioclimatic generalists.

Temperate and boreal forests

The temperate and forest belts of the Palearctic absorbed the tropical woodland species at different stages in their evolutionary history. Different lineages have had more or less time to assimilate the changing conditions of these forests. Owls and woodpeckers are very well represented in these woods, and they are the most abundant of the ancient lineages (Table 19.3) having developed adaptations that enabled year-round survival, as they are largely sedentary species. One way of dealing with the increasingly harsh winters of the north must have come easy to the owls, which were by nature nocturnal. This rare behavioural strategy of the avian world meant that these birds had extended hunting time in the winter (when nights were longest) at a time when most other birds would have their foraging time severely limited. Sedentary diurnal raptors are practically non-existent in these forests, even the accipiters migrating away from the most northerly forests in autumn.

The other long-established group, the woodpeckers, relied on diet. This diet, the seeds of broad-leaved and coniferous trees, was well-established long before the C4 grasses and the new suite of seed-eaters arrived. Seeds are designed to survive for long periods before they encounter optimal conditions in which to develop. In northern climates they have the added advantage of avoiding heat, which might increase the rate of deterioration. The added trick that the woodpeckers had was the ability to cache food and return to it at a later date (Smith and Reichman, 1984; Doherty *et al.*, 1996). This ability is also found among the boreal owls (Solheim, 1984; Korpimäki, 1987; Smith and Reichman, 1984), their prey preserving best in northern climates. So owls and woodpeckers were the long-established and hardened specialists of the temperate and boreal forests.

There is a long list of passerines also in Table 19.3, but these are birds that must have entered the northern woods much later, after their arrival in the Palearctic during the mid-Miocene: the tits, together with a few specialised corvids and a nuthatch, which entered the Palearctic using a 'launchpad' strategy from the South-east Asian forests. Those that became successful residents took on the woodpecker way of life – eating mainly seeds and developing, in some species more than others, caching behaviours (Smith and Reichman, 1984; Waite and Reeve, 1992; Carrascal and Moreno, 1993; Lanner, 1996; Štorchová *et al.*, 2010). This is a wonderful example of convergent evolution, developed in lineages of different age and different amounts of time inhabiting the northern forests.

A large number of other species of passerines, such as finches and buntings, took to seed-eating in these forests, usually adding insect food in the spring and summer. These birds show no predisposition to cache seeds and are variably migratory, from complete migrants to partial migrants in which only the northernmost populations move. They provide a contrast to the woodpeckers, corvids, tits and nuthatches.

Grouse also seem to have achieved residency in these forests, but they have done so by transferring the gamebird ground foraging strategy to these woods, so in this respect they have remained conservative with an ability to secure vegetable matter from the most difficult of situations. The extensive presence of grouse and ptarmigan in the fossil record of Pleistocene Europe tells us that these specialised birds were highly successful when dealing with the harsh climate of the Palearctic in the Pleistocene.

Species	Forest belt	L	Species	Forest belt	L	Species	Forest belt	L
Eurasian Jay*	F	1a (i)	Mistle Thrush	F	1a (iv)	Grey-headed Woodpecker*	C	1d
Siberian Jay*	B	1a (i)	Pale Thrush	C	1a (iv)	Eurasian Green Woodpecker	F	1d
Spotted Nutcracker*	C	1a (i)	Grey-backed Thrush	C	1a (iv)	Black Woodpecker*	C	1d
Golden Oriole	F	1a (i)	Redwing*	B	1a (iv)	Great Spotted Woodpecker*	F	1d
Black-naped Oriole	F	1a (i)	Eyebrowed Thrush	B	1a (iv)	White-backed Woodpecker*	C	1d
Chiffchaff*	F	1a (i)	Naumann's Thrush	B	1a (iv)	Middle Spotted Woodpecker	F	1d
Wood Warbler	C	1a (ii)	Song Thrush	B/D	1a (iv)	Lesser Spotted Woodpecker*	C	1d
Pallas's Warbler	C	1a (ii)	Red-throated Thrush	F	1a (iv)	Rufous-bellied Woodpecker	F	1d
Pale-legged Leaf Warbler	C	1a (ii)	Chinese Thrush	F	1a (iv)	Grey-capped Woodpecker	F	1d
Yellow-browed Warbler	B	1a (ii)	Eurasian Nuthatch*	F	1a (iv)	Three-toed Woodpecker*	B/D	1d
Arctic Warbler*	B	1a (ii)	Eurasian Treecreeper*	C	1a (iv)	Oriental Scops Owl	F	1e
Blackcap	F	1a (ii)	Short-toed Treecreeper	F	1a (iv)	Collared Scops Owl	F	1e
Garden Warbler	C	1a (ii)	Goldcrest*	C	1a (iv)	Tawny Owl	F	1e
Marsh Tit	C	1a (iii)	Firecrest	F	1a (iv)	Great Grey Owl*	B	1e
Willow Tit*	B	1a (iii)	Bohemian Waxwing*	B	1a (iv)	Ural Owl*	B/D	1e
Siberian Tit*	B	1a (iii)	Japanese Waxwing	C	1a (iv)	Blakiston's Fish Owl	C	1e
Coal Tit*	F	1a (iii)	Common Chaffinch	F	1a (v)	Long-eared Owl*	F	1e
Crested Tit	F	1a (iii)	Hawfinch	C	1a (v)	Brown Hawk Owl	F	1e
Great Tit*	F	1a (iii)	Japanese Grosbeak	C	1a (v)	Tengmalm's Owl*	B/D	1e
Blue Tit	F	1a (iii)	Pine Grosbeak*	B	1a (v)	Eurasian Pygmy Owl*	B	1e
European Robin	F	1a (iv)	Eurasian Bullfinch*	C	1a (v)	Northern Hawk Owl*	B/D	1e
Siberian Blue Robin	C	1a (iv)	Pallas's Rosefinch	B	1a (v)	European Honey Buzzard	C	1f
Rufous-tailed Robin	B	1a (iv)	Common Crossbill*	F	1a (v)	Oriental Honey Buzzard	C	1f
Red-breasted Flycatcher	F	1a (iv)	Two-barred Crossbill*	B	1a (v)	Eurasian Sparrowhawk*	F	1f
Collared Flycatcher	C	1a (iv)	Parrot Crossbill	B	1a (v)	Northern Goshawk*	F	1f
Mugimaki Flycatcher	C	1a (iv)	Scottish Crossbill	C	1a (v)	Common Buzzard*	F	1f
Taiga Flycatcher	B	1a (iv)	Eurasian Siskin*	B/D	1a (v)	Eurasian Woodcock*	C	2
Brown Flycatcher	C	1a (iv)	European Greenfinch	F	1a (v)	Common Cuckoo*	F	4
Dark-sided Flycatcher	C	1a (iv)	Oriental Greenfinch	F	1a (v)	Eurasian Nightjar*	F	5
Spotted Flycatcher*	F	1a (iv)	Black-faced Bunting	F	1a (v)	Wood Pigeon	F	6
Grey-streaked Flycatcher	F	1a (iv)	Pine Bunting	B	1a (v)	Siberian Grouse	C	7a
Common Redstart	F	1a (iv)	Chestnut Bunting	C	1a (v)	Western Capercaillie	B/D	7a
Daurian Redstart	F	1a (iv)	Tristram's Bunting	C	1a (v)	Black-billed Capercaillie	B	7a
Scaly Thrush	C	1a (iv)	Yellow-browed Bunting	B	1a (v)	Hazel Grouse	B	7a
Siberian Thrush	C	1a (iv)	Olive-backed Pipit	F	1a (v)			
Eurasian Blackbird	F	1a (iv)	Eurasian Wryneck*	F	1d			

Table 19.3 *Species of Palearctic birds occupying temperate-boreal forest ranges (* species with pan-Palearctic ranges). L = lineage. Forest belt refers to latitude category.*

Chats, thrushes, flycatchers and some warblers also took to the northern forests but relied heavily on migration as a survival strategy. The only successful, predominantly insect-eating resident omnivores have been the kinglets and treecreepers. They are typified by their small size and highly active way of life. It is clearly an unusual way of life in these forests but they have become among the most widespread of birds in the forests of the Palearctic.

The mid-latitude belt (MLB) as a corridor

Even though I have argued that the mid-latitude belt of the 40th parallel has acted to break up species ranges, it has acted as a corridor for species that adapted to the constant habitat of the belt – rocks. We have already seen how adaptation to breeding in rocky habitats has been a defining feature of many Palearctic species from different lineages. The genera that have used the MLB rocky habitat corridor include *Pyrrhocorax, Corvus, Ptyonoprogne, Oenanthe, Monticola, Sitta, Prunella, Montifringilla, Rhodopechys, Emberiza, Anthus, Athene, Gypaetus* and *Gyps.* The dispersal has probably been in most cases west from the Himalayan region, which makes this model a modified version of the 'launchpad'.

Aridity in the MLB and adjacent regions

The broad band covering the MLB and areas of the Sahara to the south are often regarded as a barrier preventing flow of species between Africa and the Palearctic (e.g. Snow, 1978) but I also view it as an important source of diversity of arid-adapted species whose origins are often, mistakenly, attributed to Africa or other southern regions. This belt of aridity, like the rocky part of the MLB, has a southern extension down east Africa to southern Africa. Its origins lie in the Miocene and Pliocene. A number of genera (or subsets of genera) have originated or diversified across the desert, semi-desert, shrublands and savannas of this region: *Hippolais, Sylvia,* the larks, *Cercotrichas, Passer, Petronia,* the hierofalcons, *Upupa, Merops, Coracias, Elanus, Circaetus, Aegypius, Neophron, Hieraaetus, Circus, Geronticus, Vanellus, Charadrius, Bubulcus,* the bustards, *Clamator, Caprimulgus, Aquila fasciata-spilogaster,* savanna-steppe *Aquila, Melierax, Anthropoides, Crex, Cursorius, Burhinus*, sandgrouse, and some *Strix* and *Bubo.* I have only listed genera with Palearctic representatives. The species which adapted to the expansion of the C4 grasses are embedded within this group.

The emergence of the C4 grasses

We saw earlier (see p. 18) how a new world of grasses came to the fore in the Miocene and Pliocene. It opened up new kinds of habitats, most notably grasslands and savannas, and new opportunities for birds able to exploit them. Grass-seed eaters became prevalent, but there were other birds that took to these habitats to exploit other food sources, like the insects. Groups that emerged or diversified as a result included the cisticolas, larks, sparrows, finches, buntings, doves, quails, harriers, falcons and the corncrake. The harriers were, in fact, a lineage of accipiters that took to new habitats as exploiters of the mixed carnivore strategy in these habitats, alongside the bustards, storks and some falcons. Not surprisingly they were mainly passerines, birds of the modern world. But other lineages exploited the grasslands as extensions of other, established, habitats, like wetlands or semi-arid open habitats. They included the storks, herons, bustards, cranes, pigeons, sandgrouse, partridges, francolins, rollers, bee-eaters and ostriches. When these habitats flooded seasonally they became extensions of wetland habitats for many more species.

Wetlands of the mid-latitude belt

The MLB is also characterised by a broken series of water bodies and wetlands that stretch from the Mediterranean Sea in the west, via Black, Caspian and Aral Seas eastwards. This chain also includes a number of smaller wetlands and is a relict of a wetter past. Even in recent times, when the Sahara was a land of lakes, or when southern Siberia held large inland freshwater seas, this belt's wetlands must have been more continuous and of greater extent. A large number of species associated with wetlands and enclosed seas have fragmented ranges across this belt or in parts of it (Table 19.4). It is of particular interest that many of the species come from ancient lineages that today have highly fragmented ranges: herons, ibises, stilts, avocets, pratincoles, gulls and terns. The lineages that, on the other hand, adapted to tundra wetlands are absent. It is also notable that there are only three passerines in this group.

Species	Lineage	Species	Lineage
Moustached Warbler	1a (ii)	Caspian Tern	2
Clamorous Reed Warbler	1a (ii)	Sandwich Tern ('southern' pop.)	2
Penduline Tit*	1a (iii)	Gull-billed Tern	2
Pallas's Fish Eagle	1f	Black-crowned Night Heron	3
Black-winged Stilt	2	Squacco Heron	3
Avocet	2	Little Egret	3
Collared Pratincole	2	Great White Egret	3
Black-winged Pratincole	2	Purple Heron	3
Kentish Plover	2	Eurasian Spoonbill	3
Greater Sand Plover	2	Greater Flamingo	3
Caspian Plover	2	Dalmatian Pelican	3
Spur-winged Plover	2	Great White Pelican	3
White-tailed Plover	2	Baillon's Crake	4
Northern Lapwing*	2	Ruddy Shelduck	7b
Pallas's Gull	2	Shelduck ('southern' pop.)	7b
Relict Gull	2	Greylag Goose*	7b
Mediterranean Gull	2	Gadwall	7b
Yellow-legged Gull	2	Garganey	7b
Armenian Gull	2	Shoveler	7b
Slender-billed Gull	2	Marbled Duck	7b
Audouin's Gull	2	Red-crested Pochard	7b
Black Tern	2	Pochard*	7b
White-winged Black Tern	2	Ferruginous Duck	7b
Whiskered Tern	2	White-headed Duck	7b
Little Tern	2		

Table 19.4 *Species of Palearctic birds occupying mid-latitude wetlands; * = species with range extending north of the mid-latitude belt.*

Range shifts in the Pleistocene

I have argued in this book that the Pleistocene, with its abrupt and rapid climate oscillations, was typified by the southward displacement of species during glacials. This displacement was often accompanied by westward shifts of steppe and cold desert species. There is a large body of supporting fossil evidence for such displacements in mammals (Finlayson, 2004). The data for birds are limited but strongly indicative, as I have described in the corresponding chapters.

Evidence of displacements south into areas not occupied today is found from the following fossil species: Lapland Longspur, Snow Bunting, Gyrfalcon, Merlin (possible), Snowy Owl, Rough-legged Buzzard, Grey Plover (possible), Glaucous Gull, Little Auk, Red-throated Diver, Long-tailed Duck and Velvet Scoter. To this record we should add a high representation of tundra geese, wetland ducks and open habitat gamebirds (especially grouse and ptarmigan). In fact, evidence of significant tracts of open, treeless landscapes also comes from the presence of bustards, corncrake and, particularly, quail and thrushes in many localities.

South-west displacements of boreal birds are recorded in a number of fossil species: Spotted Nutcracker, Siberian Jay, Grey-headed Woodpecker, White-backed Woodpecker, Ural Owl, Great Grey Owl, Northern Hawk Owl, Eurasian Pygmy Owl, Tengmalm's Owl and Greater Spotted Eagle. Westward displacements of steppe and savanna birds are recorded in Red-footed Falcon, Saker, Imperial Eagle, Steppe Eagle, Pallid Harrier and Pallas's Gull. To this we may add (with caution) ground jays and White-winged Lark.

Finally, there are species that spread from high mountain habitats to lower ground and became more abundant and widespread than they are today. The Alpine Chough is the most striking example. Others are the Shorelark, Alpine Accentor and White-winged Snowfinch.

Of great interest is a different and unusual form of expansion. It involved Bearded, Black and Griffon Vultures, which seem to have ranged more widely and to the north across the Eurasian Plain. This atypical pattern would seem to be due to the vast herds of reindeer, mammoths and other large ungulates that swarmed on the plains. They were tracked by lions, hyenas, wolves and people and it seems also by the vultures, which reached large dimensions.

Importantly, bar island endemics, there is *no* fossil evidence of the emergence of new species or extinction of existing ones in the Pleistocene, a time when a number of mammal species *did* disappear from the planet.

Species	Latitudes occupied	Lineage
Greenish Warbler	A/D	1a (ii)
Bluethroat	B/D	1a (iv)
Siberian Rubythroat	B/D	1a (iv)
Red-flanked Bluetail	B/D	1a (iv)
Pied Flycatcher	B/D	1a (iv)
Whinchat	B/D	1a (iv)
Song Thrush	B/D	1a (iv)
Ring Ouzel	B/D	1a (iv)
White-throated Dipper	A/D	1a (iv)
Dunnock	B/D	1a (v)
Siberian Accentor	A/C	1a (v)
Black-throated Accentor	A/D	1a (v)
Eurasian Siskin	B/D	1a (v)
Twite	B/D	1a (v)
Ortolan Bunting	B/D	1a (v)
Water Pipit	B/D	1a (v)
Tree Pipit	B/D	1a (v)
Three-toed Woodpecker	B/D	1d
Ural Owl	B/D	1e
Tengmalm's Owl	B/D	1e
Northern Hawk Owl	B/D	1e
Eurasian Dotterel	A/D	2
Rock Ptarmigan	B/D	7a
Western Capercaillie	B/D	7a

Table 19.5 *Species of Palearctic birds occupying split high/mid-latitude ranges.*

Split-category birds and the 'mountain trap' effect

If the Himalayas acted as a launchpad during the Tertiary, the mid-latitude belt mountains were a trap during the Pleistocene. As many boreal species shifted range southwards with the cold, they reached the mid-latitudes. Subsequent global warming pushed these populations back north but others tracked bio-climatic belts up mountains. The result is observable today in split-range species that are mainly derived from muscicapoid and passeroid lineages, with some owls and gamebirds (Table 19.5).

Split-category waterbirds

A small number of waterbirds seem to have developed split ranges too, even reaching population separation at subspecific level (Great Cormorant, for example). These birds are part of the mid-latitude waterbird fauna, but they seem to have succeeded also along the temperate and boreal coasts of north-western Europe, where the mildness of the climate probably equates to lower latitude representation in more continental locations. Invariably, the continental populations are more migratory than the more northerly, oceanic ones. The species are Mediterranean Gull (C/D), Gull-billed Tern (C/D), Caspian Tern (B/D), Sandwich Tern (C/D), Great Cormorant (B/D), Common Crane (B/D) and Common Shelduck (B/D). There is no equivalent pattern of distribution in the Eastern Palearctic.

'Tethyia' and the African connection

We have seen how the broad arid band that includes the Sahara, the Arabian and Central Asian Deserts has, along with the mid-latitude belt's mountains and seas, been a major barrier to bird dispersal. Moreau's (1972) analysis of its effect on migratory birds gives us a clear indication of its extent and dominating influence. It is not surprising to find, then, that Africa's avian contribution to the Palearctic has been minimal (Snow, 1978). I would go even further than Snow. If we consider the mid-latitude belt and its associated arid and mountain habitats as a unit in its own right, with parts of Africa and the Palearctic within it, then what we find is that the perceived African influence is reduced practically to nil. I realise that this argument may seem to be a semantic one but attributing species to Africa and not the Palearctic within a region that has its own internal character (and influences from both) prevents us from understanding the nature of this belt and the history of its birds. Dennell and Roebroeks (2005) called a large part of this region 'Savannahstan' but this belt also included wetlands, seas, mountains and deserts. It was a new world that emerged after the Miocene. To give this region its own identity I propose the name Tethyia, as it embraces the lands once lapped by the ancient Tethys Ocean.

The Palearctic-Nearctic interchange

As recently as 100,000 years ago, Barn Swallows entered North America from the Palearctic across the Bering Strait. Soon after, a North American population re-entered the Palearctic and colonised the Baikal area (Zink *et al.*, 2006). The connection between the Palearctic and Nearctic has been largely through Beringia for much of the Tertiary, once the ancient connection via Greenland was severed (see p. 15). Given its high-latitude position, Beringia's permeability to birds has depended heavily on climate. In many instances, particularly during warm periods like the Early Miocene and others before that, it allowed the entry of temperate and warm forest birds into the Nearctic. It was, in effect, part of the 'launchpad' process that catapulted species across the Palearctic. *Perisoreus, Nucifraga, Sitta* and *Athene* are examples. Once in the Nearctic, species behaved differently on either side of Bering Strait, and became isolated when climate severed access. When the time of separation was ancient and long the splits became significant, for example the separation of *Otus* and *Megascops* or *Aegithalos* and *Psaltriparus.*

Entry into the Palearctic from the Nearctic has happened on a number of occasions and it has typically involved the spread of a single species, a situation akin to the launchpad model, most often across the Arctic and Boreal belts – *Troglodytes, Calcarius, Plectrophenax, Picoides, Aegolius* and *Glaucidium.* For a number of genera and species, with Arctic and Boreal character, the Nearctic and Palearctic have been parts of a single unit. This is most obvious for Arctic birds – geese, sandpipers, auks and the like – but also for forest dwellers like *Regulus, Bombycilla* and *Spinus.* In some cases the entry from North America has been followed by Palearctic radiation. This seems to have been the case in *Falco* and, earlier, with the entry of the swifts. Much more recent, possibly even during the Early Pleistocene, was the entry of *Buteo* with subsequent differentiation into

(a)

Macaronesian endemics	Lineage
Canary Islands Grey Shrike	1a (i)
Cape Verde Swamp Warbler	1a (ii)
Canary Islands Chiffchaff	1a (ii)
African Blue Tit	1a (iii)
Canary Islands Chat	1a (iv)
Madeira Firecrest	1a (iv)
Cape Verde Sparrow	1a (v)
Blue Chaffinch	1a (v)
Azores Bullfinch	1a (v)
Canary	1a (v)
Berthelot's Pipit	1a (v)
Canary Islands Oystercatcher	2
Plain Swift	5
Trocaz Pigeon	6
Bolle's Pigeon	6
Laurel Pigeon	6

(b)

Other Macaronesian species	Lineage
Red-billed Chough	1a (i)
Raven	1a (i)
Barn Swallow	1a (ii)
Blackcap	1a (ii)
Spectacled Warbler	1a (ii)
Sardinian Warbler	1a (ii)
Lesser Short-toed Lark	1a (ii)
Blue Tit	1a (iii)
European Robin	1a (iv)
Eurasian Blackbird	1a (iv)
Eurasian Starling	1a (iv)
Goldcrest	1a (iv)
Spanish Sparrow	1a (v)
Eurasian Tree Sparrow	1a (v)
Rock Sparrow	1a (v)
Chaffinch	1a (v)
Serin	1a (v)
Greenfinch	1a (v)
Goldfinch	1a (v)
Linnet	1a (v)
Trumpeter Finch	1a (v)
Corn Bunting	1a (v)
Grey Wagtail	1a (v)
Barbary Falcon	1c
Common Kestrel	1c
Eurasian Hoopoe	1d
Great Spotted Woodpecker	1d
Barn Owl	1e
Long-eared Owl	1e
Egyptian Vulture	1f
Eurasian Sparrowhawk	1f
Common Buzzard	1f
Stone Curlew	2
Cream-coloured Courser	2
Eurasian Woodcock	2
Houbara Bustard	4
Pallid Swift	5
Rock Dove	6
Wood Pigeon	6
Collared Dove	6
Eurasian Turtle Dove	6
Black-bellied Sandgrouse	6

(c)

Ibero-Mauritanian endemics	Lineage
Iberian Azure-winged Magpie	1a (i)
Southern Grey Shrike	1a (i)
Algerian Grey Shrike	1a (i)
Western Olivaceous Warbler	1a (ii)
Iberian Chiffchaff	1a (ii)
Tristram's Warbler	1a (ii)
Dupont's Lark	1a (ii)
Atlas Flycatcher	1a (iv)
Moussier's Redstart	1a (iv)
Black Wheatear	1a (iv)
Algerian Nuthatch	1a (iv)
Levaillant's Woodpecker	1d
Spanish Imperial Eagle	1f
Red-necked Nightjar	5
Barbary Partridge	7a

Table 19.6 *List of endemic (a) and non-endemic (b; some are endemic subspecies) terrestrial species of Macaronesia, and of endemic birds of the Iberian peninsula and the Maghreb (c).*

the closely related Palearctic species and subspecies. If South-east Asia was the main supplier of species to the Palearctic, then the Nearctic was next, making a much greater contribution that sub-Saharan Africa.

Pelagic influence from the southern hemisphere

Species from Group 3, notably shearwaters and storm-petrels, of southern-hemisphere origins have had success at colonising the Atlantic and Pacific waters of the Palearctic. They include the genera *Morus, Fulmarus, Pterodroma, Oceanodroma* and *Pelagodroma*. The *Puffinus* shearwaters include a branch that seems to have radiated in the northern hemisphere and re-entered the southern hemisphere. A more recent example is the colonisation of the Antarctic region by the Great Skuas *Catharacta*. Northern-southern hemisphere exchanges of bird taxa seems largely limited to these seabirds. The Palearctic pelagic avifauna is thus a composite of southern immigrants (gannets, shearwaters and storm-petrels), with subsequent *in situ* radiation, and locally-derived species (including the auks and gulls).

Island endemics

We have not discussed the Western Palearctic's island endemics until now, but they deserve consideration. Most of the island endemics are on the Atlantic seaboard, largely in the south-western Canaries, Madeira, Azores and Cape Verde (collectively termed Macaronesia). The Scottish Crossbill is an exceptional Atlantic island endemic of the north-west as is the British race *scoticus* of the Willow Grouse. The Mediterranean islands, close to the coast, have produced few endemics: Cyprus Warbler and Wheatear, Marmora's and Balearic Warblers, Corsican Nuthatch and Corsican Finch (Appendix 1). In the Pleistocene Malta harboured a unique fauna that included a large swan, *Cygnus falconeri*. But it is the Atlantic south-west that has offered most by way of island species diversity. In Table 19.6 I list the Macaronesian island endemics and other terrestrial bird species that breed on the islands.

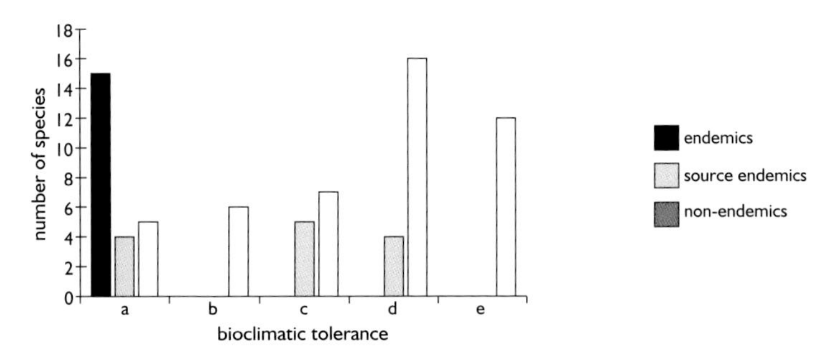

Figure 19.1 *Bioclimatic tolerance of Western Palearctic island endemics compared to closest mainland species (source endemics) and non-endemic species that also breed on these islands.*

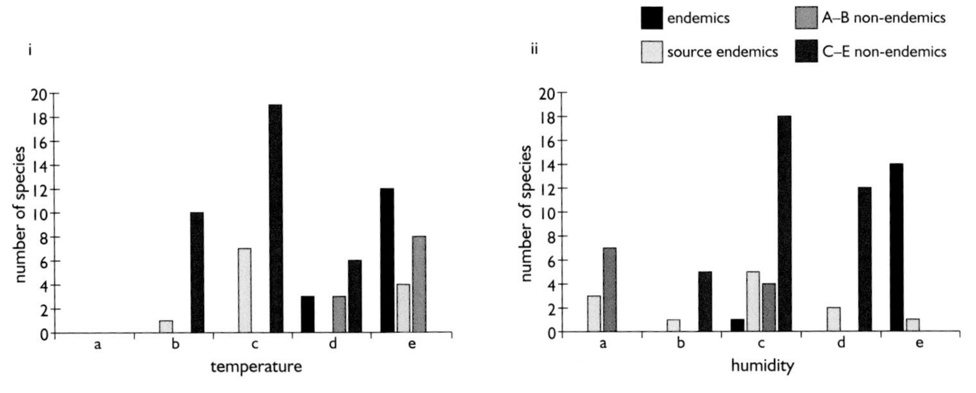

Figure 19.2 *Temperature (i) and humidity (ii) tolerance of Western Palearctic island endemics compared to closest mainland species (source endemics) and non-endemic species that also breed on these islands. The latter are divided into low- (A–B) and high- (C–E) tolerance bioclimatic groups.*

I tested their bioclimatic characteristics to see if there was any trend among the species that have successfully colonised these islands. The island endemics are all, not suprisingly given their restricted ranges, bioclimatic specialists of warm, humid bioclimates (Figures 19.1–19.2). When compared to the closest mainland species (source endemics) we find that they are derived from species with a wide range of tolerances, temperatures (except the coldest) and humidities. They are, in effect, a specialised subset of their mainland counterparts. The other island species (non-endemics) are strikingly different in that they show a range of bioclimatic tolerances (many high, Figure 19.1). These birds seem to separate into two groups, those with low tolerance (A–B) and those with moderate to high tolerance (C–E). The low-tolerance species are birds of warm, mainly dry, bioclimates and seem to be derived mainly from the Iberian or North African

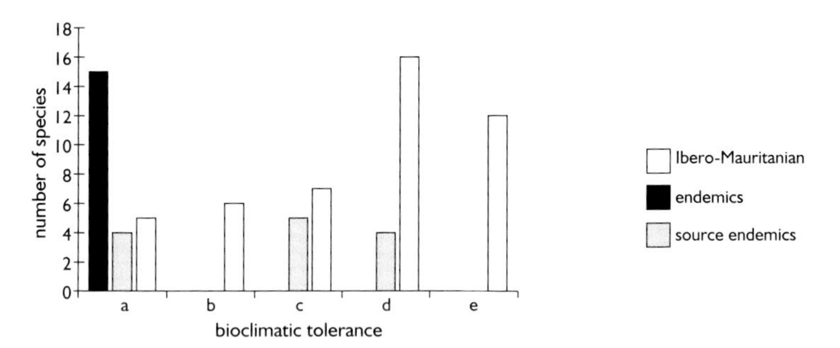

Figure 19.3 *Bioclimatic tolerance of Western Palearctic island endemics compared to Ibero-Mauritanian endemics and non-endemic species that also breed on these islands.*

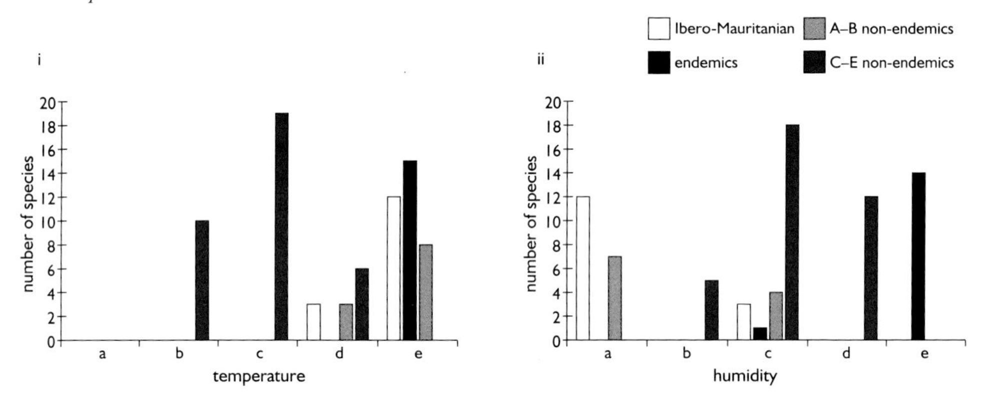

Figure 19.4 *Temperature (i) and humidity (ii) tolerance of Western Palearctic island endemics compared Ibero-Mauritanian endemics and non-endemic species that also breed on these islands. The latter are divided into low- (A–B) and high- (C–E) tolerance bioclimatic groups.*

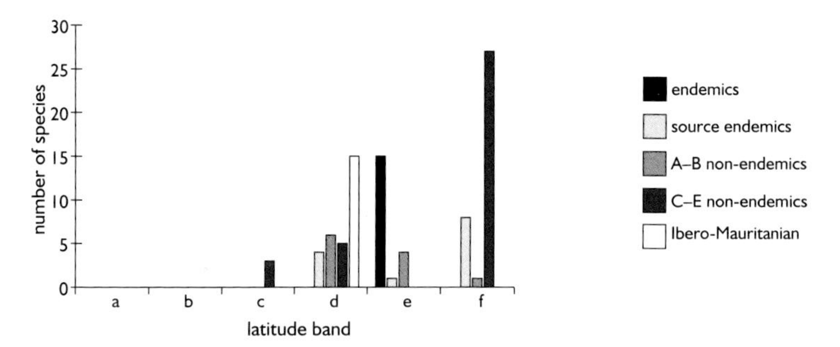

Figure 19.5 *Main latitude bands from which Western Palearctic island species are derived.*

mainland (Figure 19.2). The tolerant group is very different. These are birds of cooler climates and a range of humidity tolerances.

When we add the Ibero-Mauritanian mainland endemics we observe that they are also bioclimatically specialised (Figure 19.3, though less so than the island endemics). They are birds of warm and largely dry bioclimates (Figure 19.4). They thus appear to be related to the low-tolerance non-endemics that we have identified on the islands.

Proximity to the mainland coast has probably prevented these island birds from becoming full endemics. Figure 19.5 shows the main latitude bands from which the island species are derived. I have included the island and Iberian endemics as reference markers as they are all from the respective latitude locations of the islands and mainland. The low-tolerance non-endemics are clearly derived from latitudes opposite the islands, while the tolerant non-endemics are overwhelmingly multi-latitude species, with some from the warm and temperate belts. The source endemics are mainly multi-latitude species, with a few warm belt and very few subtropical belt species. I conclude from this that the island endemics are largely derived from high-tolerance species with wide distributions, probably species that are not derived from the immediate mainland.

The island endemics are sedentary species, but there is no indication that they are derived from migratory species that may have accidentally drifted to the islands (Figure 19.6). The source endemics exhibit a range of behaviours and full migrants are few. Trans-Saharan migrants are certainly not major contributors to the colonisation of these islands (Table 19.6). We observe a similar pattern for the island non-endemics, which is curiously in contrast with that of the mainland endemics that are sedentary or migratory (in fewer cases) but never partial migrants. The successful long-term (measured by species endemism) colonisation of the Macaronesian islands has been largely achieved by sedentary or partially migratory non trans-Saharan species, which have broad latitudinal ranges and wide bioclimatic tolerances and are largely of temperate origin. It may be a reflection of refugium survival by these species, which also escaped extinction on the south-western mainland of Iberia (Finlayson *et al.*, 2006) and probably the wetter parts of North Africa too. A feature of these island colonists is their wide occupation of habitats and altitudes on the Macaronesian

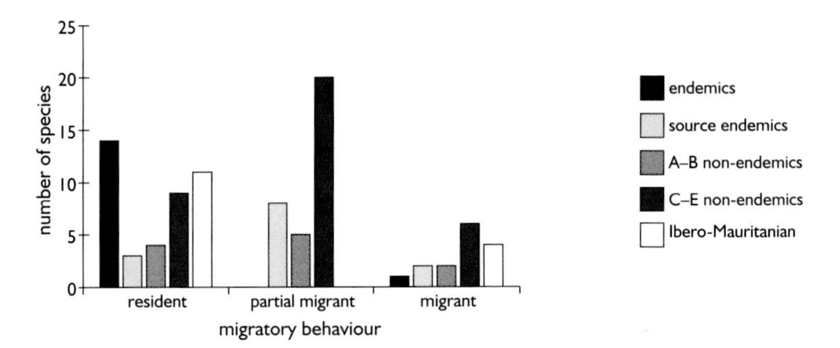

Figure 19.6 *Residency-migratory status of Western Palearctic island birds.*

islands. They may well have been better predisposed to expand into all available space on islands than bioclimatic specialists and the ability to colonise all available space would have permitted the build-up of significant populations that became resistant to extinction.

The available evidence points to much of the speciation having occurred during the Pleistocene, with exceptional colonisations much earlier. For example, Gonzalez *et al.* (2009) have shown that the Laurel Pigeon is derived from a lineage that colonised the Canary Islands around 20 mya, and Bolle's Pigeon around 5 mya. By contrast the Goldcrest's colonisation seems to have involved several steps between 2.3 and 1.3 mya (Päckert *et al.*, 2006) and we have already seen that the Chaffinch colonisations took place during the Pleistocene. The Macaronesian island endemics are therefore relicts of the glaciations that remained trapped on these islands, much like others were on the mountain peaks of the mid-latitude belt. They seem to have

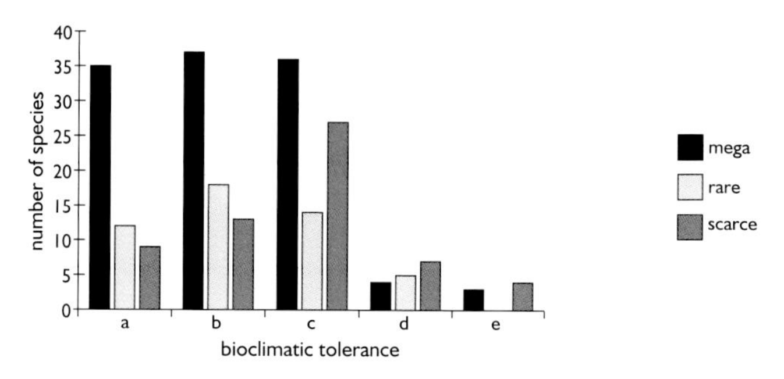

Figure 19.7 *Bioclimatic tolerance of rarities on the British list (following www.birdguides.com).*

been derived from bioclimatically tolerant, not very highly migratory, species – in other words, the species would have been among the best survivors.

Are rarities good colonisers?

To try and answer this question I have examined the bioclimatic and related profiles of all Palearctic species described as 'mega-rare', rare or scarce on the British list (according to www.birdguides.com). High bioclimatic tolerance was a characteristic feature of the island endemics. How do the rarities compare? Figure 20.9 reveals that there are comparatively few bioclimatically tolerant species among the rarities, with most species derived from moderate or specialist categories.

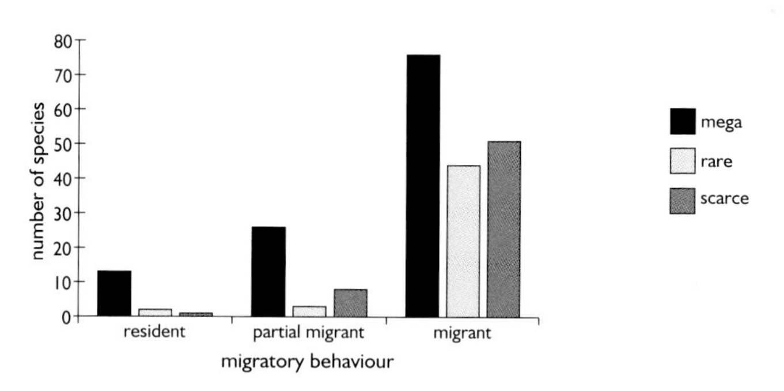

Figure 19.8 *Migratory-residency status of rarities on the British list (following www.birdguides.com).*

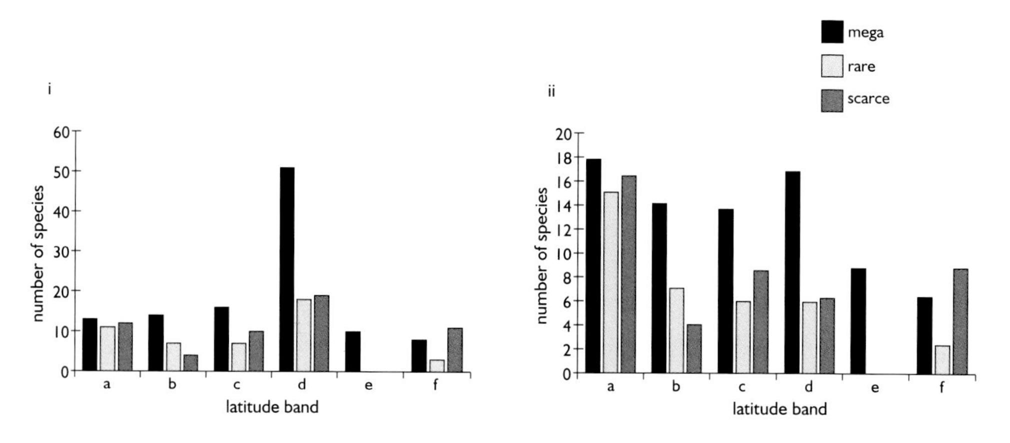

Figure 19.9 *Latitude sources of rarities on the British list (following www.birdguides.com) (i); latitude sources of rarities on the British list corrected to take overall number of species present in each latitude band into account (ii).*

A second feature was that few were full migrants. Figure 19.8 analyses the migratory status of British rarities, which are clearly derived from migratory birds, contrasting once again with the picture for the island endemics. Most island endemics were derived from multi-latitude species. This category scores low among the British rarities (Figure 19.9i), even after we take the available numbers in the source pool (the entire Palearctic) into account (Figure 19.9ii). Instead, the rarities are largely birds of the mid-latitude belt, although there is a major proportional contribution (relative to available number of species) among birds from higher latitudes. In conclusion, island colonisation has clearly been a process distinct from the regular fall-out of rare birds away from their typical home ranges and migratory routes. These rarities do not fit the profile of good colonisers, which are instead derived from highly bioclimatically tolerant species, which occupy wide latitude ranges, and tend not to be highly migratory.

The Palearctic avifauna

The present Palearctic avifauna, of which the Western Palearctic's is a subset, is a composite. It includes ancient lineages that have remained conservative and kept to those environments and climates that suited them best. In many cases they display highly fragmented ranges today. Then there are groups that adapted to changing conditions, over long periods (shorebirds) or after a relatively recent arrival (passerines). These birds took on the tundra, cold seas, grasslands and savannas, rocky habitats, arid lands and temperate and boreal forest belts. Their success has depended on the ability to tolerate a wide range of bioclimates, habitats that (in many cases) differed from their native tropical forests, and to migrate seasonally. Diet, often conservatively retained within lineages, has had an important bearing on survival in the Palearctic – the sedentary seed-eaters of the temperate and boreal forests are examplars.

Dispersal across the Palearctic, followed by radiation into different species, has tended to follow the South-east Asian 'launchpad' route in many cases, the Nearctic and southern hemisphere pelagic in others, and rarely the African channel. Periods of climatic upheaval, especially during the Pleistocene, has forced many species into glacial refugia. Subsequent retreat of the ice has left many species trapped on mountain 'islands'. The mid-latitude belt and its associated desert, rocky and wetlands – Tethyia – has had a major influence on the diversity, endemism and distribution of the birds of the Palearctic.

CHAPTER 20

Surviving climate change – characteristics of survivors

BIOCLIMACTIC TOLERANCE

Having looked at the lineages of Palearctic species and the key features of their history, it is now time to look at the characteristics that have allowed these species to survive and be successful in the region. It should be clear that I regard the present set of species as those that emerged during the Miocene and Pliocene, survived the great upheavals of these periods, and then managed to pull through the new kinds of stresses imposed upon them by the glaciations of the Pleistocene. The present Palearctic species have to be regarded as the survivors of the Pliocene aridification and the Pleistocene glaciations.

Lineages, bioclimatic tolerance and migration

A glance at Table 20.1 immediately shows the importance of the Miocene as a period of transition between the old avifauna of the Lower Tertiary and the new one that would survive to today. The absence of passerines from the Eocene and Oligocene fossil record of Europe is evident, as is their emergence during the Miocene and subsequent dominance. Harrison (1979) suggested that small non-passerines were prevalent during the Lower Tertiary and were later replaced by the passerines. This suggestion implied a competitive superiority on the part of passerines, but this is something that cannot be confirmed as the alternative possibility, that passerines only radiated once the small non-passerines had gone, cannot be excluded. It is clear, however, that terrestrial non-passerines (our Group 1d) lost many genera by the Miocene and is, today, an impoverished lineage (Table 20.1). So it does seem as though the passerines have been the inheritors of the small terrestrial bird nice, and we should look at the characteristics that made them successful.

Nine other groups have been successful in the Palearctic since the Miocene: Falcons (Group 1c), owls (1e), diurnal raptors (1f), shorebirds (Group 2), waterbirds (3), the crane group (4), the pigeon group (6), gamebirds (7a) and the waterfowl (7b). They show a significant turnover of genera in the Miocene, with loss of genera compensated by a number of new genera (Table 20.1). By contrast, the swifts and nightjars (5) reveal a similar pattern of losses as the terrestrial non-passerines.

In Table 20.2, I analyse two features that appear to have been important in the success of the different groups in the Palearctic – the ability to become bioclimatic generalists, and the ability to migrate. It is clear from Table 20.2 (a) that four of the lineages that appear to have done particularly well after the Miocene, the shorebirds, waterfowl, waterbirds and the crane group, have done so relying heavily on the ability to migrate. This is reflected in the low proportion of sedentary species and bioclimatic generalists. It may not be coincidental that, after the passerines, three (shorebirds, waterfowl, waterbirds) of the four groups contribute the most species to the Palearctic avifauna today.

Three groups – falcons, owls and diurnal raptors – have succeeded by becoming bioclimatic generalists, which seems to have allowed a higher proportion of birds to remain sedentary compared to the previous

Lineage		Eocene	Oligocene	Miocene	Pliocene	Pleistocene
1a (i)	Shrikes, corvids, orioles			(1)/2	4	
1a (ii)	Hirundines, warblers, larks			6	3	6
1a (iii)	Tits				2	
1a (iv)	Chats and thrushes			2	8	4
1a (v)	Sparrows, finches, buntings and pipits			4	7	4
1c	Falcons			1		
1d	Terrestrial non-passerines	(17)	(3)/1	(9)/3	1	2
1e	Owls	(6)	(1)	(4)/3	5	
1f	Diurnal Raptors	(1)	(2)	6	(1)/1	6
2	Shorebirds, gulls, terns and auks	1	(4)	(1)/7	6	11
3	Waterbirds	(10)	(3)/1	(9)/10	1	4
4	Cranes, rails, bustards and cuckoos	(12)	(1)	(7)/4	(1)/5	1
5	Swifts and nightjars	(12)		2		2
6	Pigeons, flamingoes and grebes	(2)	(1)/1	(2)/1	4	3
7a	Gamebirds	(5)/1		(2)/1	4	3
7b	Wildfowl	(5)	(3)	(3)/8	2	5
8	Ratites	(2)		1		

Table 20.1 *First presence of genera by epochs. Numbers in brackets indicate extinct genera.*

groups of migrants (Table 20.2). This alternative strategy of survival would seem to have been specific to a particular ecological way of life (raptorial), and was derived independently on three occasions.

The pigeon and gamebird groups defy the logic of the above as they represent non-migratory and bioclimatically intolerant strategies (Table 20.2). Their success may lie in their diet, many being herbivores or omnivores that consume a high proportion of seeds (Appendix 1).

The swifts and nightjars have not been as successful as the shorebirds, waterfowl, waterbirds or crane group, either in terms of surviving genera or in number of species present in the Palearctic (Table 20.2a), despite the lack of generalists and universally migratory strategy. One possibility is that their role has been taken over by the hirundines (bioclimatically tolerant *and* migratory), which would represent an example of a body plan and life history strategy better able to cope with the increased aridity and cold of the Palearctic from the Miocene onwards.

Curiously, the terrestrial non-passerines and the passerines resemble each other in having low levels of bioclimatic tolerance but also high levels of residency (especially among non-passerines). If we separate these groups into lower taxonomic categories (Table 20.2b) we observe two patterns. One is of three lineages that are dominated by sedentary species and which have low to intermediate levels of bioclimatic tolerance (higher than the first cluster but lower than the raptors). These lineages are the corvids, the tits and the woodpeckers, and they seem closest to the pattern we have described for the pigeon and gamebird groups. It is interesting that all three are omnivores that rely heavily on seeds. The other lineages – sylvioids, muscicapoids, passeroids and the remaining terrestrial non-passerines – exhibit low levels of bioclimatic tolerance but are more migratory than the previous group without reaching the proportions of the first cluster (Table 20.2). It is also interesting that the few terrestrial non-passerines that have succeeded (hoopoes, bee-eaters,

(a)

	1	2	3	4	5	6	7	8	9	10	11	12
% generalists	6.16	4.17	1.92	15.91	18.75	17.39	3.45	0	0	3.45	0	7.69
% sedentary	33.17	4.86	3.85	27.27	18.75	69.57	44.83	91.67	12.28	10.34	0	65.39
n	406	144	52	44	16	23	29	24	57	29	13	26

(b)

	A	B	C	D	E	F	G
% generalists	10.81	3.94	11.11	5.67	5.17	6.67	9.09
% sedentary	59.46	24.41	94.44	28.3	29.31	93.35	22.27
n	37	127	18	106	116	15	11

Table 20.2 *Proportion of Palearctic bioclimatic generalists and sedentary species among major avian lineages (a) (in which 1 = passerines, 2 = shorebirds, 3 = wildfowl, 4 = raptors, 5 = falcons, 6 = owls, 7 = pigeons, 8 = gamebirds, 9 = waterbirds, 10 = cranes, 11 = swifts and 12 = terrestrial non-passerines) and subsets of passerines and terrestrial non-passerines (b) (in which A = Corvoidea, B = Sylvioidea, C = Paroidea, D = Muscicapoidea, E = Passeroidea, F = Picidae and G = other terrestrial non-passerines).*

rollers etc.) have a similar profile, but are larger than their passerine counterparts (warblers, flycatchers, chats etc.), which may indicate that survival has been dependent on the exploitation of different resources from the passerines.

If we look at the families within the sylvioids, muscicapoids and passeroids (Table 20.3) we see that the patterns repeat themselves at this lower taxonomic scale. Here we have five families with a low level of bioclimatic tolerance and sedentary behaviour, just like the shorebird, waterfowl and other major lineages – the Sylviidae, Muscicapidae, Turdidae, Emberizidae and Motacillidae. These are largely omnivorous families that rely on insects during the breeding season. Four other families (Alaudidae, Prunellidae, Passeridae, Fringillidae), all relying to a large degree on seeds for food, behave like the pigeon and gamebird lineages and also the crows, tits and woodpeckers, that is they have low bioclimatic tolerance but include a high proportion of sedentary species. There is one family, Hirundinidae, which departs from the rest in being highly bioclimatically tolerant and highly migratory (Table 20.3). This may be a difficult strategy to follow (the number of Palearctic hirundines is low), and may also explain the poverty of swift and nightjar species (Table 20.2).

In summary, four strategies have been successful among major- and lower-order lineages in the Palearctic. On the one hand is the rare strategy of high bioclimatic tolerance coupled with migration. The only lineage to have achieved this is the hirundines. Second, we observe a strategy that consists of low bioclimatic tolerance and migration. We find it among shorebirds (Group 2), wildfowl (7b), waterbirds (3) and cranes (4), then at another level among sylviids, muscicapids, turdids, emberizids and motacillids. The third strategy involves low bioclimatic tolerance and sedentary behaviour and seems linked with mainly herbivorous and seed-eating omnivory – pigeons (Group 6), gamebirds (7a), at a second level crows, tits

	1	2	3	4	5	6	7	8	9	10
% generalists	50	0	4	3.33	5.26	0	7.69	5	3.125	11.11
% sedentary	12.5	8.33	64	18.33	5.26	40	53.85	45	9.375	18.33
n	8	84	25	60	19	10	13	40	32	18

Table 20.3 *Proportion of Palearctic bioclimatic generalists and sedentary species among major families of Sylvioidea, Muscicapoidea and Passeroidea (1 = Hirundinidae, 2 = Sylviidae, 3 = Alaudidae, 4 = Muscicapidae, 5 = Turdidae, 6 = Prunellidae, 7 = Passeridae, 8 = Fringillidae, 9 = Emberizidae, 10 = Motacillidae).*

and woodpeckers, and at a lower taxonomic level among larks, accentors, sparrows and finches. The fourth strategy involves higher levels of bioclimatic tolerance and sedentary behaviour and is found among diurnal raptors and owls, falcons being intermediate between this and the hirundine strategy. Since bioclimatic generalisation and migration seem two key ingredients in the survival of Palearctic birds, I shall now examine the various possible strategies at the level of species. It will be especially interesting to see where individual species deviate from main-lineage strategies that we have identified here.

Bioclimatic tolerance and sedentary behaviour

Strategy 1a: Generalist residents

To be able to survive in a region like the Palearctic without having to perform seasonal migratory movements must be regarded as the hallmark of survival. Only 15 species (1.74%) have managed it (Table 20.4). Not surprisingly this is not a random group of species, but several lineages dominate – corvids (4 species), tits (2), muscicapids (3), finches (2), woodpeckers (1), owls (2) and diurnal raptors (1).

All of these species have pan-palearctic ranges, nine of which are multi-latitude and five temperate. Their bioclimatic mid-points are centred around the cool-temperate, humid parts of the ranges. The only exception is the Red-billed Chough, which occupies a mid-latitude belt position as an omnivore of rocky habitats. Forest-dwelling omnivores dominate (Group 7) but Table 20.4 shows that there are alternative ways of achieving this strategy. All the species are derived from one major lineage (Group 1) and from four subgroups (1a, 1d, 1e and 1f). Four of the passerine (Group 1a) sub-lineages are represented, with only the sylvioids missing.

Strategy 1b: Semi-generalist residents

This is close to strategy 1a but involves species with slightly lower tolerance. 13 species (1.51%) have achieved it (Table 20.5) and the lineages represented are familiar when compared with strategy 1a – corvids (1 species), tits (1), muscicapids (1), woodpecker (2), owls (3) and diurnal raptors (1). Novelties are pigeons (2 species) and cormorants (2).

Species	TOL	LAT	TEM	HUM	MON	WP	CP	EP	HF	HN	DT	MIG
Northern Goshawk	E	C	B	B		+	+	+	F	F	E	S
Spotted Nutcracker	E	C	B	D		(+)	+	+	F	F	O	S
Red-billed Chough	E	D	C	D	M	(+)	(+)	+	O	R	O	S
Black-billed Magpie	E	C	B	B		+	+	+	O	O	M	S
Common Raven	E	F	B	B		(+)	+	+	M	R	M	S
Goldcrest	E	C	B	C		+	+	(+)	F	F	O	S
Eurasian Nuthatch	E	F	C	D		+	+	+	F	F	O	S
Wren	E	F	C	C		+	(+)	+	F	R	O	S
Coal Tit	E	F	B	D		+	+	+	F	F	O	S
Great Tit	E	F	C	D		+	+	+	F	F	O	S
Eurasian Tree Sparrow	E	F	B	D		+	+	+	M	M	O	S
Common Crossbill	E	F	B	D		(+)	+	+	F	F	O	S
Tawny Owl	E	F	C	D		+	(+)	(+)	F	F	M	S
Eagle Owl	E	F	B	D		(+)	+	+	M	R	E	S
Grey-headed Woodpecker	E	C	B	D		(+)	+	+	F	F	O	S

Table 20.4 *The fifteen resident bioclimatic generalists of the Palearctic. For key see Appendix 1 (p. 242).*

Species	TOL	LAT	TEM	HUM	MON	WP	CP	EP	HF	HN	DT	MIG
Bonelli's Eagle	D	D	D	C		(+)	(+)	(+)	M	R	E	S
Eurasian Collared Dove	D	F	C	D		+	(+)	(+)	M	M	O	S
Rock Dove	D	F	D	D		+	+	(+)	M	R	O	S
Eurasian Jay	D	F	C	D		+	+	+	F	F	O	S
Eurasian Treecreeper	D	C	B	D		+	+	+	F	F	O	S
Blue Tit	D	F	C	C		+	(+)	-	F	F	O	S
Tengmalm's Owl	D	B/D	B	B		(+)/(+)	+/+	+/+	F	F	M	S
Barn Owl	D	F	C	C		+	-	-	O	R	M	S
Little Owl	D	D	C	C		+	+	+	M	R	M	S
White-backed Woodpecker	D	C	B	D		(+)	+	+	F	F	O	S
Great Spotted Woodpecker	D	F	C	B		+	+	+	F	F	O	S
European Shag	D	F	B	B		+	-	-	Ma	R	F	S
Pelagic Cormorant	D	B	B	E		-	-	+	Ma	R	F	S

Table 20.5 *The thirteen resident bioclimatic semi-generalists of the Palearctic. For key see Appendix 1 (p. 242).*

Species	TOL	LAT	TEM	HUM	MON	WP	CP	EP	HF	HN	DT	MIG
Common Buzzard	E	F	B	C		+	+	+	M	F	M	P
Golden Eagle	E	F	C	B		+	+	+	M	R	E	P
Eurasian Sparrowhawk	E	F	C	D		+	+	+	F	F	E	P
Black-headed Gull	E	F	B	C		+	+	+	W	W	M	P
Common Snipe	E	B	B	B		(+)	+	+	W	W	M	P
Sandwich Tern	E	C/D	C	C		+/(+)	+/(+)	-	W	W	M	P
Common Kestrel	E	F	C	D		+	+	+	M	R	M	P
Peregrine Falcon	E	F	D	C		+	+	+	A	R	E	P
Eurasian Blackbird	E	F	C	C		+	(+)	-	F	F	O	P
Hawfinch	E	C	B	C		+	+	+	F	F	O	P
Reed Bunting	E	F	B	C		+	+	+	W	W	O	P
White Wagtail	E	F	B	C		+	+	+	W	W	I	P
Grey Wagtail	E	F	B	C		+	(+)	+	W	W	I	P
Long-eared Owl	E	F	B	C		(+)	+	+	F	F	M	P
Common Kingfisher	E	F	C	D		(+)	(+)	(+)	W	W	F	P
Great Bittern	E	C	C	D		(+)	+	+	W	W	M	P
Mallard	E	F	B	C		+	+	+	W	W	M	P

Table 20.6 *The seventeen partly migratory bioclimatic generalists of the Palearctic. For key see Appendix 1 (p. 242).*

Nine species have pan-palearctic ranges that seem to indicate that reduced tolerance is sufficient to restrict distribution. Seven species are multi-latitude, two temperate and two are mid-latitude. The mid-latitude species – Bonelli's Eagle and Little Owl – share the rocky habitat character of the Red-billed Chough of strategy 1a. Forest omnivores are still dominant (Group 5) but other alternatives are apparent in the table. The cormorants represent the first presence of a lineage (waterbirds, Group 3) outside Group 1. The remaining species belong to Group 1 (subgroups a (i), (iii), (iv), d, e, f), with sylvioids and passeroids missing.

Strategy 1c: Generalist partial-migrants

This strategy is similar to 1a but some populations of each species, usually the northern or continental ones, migrate. 17 species (1.97%) have achieved it (Table 20.6) and are from familiar lineages – muscicapids (1 species), finches and allies (4), owls (1) and diurnal raptors (3). Corvids, tits and woodpeckers, strict residents with similar tactics as we have seen (Table 21.2b), are missing. Novelties are falcons (2 species), shorebirds (3) and waterfowl (1). Waterbirds (1) and terrestrial non-passerines (1) have already appeared but the species represented here, a kingfisher and a heron, come from new families.

All species have pan-palearctic ranges, which indicate that the partially migratory strategy, in combination with bioclimatic tolerance, has been a successful solution to the maintenance of wide ranges. Thirteen species have multi-latitude ranges. Forest omnivores are represented by only two species, suggesting that this way of life allows for strict residency. Instead wetland species are dominant (9) and these include mixed-strategy carnivores, fish-eaters, insectivores and omnivores.

Species	TOL	LAT	TEM	HUM	MON	WP	CP	EP	HF	HN	DT	MIG
Hen Harrier	E	B	B	B		(+)	+	+	O	O	M	M
Black Kite	E	F	C	C		+	+	+	M	M	M	M
Osprey	E	F	B	D		(+)	+	+	W	W	F	M
Common Tern	E	F	B	D		(+)	+	+	W	W	M	M
Eurasian Woodcock	E	C	B	D		(+)	+	+	F	F	M	M
Little Ringed Plover	E	F	B	D		(+)	+	+	W	W	M	M
Oriental Turtle Dove	E	F	B	D		-	(+)	+	M	M	O	M
Eurasian Hobby	E	F	B	D		(+)	+	+	A	M	M	M
Common Cuckoo	E	F	C	D		+	+	+	F	F	I	M
Siberian Stonechat	E	F	C	C		(+)	+	+	O	O	O	M
Brown Flycatcher	E	C	B	C	M	-	-	+	F	F	I	M
Short-eared Owl	E	B	C	C		(+)	+	+	O	O	M	M
Barn Swallow	E	F	B	B		+	+	+	A	R	I	M
House Martin	E	F	B	D		+	+	+	A	R	I	M
Sand Martin	E	F	B	C		(+)	+	+	A	R	I	M
Red-rumped Swallow	E	D	D	D		(+)	+	+	A	R	I	M
Black-crowned Night Heron	E	D	C	D		(+)	(+)	(+)	W	W	M	M
Great White Egret	E	D	D	C		(+)	(+)	+	W	W	M	M
Shore Lark	E	A/D	B	B	-/M	+/(+)	+/+	+/(+)	O	O	O	M/S

Table 20.7 *The nineteen migrant bioclimatic generalists of the Palearctic. For key see Appendix 1 (p. 242).*

Species	TOL	LAT	TEM	HUM	MON	WP	CP	EP	HF	HN	DT	MIG
Egyptian Vulture	D	D	D	B	M	(+)	+	-	O	R	N	M
White-tailed Eagle	D	F	B	B		(+)	+	+	W	W	M	M
Fork-tailed Swift	D	F	B	C		-	-	+	A	R	I	M
Common Sandpiper	D	F	B	C		(+)	+	+	W	W	M	M
Common Redshank	D	F	B	C		(+)	+	(+)	W	W	M	M
Caspian Tern	D	B/D	C	B		+/(+)	-/(+)	-/(+)	W	W	M	M
Little Tern	D	F	D	D		(+)	+	(+)	W	W	M	M
Eurasian Turtle Dove	D	F	C	C		+	(+)	-	M	M	O	M
Golden Oriole	D	F	C	B		+	+	-	F	F	O	M
Common Quail	D	F	D	D		+	+	-	O	O	O	M
Baillon's Crake	D	D	C	D		(+)	+	+	W	W	O	M
Oriental Cuckoo	D	E	B	C		-	+	+	F	F	I	M
Daurian Redstart	D	F	B	B	M	-	-	+	F	F	O	M
Northern Wheatear	D	F	B	B	M	+	+	+	O	O	O	M
Common Redstart	D	F	C	B		+	+	-	F	F	O	M
Dark-sided Flycatcher	D	C	B	B	M	-	-	+	A	F	I	M
Common Rosefinch	D	F	B	B	M	(+)	+	+	M	M	O	M
Olive-backed Pipit	D	F	B	B	M	(+)	+	+	F	F	O	M
Richard's Pipit	D	C	B	C		-	-	+	O	O	O	M
Western Yellow Wagtail	D	F	C	B		(+)	(+)	-	W	W	I	M
Lanceolated Warbler	D	B	B	C		-	(+)	+	W	W	O	M
Sedge Warbler	D	C	C	B		+	+	-	W	W	O	M
Eurasian Hoopoe	D	F	C	D		+	+	+	M	M	M	M
Dollarbird	D	F	C	D		-	-	+	M	M	M	M
Black Stork	D	C	C	B		(+)	+	+	W	W	M	M
Little Bittern	D	D	D	D		(+)	+	-	W	W	M	M
Purple Heron	D	D	D	D		(+)	(+)	(+)	W	W	M	M

Table 20.8 *The twenty-seven migrant bioclimatic semi-generalists of the Palearctic For key see Appendix 1 (p. 242).*

Bioclimatic tolerance and migratory behaviour

Strategy 2a: Generalist migrants

Bioclimatic tolerance in the breeding season and an ability to move somewhere else the rest of the time confers great advantages, and it is somewhat surprising that only 18 species (2.09%) are in this category (Table 20.7). Some lineages from strategy 1 are represented – muscicapids (2 species), owls (1), diurnal raptors (3) and pigeons (1) from among the residents and also falcons (1), shorebirds (3) and waterbirds (2) from among the partial migrants. Novelties are the hirundines with 4 species, a lark and a cuckoo. Corvids, tits and woodpeckers are conspicuously absent.

Sixteen of these 18 species have pan-palearctic ranges confirming our observation with partial migrants, which is that migration assists in the maintainance of wide geographical ranges among generalists. Eleven species have multi-latitude ranges and there are some from the mid-latitude (4 species) and temperate (2) belts. The former includes a split-category species – Shore Lark – which has migrants in the boreal parts of the range and residents in the southern mountains. Dietary constraints are lifted by migration and birds in this category include aerial and forest insectivores. These birds also occupy a wide range of habitats.

Strategy 2b: Semi-generalist migrants

Twenty-seven species (3.13%) are in this category (Table 20.8). They are orioles (1 species), warblers (2), muscicapoids (4), finches and allies (4), terrestrial non-passerines (2), diurnal raptors (2), shorebirds (4), waterbirds (3), rails (1), cuckoos (1), swifts (1), pigeons (1) and game birds (1). Orioles, rails and swifts are novelties.

Twelve species have pan-palearctic ranges, illustrating the combined power of generalisation and migration in producing wide geographic ranges. Among semi-generalists wide ranges are reduced, and many are fragmented (Table 20.8) in spite of migration. There is a predominance of multi-latitude species and there is no dominance of a particular dietary category. A range of habitats are occupied but wetland birds, derived from four separate lineages (Table 20.8), stand out.

Bioclimatic tolerance and genera

Seventy-three genera are represented in the above analysis (Tables 20.4–20.8). The most striking feature to emerge is the paucity of genera that contribute more than one species. The exceptions are helpful. There are only five cases of two or more generalists in the same genus: (a) Great Tit–Coal Tit; (b) Grey–White Wagtail; (c) Peregrine Falcon–Eurasian Hobby–Common Kestrel; (d) Long-eared–Short-eared Owl; and (e) Eurasian Sparrowhawk–Northern Goshawk. Three of these cases – tits, falcons and accipiters – involve species of significantly different sizes, indicating major differences in food. The two owls differ in habitat in a major way, one in woodland, the other avoiding trees altogether. The wagtails are more difficult to separate as clearly; they occupy different habitats but also overlap to some degree, and they are similar in size. The first four examples are of pairs or groups of species whose scale of ecological difference is such that it is equivalent to species in different genera. Only the issue in the wagtails remain unresolved. This suggests that bioclimatic tolerance is achievable, usually, only once in a genus; the species that achieves it first becomes dominant and widespread, leaving little ecological space for others to achieve the same degree of bioclimatic tolerance.

Pairs or groups of species with generalist and semi-generalist pairs are also infrequent, suggesting that this degree of difference in tolerance is too small to allow multi-species groups within genera. The few cases are the three *Streptopelia* doves, which are separated by habitat and geographical range, and the Western Yellow Wagtail with the other two wagtails, again separated by habitat. Two pipits – Olive-backed and Richard's - are semi-generalists with overlapping ranges, but they differ significantly in size and habitat. The two semi-generalist *Phalacrocorax* cormorants are at either end of the Palearctic and do not meet. Finally, the White-backed and Great Spotted Woodpeckers differ in size and, to some degree, in forest types occupied.

Overall, we can conclude that achieving bioclimatic tolerance, especially generalisation, has not been difficult, given that it appears in so many lineages and genera. But its achievement by a species seems to close the door for others in the same genus, which are consequently constrained and limited to specialised portions of the bioclimatic range.

MIGRATION, DIET AND HABITAT

The origins of migration

It is clear that migration is an important component, along with bioclimatic tolerance, in the success of species in the Palearctic. In the same manner that bioclimatic tolerance and other ecological attributes have evolved many times independently, during the Miocene and Pliocene in particular, migration must also have its origins, or have received a major impetus, during the climatic stresses of these epochs. This view has received recent support (Louchart, 2008) and supplants earlier interpretations of the origins of migration that disregarded past changes, and looked at species exclusively from the recent perspective of the climatic conditions of the Pleistocene and Holocene.

One prevalent interpretation has been the 'southern origin model', in which resident species in tropical areas took advantage of seasonal bursts of food resources to the north, which they exploited by migrating to these areas in the spring and returning south in autumn (Alerstam and Enckell, 1979; Rappole, 1995, 2005; Safriel, 1995; Rappole and Jones, 2002; Böhning-Gaese and Oberrath, 2003; Helbig, 2003). This idea is flawed because it fails to explain the presence of many migratory species with no resident counterpart populations in the tropics, and relies on successful initial random dispersal towards unknown areas, which happen to be good. A second model looked at northern origins of migrations, from species living at high latitudes that gradually shifted their winter ranges southwards (Gauthreaux, 1982). Cox (1985) attempted a 'middle' position, which combined northern and southern origins models. In some sense, the northern origins model is closer to the ideas I put forward here, except that it does not take account of climatic and ecological changes that, as we have seen throughout this book, have taken place throughout the Tertiary, particularly after the Miocene. These changes have been the driving force behind the origin and evolution of Palearctic-African migration systems.

Lineages that today include migrants were present in the Late Cretaceous (e.g. buttonquails, see p. 149) so the potential for migration is ancient. Modern-type migrations probably emerged during the Miocene, and the Late Pliocene/Early Pleistocene Olduvai (East Africa) avifauna gives a very clear picture of migration at that point. A vast number of fossils (around 5,000) reveals the presence of adult scolopacid waders, of species that do not breed in the tropics today. The absence of juveniles precludes breeding. *Calidris, Numenius, Limosa, Tringa, Lymnocryptes, Arenaria, Phalaropus* and *Limicola* – all taiga and tundra breeding species – were represented (Matthiesen, 1990). There is little reason to suppose that this image is exceptional. We can glean further insights by looking at the first appearance of bird genera that today have migratory, and also trans-Saharan, species (Table 20.9). These results confirm the presence of genera with potential migrants from the Eocene, with a sharp rise from the Miocene.

So, in a nutshell, I propose that migration evolved as one of a suite of responses by birds living within the Palearctic as its climate became increasingly seasonal, arid and cold. Early migration probably existed in the tropical Palearctic world of the Palaeocene and Eocene as responses to latitudinal shifts of daylength. It would have taken a new dimension with the significant climate cooling of the Oligocene, and may have lost importance during the Early Miocene warming, only to intensify with the cooling and aridification of the later Miocene and Pliocene. The early migrations would have been short-distance, to the nearest available wintering area, but the expansion of the Central Asian deserts and the Sahara, the flooding of the Mediterranean Sea and the continuing uplift of the mountains of the mid-latitude belt severed the continuous gradient between temperate and tropical zones, except in the Far East. Trans-Saharan and trans-Central Asian migrations probably had early origins in the Miocene and Pliocene, but their intensity would have waxed and waned with climate change.

Epoch	Genera present	Migratory genera	Trans-Saharan genera
Eocene	76	2 (2.63%)	1 (1.32%)
Oligocene	43	4 (9.3%)	1 (1.32%)
Miocene	105	44 (41.91%)	22 (51.16%)
Pliocene	120	68 (56.67%)	37 (30.83%)
Pleistocene	164	106 (64.63%)	57 (34.76%)

Table 20.9 *Distribution of fossil migrant genera by epoch.*

Even as recently as 5,000 years ago, when the Sahara was a land of megalakes and savannas, the Trans-Saharan migration systems would have had a very different profile. So the Palearctic-African migration systems described by Moreau (1972) are only 5,000 years old but they are embedded within a long history of adaptation to climate change that started in the Miocene. The birds that survived to the Pleistocene and through to the present came from lineages that had repeatedly and independently evolved migration. The kinds of short-term genetic responses of migratory birds to rapid climate change that are now well-documented (e.g. Berthold and Helbig, 1992; Berthold *et al.*, 1992; Helbig, 1996; Pulido and Berthold, 2010) are the legacy of a flexibility that evolved in the Tertiary.

In Table 20.9 and Figure 20.1 I sort out migratory, partially migratory and sedentary birds in relation to bioclimatic tolerance. It is clear that migrants, both in absolute and relative terms, are scarcest among the bioclimatic generalists and semi-generalists. The latter, which as we already saw are few in number (see pp. 223–227), are balanced between migratory, partially migratory and sedentary species. Bioclimatic tolerance is difficult to achieve, as we saw in the previous chapter, and migratory behaviour is limited within the subset of those that achieve it. Migratory behaviour is high among species of low bioclimatic tolerance (Table 20.9; Figure 20.1) and is a complement that permits survival. It reaches a peak among bioclimatic moderates, a category that we will examine in more detail below. Bioclimatic moderates have succeeded by combining an intermediate level of tolerance with highly developed migratory abilities.

Latitudinal distribution of migrants in relation to bioclimatic tolerance

Bioclimatic specialists and semi-specialists are very similar to each other (Figures 20.2 and 20.3). Migrants are distributed mainly in the mid-latitude belt, with a second peak in the Arctic. This second peak seems to shift towards the Boreal belt among semi-specialists. Multi-latitude species are virtually absent among these birds and the representation of temperate and subtropical species is low. The huge presence of migratory

Tolerance	Migratory	Part-migratory	Sedentary
A	136 (49.82%)	28 (10.26%)	109 (39.93%)
B	125 (53.65%)	37 (15.88%)	71 (30.47%)
C	141 (58.51%)	71 (29.46%)	29 (12.03%)
D	27 (42.19%)	24 (37.5%)	13 (20.31%)
E	19 (36.54%)	17 (32.69%)	16 (30.77%)

Table 20.10 *Distribution of migratory, partially migratory and sedentary species by bioclimatic tolerance.*

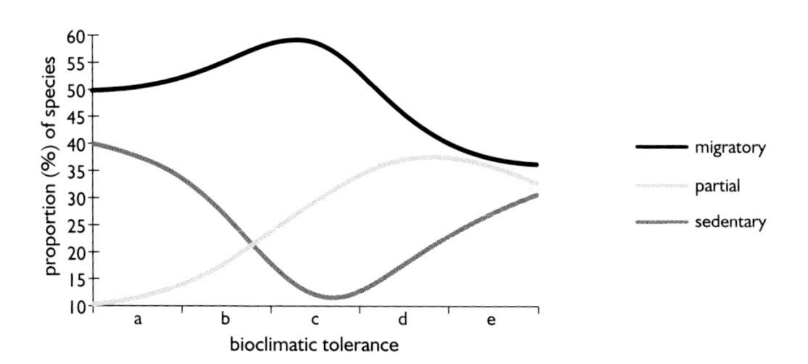

Figure 20.1 *Relationship between bioclimatic tolerance and residency-migratory status among Palearctic birds.*

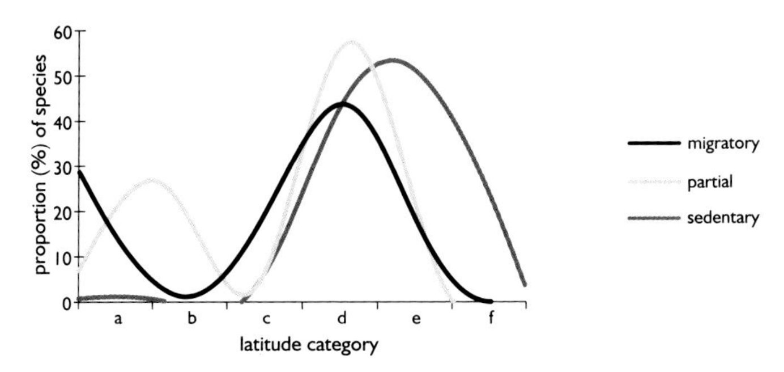

Figure 20.2 *Distribution of residency-migratory status by latitude among bioclimatic specialists.*

species in the mid-latitude belt disrupts an otherwise simple picture of a latitudinal gradient of migratory species. The reason for this peak of migratory species, at a relatively low latitude, seems to be aridity. The mid-latitude belt includes the northern part of the Sahara, the Arabian desert and much of the Cantral Asian desert system as well as highly seasonal continental steppes.

Table 20.10 supports the view that aridity and continentality are the drivers of mid-latitude belt migration. The species that occupy this belt are predominantly bioclimatic specialists and semi-specialists (71.28% of all MLB species) and bioclimatically tolerant species are rare (4.29%). The specialists are predominantly species with arid bioclimatic mid-points (Categories A and B combined – 55.09% of all MLB species). The proportion of migrants in the MLB is highest among the species with humid bioclimatic positions (Categories D and E), which presumably are the ones that resist aridity most poorly. But the proportion of migrants is high even among those specialised for these conditions (arid bioclimatic specialists). The resulting combination is a high level of migration among birds of the MLB, which is related to a combination of specialisation and the aridity of the belt. The seasonality is so intense that the proportion of migrants exceeds that of the high latitudes (Figures 20.2 and 20.3).

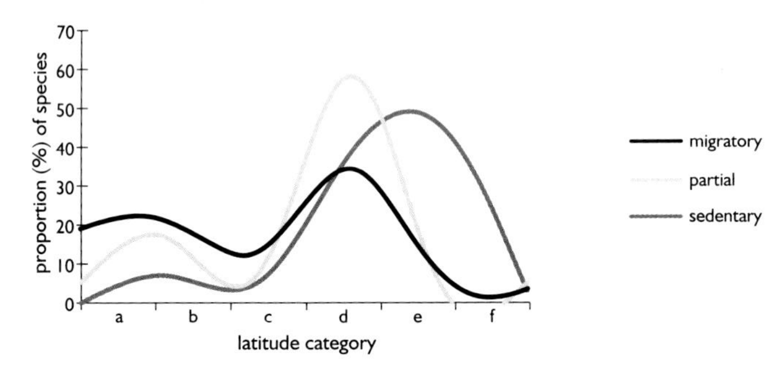

Figure 20.3 *Distribution of residency-migratory status by latitude among bioclimatic semi-specialists.*

The distribution of partial migrants among specialists and semi-specialists resembles that of the full migrants, with a clear peak in the mid-latitude belt, but the second peak is clearly in the Boreal, not Arctic, belt (Figures 20.2 and 20.3). The pattern of sedentary bird distribution follows a latitudinal gradient and peaks in the subtropical belt as would be expected.

The pattern of high migratory levels in the mid-latitude belt persists among bioclimatic categories that represent higher tolerance (Figure 20.4). The difference lies in a drop in migratory species among high-latitude species and a sharp rise among multi-latitude species. At the same time the sedentary species pattern changes with a large proportion among multi-latitude species, and a second peak that shifts towards the mid-latitude belt among moderates and eventually the temperate belt among generalists (Figure 20.4). Thus migratory and sedentary behaviours are dependent on tolerance and, as this changes, with latitude.

HUMIDITY					
Latitude	A	B	C	D	E
All species	44.15	10.94	20.37	7.92	16.60
Migrant	47.00	55.17	57.40	52.38	63.63
Partial migrant	19.65	34.48	22.22	33.33	11.36
Sedentary	33.33	10.34	20.37	14.28	25
TOLERANCE					
All species	41.25	30.03	24.42	2.97	1.32

Table 20.11 *Distribution of mid-latitude (Latitude Category D) birds by position in humidity gradient, migration and bioclimatic tolerance; n = 265.*

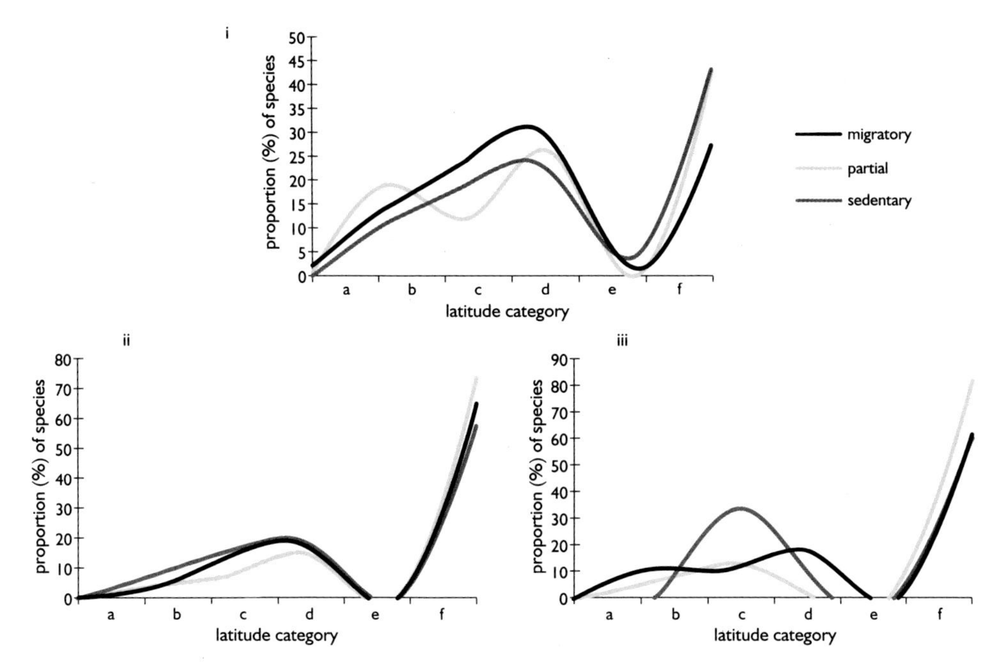

Figure 20.4 *Distribution of residency-migratory status by latitude among bioclimatic moderates (i), semi-generalists (ii) and generalists (iii).*

Diet

Diet is heavily lineage-dependent (Table 20.12), which is not surprising as it is a reflection of deep-rooted body plans. The most successful strategies, in terms of number of species represented, are omnivory, mixed-strategy carnivory and insectivory (Figure 20.5). The latter has been heavily dependent on the migratory habit for its success while a number of other species, capable of adding plant matter at particular times of the year (e.g. *Sylvia* warblers consuming fruit) or in variable proportions (e.g. buntings and finches consume seeds in different quantities relative to insects depending on the season), have been able to reduce the need to migrate. Similarly, many carnivores have succeeded by adopting a mixed diet (for example corvids or many diurnal raptors). This particular (mixed-category) strategy has been much more successful than the more specialised carnivorous strategies of consuming only warm-blooded animals, fish or carrion. Herbivory is also rare among Palearctic birds and is limited to two lineages – Anseriformes (Group 7a) and Columbiformes (Group 6). The sandgrouse of this latter group and the geese of the former provide a

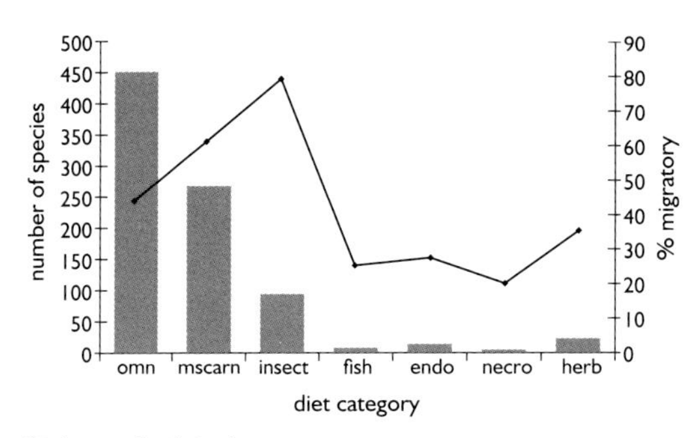

Figure 20.5 *Distribution of Palearctic birds by diet.*

contrast. Sandgrouse occupy low latitudes and are sedentary; geese occupy high latititudes and are highly migratory. The low frequency of herbivores and carnivores of warm-blooded animals is perhaps a reflection of the dominance of these dietary strategies in mammals.

Omnivory has been adopted successfully by most lineages with the exception of the falcons (Group 1c), owls (1e), diurnal raptors (1f) and swifts and nightjars (Group 5) (Table 20.12). This suggests that the specialised morphology of these lineages has prevented this dietary strategy, which may have been an added impediment to success in the Palearctic. The first three groups have compensated it with mixed-strategy carnivory, but the last group has been restricted to insectivory (which requires migration, Figure 20.5) and has remained marginal in the Palearctic. Mixed-strategy carnivory has been successfully adopted by a number of lineages, most notably among shorebirds (Group 2), waterbirds (3), the three raptor categories and also corvoids (Group 1a i) (Table 20.12). Insectivory, the third most important diet category, has been adopted by several lineages among which the sylvioids (1a ii), muscicapoids (1a iv) and the swifts and nightjars (5) stand out.

Lineage	Omnivory	M-s carn	Insectivore	Fish	Endo-carn	Necrophyte	Herbivore
1a (i)	18	19					
1a (ii)	86	2	40				
1a (iii)	18						
1a (iv)	87	3	16				
1a (v)	101	4	9				
1c		10			3		
1d	14	7	4				
1e		20	1		2		
1f		29		1	9	5	
2	31	105	6	2			
3	1	55		4			
4	23	1	1				
5			13				
6	14	6					9
7a	22						
7b	35	7					10

Table 20.12 *Distribution of dietary categories by lineage. M-s carn = mixed-strategy carnivore; endo-carn = endothermic carnivore.*

Habitat

The occupation of habitats has been less restricted by lineages than diet (Table 20.13). The exploitation of aerial and marine habitats for feeding has required specialised morphology and has evolved in particular lineages. Aerial feeding has emerged in two passerine subgroups (sylvioids, Group 1a ii and muscicapoids, 1a iv), in falcons (1c), terrestrial non-passerines (1d), shorebirds (pratincoles, Group 2) and swifts and nightjars (5). Marine habitat exploitation has been limited to three groups: shorebirds (Group 2), waterbirds (3) and tropicbirds (of Group 6). The success of many groups in mixed and open terrestrial habitats as well as wetlands (Figures 20.6 and 20.7) has been a key to the success in the Palearctic where forests were lost frequently at the expense of these habitats. Most forest lineages have survived by tracking the movement of the forests and those that have adjusted to the vast tracts of boreal and temperate forests (reflected in Latitude Categories B, C and F) are best represented (Figure 20.8).

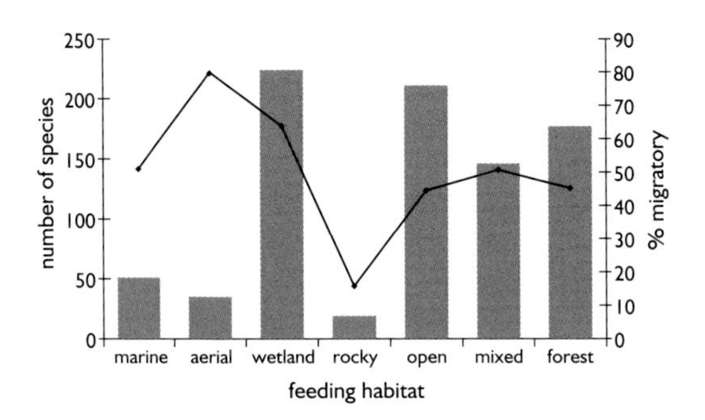

Figure 20.6 *Distribution of Palearctic birds by feeding habitat.*

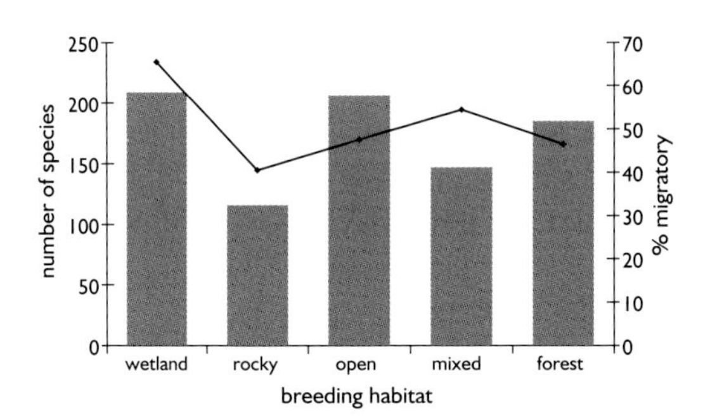

Figure 20.7 *Distribution of Palearctic birds by breeding habitat.*

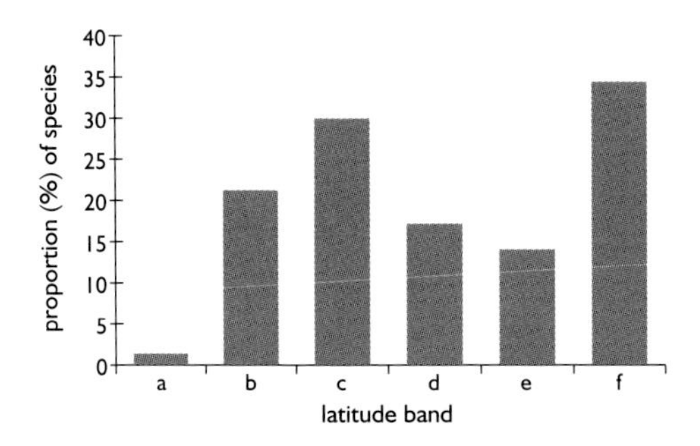

Figure 20.8 *Distribution of forest species by latitude band.*

Lineage	Aerial	Marine	Rocky	Wetlands	Open	Mixed	Forests
1a (i)			0/4		17/14	14/13	6/6
1a (ii)	8/0		0/8	22/22	31/31	44/44	22/22
1a (iii)				3/3		1/1	14/14
1a (iv)	3/0		9/10	2/2	24/24	13/13	55/57
1a (v)			7/8	8/8	42/41	31/31	28/28
1c	5/0		0/6		6/5	2/2	
1d	3/0			4/3	1/2	2/4	16/17
1e			2/6	3/2		4/2	11/12
1f			1/7	7/6	12/8	17/15	7/8
2	3/0	28/0	0/34	74/66	35/38	3/3	1/3
3		22/0	0/22	36/35	2/1	2/0	
4				17/17	6/6	1/1	5/5
5	13/0		0/8			0/2	0/1
6		1/0	0/3	6/6	11/9	5/5	6/6
7a					10/10	9/9	5/5
7b				42/40	9/11		1/1

Table 20.13 *Distribution of feeding and breeding habitat by lineage.*

The occupation of rocky habitats for breeding has been highlighted in the preceding chapters as unusual in many lineages. Nevertheless it has been adopted independently by many Palearctic lineages and reflects the large tracts of open rocky habitat and cliffs that were created during periods of mountain building, especially during the Tertiary. The exploitation of rocky habitats, a speciality of the Palearctic, has been taken up by all five passerine lineages (if we include Hume's Groundpecker *Pseudopodoces humilis*, an aberrant tit of the Tibetan massif), falcons (Group 1c), owls (1e), diurnal raptors (1f), shorebirds (2), water birds (3), swifts (5) and pigeons (6).

CHAPTER 21

The Palearctic avifauna of yesterday, today and tomorrow

At the start of this book I wrote about the Common and Pallid Swifts of Gibraltar, and how I had tried to understand how, appearing so similar, they were able to coexist. I also remarked how it had taken me some time to assimilate what I was really looking at, and how much I was able to glean from studying what was in effect a 'micro-slice' of the evolutionary history of the two species. Today, I read the story of these these swifts very differently. Certainly at the smallest of scales these birds lead their lives, reproduce and leave offspring, and it all happens under the watchful eye of natural selection. But what is happening in Gibraltar need not be the same as what is happening in other places where the two swifts' ranges overlap, or indeed in places where they do not. Multiply those individual histories and scale up in space and time, and what emerges is a picture of fluctuating ranges and population sizes of the two species across the immensity of geological time. Our snapshot is just that. These birds are where they are because of a protracted history, which included contingent events. Pallid Swifts may have been commoner once, rarer at other, with the same applying to Common Swifts. They may be competing with each other today, or the differences we observed may be keeping them apart. Most likely the answer is somewhere in the middle. The outcome, in Gibraltar today, is two populations in the same place at this moment in time. But that picture is the result of the histories of many individuals. In trying to understand the ecology and distribution of species we simply cannot ignore history.

The past

We have seen that the avifauna of the Palearctic, and its sub-region the Western Palearctic, has a long history which started while dinosaurs still roamed the planet. It has been a history dominated by seven of the eight major lineages of birds. Some of the families and genera that emerged at an early stage, during the Eocene and Oligocene, are still with us today but they are a minority. The architect of the modern avifauna has been climate, but its choreographers have been the Earth's landmasses, which were responsible for the cooling and drying of the climate. Orbital cycles also contributed by generating climatic oscillations that re-arranged where species lived over and over again.

The big transition into the world of modern birds began during the Miocene and was probably directly linked to the cooling that followed the Early Miocene warm period. It started around 15 mya. So even though today's avifauna has pre-15 mya components, it has largely built up after this. This was also around the time when the passerines entered the region from the south-east, having originated in an early Gondwanaland and remained in the fragmented southern land masses (Australia, New Zealand, South America) without an opportunity to move north before then. This was also the time when some shorebirds started to adapt to the cooling world of northern coasts.

Much of the history of the Palearctic avifauna has been determined by aridity; in my view aridity has been far more important than temperature. Aridity opened up ancient forests, created savannas, killed off vegetation in places that were taken over by deserts, and introduced seasonal wetlands. The Pliocene was the stage for the world of arid lands to emerge, and it is during this time that modern lineages branched off into many new lines and adapted to a new world that was so unlike the ancient warm and wet world of forests.

It follows that much of what went on happened *in situ*. Species that had arrived in the Palearctic at various stages in the history of its birds adapted to the new conditions as they arose. This included migration, which has to have been independently derived on many occasions by Palearctic populations experiencing changes to the climate and its seasonality. We need look no further than this for the origins of this phenomenon. Migration did not emerge in response to the Pleistocene glaciations, and southern species did not 'discover' untapped seasonal ecologies to the north with global warming. The intensity of migration must have waxed and waned many times, and the lineages capable of switching from residency to migration and back fared best. It is not surprising that species like the Blackcap can switch with such incredible speed. The changes continued to the present day, and continue still. The present trans-Saharan migrations are only around 5,000 years old. The birds that perform wonderful migrations that have amazing abilities for coping with overflying the Sahara have adapted to the new regime in a matter of generations, but they had a genetic legacy that had helped their lineages cope with similar changes since the Miocene. In this, and in many other ways, the past achieved one thing above all others. It weeded out the body plans that could not live in this kind of dry, seasonally fluctuating world. The survivors included only the toughest.

The glaciations brought a new kind of pressure, the influence of cold. But it did not affect the Palearctic's birds in the way the Pliocene droughts had, because it introduced little by way of novel pressures. Cold seasonality could be dealt with through migration, while sedentary species simply shifted geographical position. Among those that found refuge on the southern peninsulas and other cryptic refugia were the sedentary and partially migratory bioclimatic tolerants; they got into oceanic islands, where they remained trapped following deglaciation. The boreal birds returned north too but also left prisoners, this time trapped around the mountains of the mid-latitude belt.

So, with the arrival of renewed global warming 10,000 years ago, the present pattern of bird distribution emerged. The species that greeted the ice-free Palearctic had practically all made it through two million years of glaciations. These birds represent the avifauna of today.

The present

So today's Palearctic avifauna can be defined according to characteristics that have evolved over the past 15 mya. They include a proportion of species that are highly bioclimatically tolerant (Figure 21.1). These species are characterised by having multi-latitude ranges, contrasting with more specialised species that are distributed along particular latitude belts (Figure 21.2). The bioclimatic tolerants are characterised by having taken up central positions on the temperature and humidity gradients (Figures 21.3–21.4); with a wide tolerance, this has permitted expansion in either direction along these gradients. It is noteworthy that this central position is in actual fact slightly towards cool and wet. In contrast the intolerant species are clustered at either end of the spectra, with more species at the warm than cold end (Figure 21.3); there is an even more marked bias

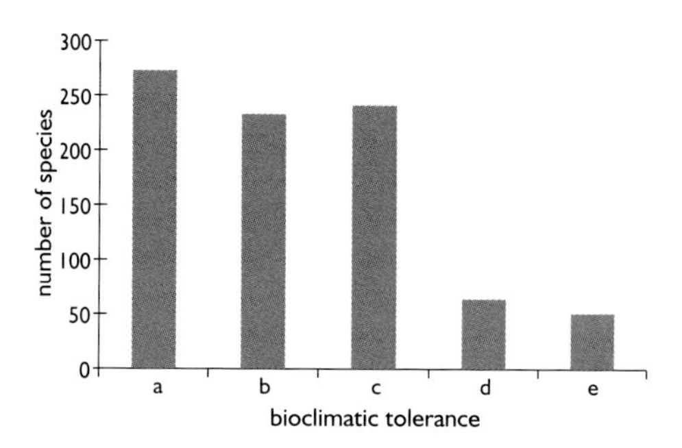

Figure 21.1 *Distribution of Palearctic birds by bioclimatic tolerance.*

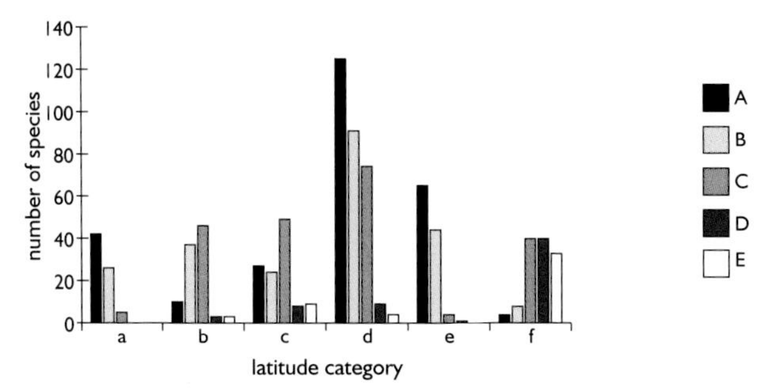

Figure 21.2 *Distribution of Palearctic birds of different bioclimatic tolerance by latitude band.*

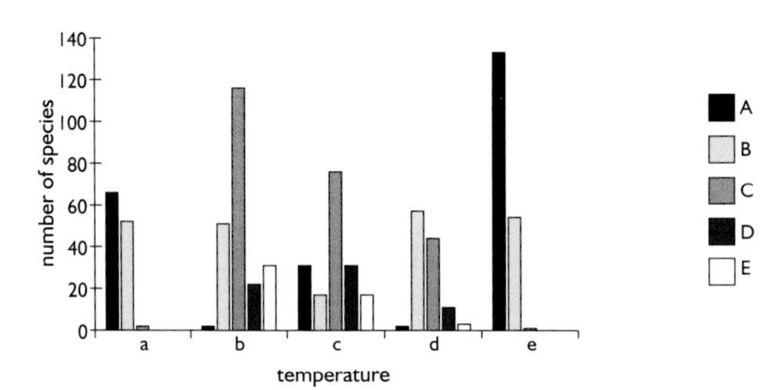

Figure 21.3 *Distribution of Palearctic birds of different bioclimatic tolerance by temperature mid-point.*

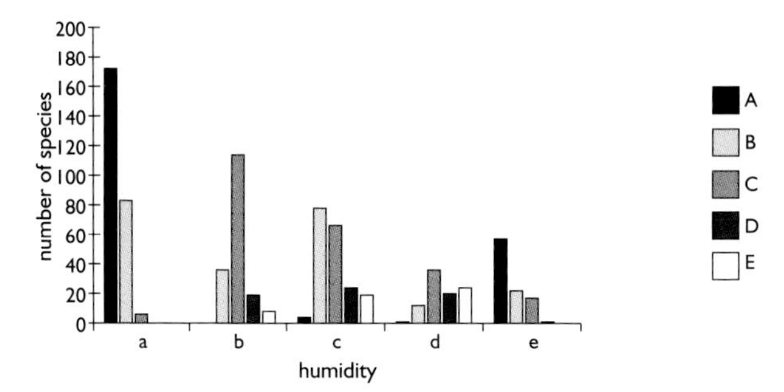

Figure 21.4 *Distribution of Palearctic birds of different bioclimatic tolerance by humidity mid-point.*

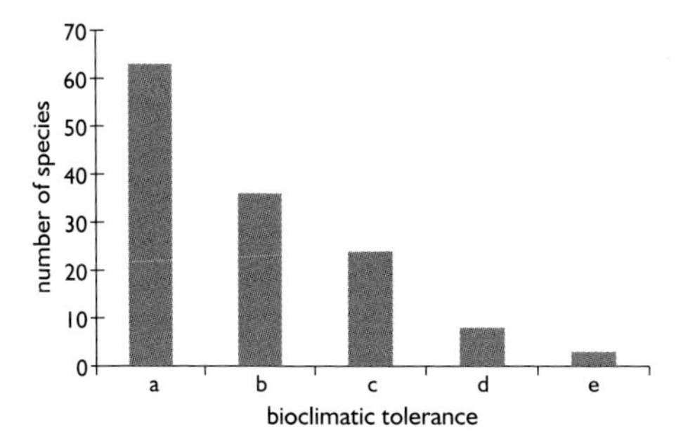

Figure 21.5 *Bioclimatic tolerance levels of montane species.*

towards arid climates (Figure 21.4). We have seen the consequences that these distributions have had on the migratory behaviour of species (see pp. 229–231). These observations are strikingly reproduced in montane birds, most of which are specialists, in equivalence to latitude specialisation (Figure 21.5).

One obvious result of bioclimatic tolerance is that it is among those species that pan-Palearctic ranges have been achieved (Figure 21.6i), whereas intolerant species are characterised by localised or fragmented ranges. The pan-palearctic species tend to have cool, mainly dry bioclimatic mid-points (Figures 21.6 ii–iii). Figure 21.7 summarises the situation. The species that have achieved pan-palearctic ranges are those that have also achieved multi-latitude ranges, the most bioclimatically tolerant species. Next with pan-palearctic ranges are the species of the Boreal belt, which has continued uninterrupted for much of the history of

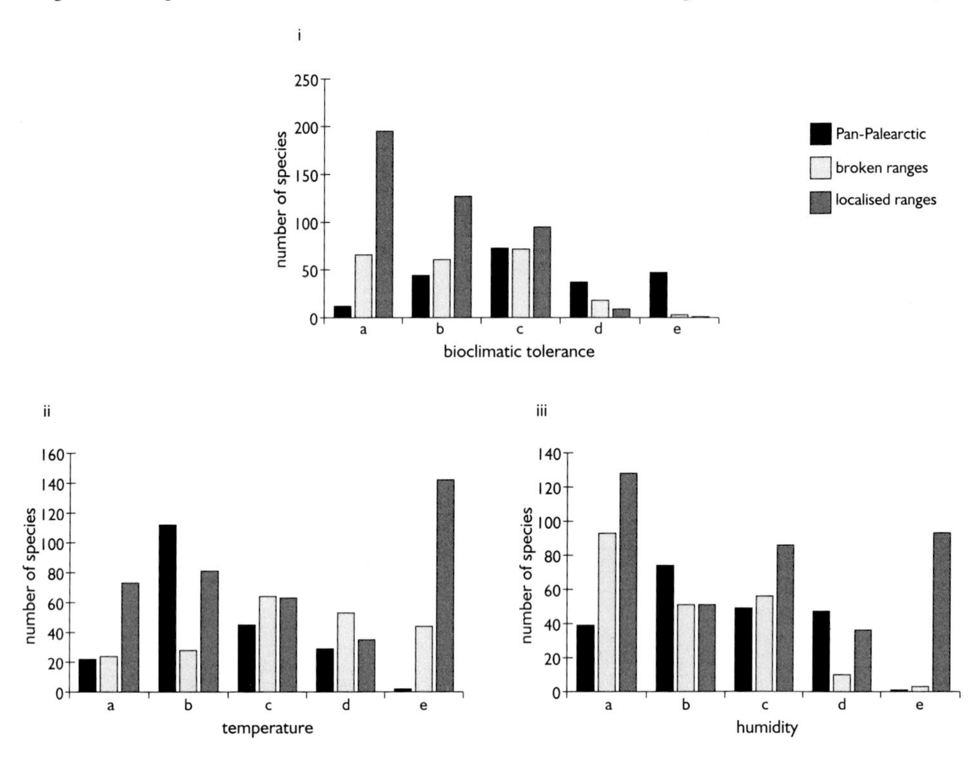

Figure 21.6 *Range continuity and fragmentation in relation to bioclimatic tolerance (i), position on the temperature gradient (ii) and position on the humidity gradient (iii).*

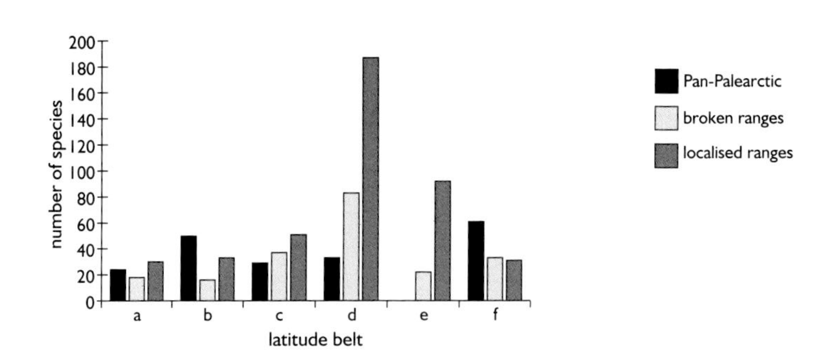

Figure 21.7 *Range continuity and fragmentation in relation to position on the latitude gradient.*

Palearctic birds. But the belts to the south, especially the mid-latitude and subtropical belts, have been severed by deserts and mountains. The result is a huge proportion of species with broken or localised ranges (Figure 21.7). There are no pan-palearctic range species in the subtropical belt.

What can we say about habitat and diet? The dominant categories of feeding habitat are open and wetland which exceed forest and mixed habitats in importance (Appendix 1). Given the importance of forests elsewhere this result is a clear indication of the way in which Palearctic species have adapted to unforested habitats in response to aridity-generated fragmentation. These species are bioclimatically intolerant and combine the use of these habitats with migration. The subset of bioclimatically tolerant species is dominated by birds that use forest and wetlands, with some open and mixed habitats (Appendix 1). There are important differences when breeding habitat is considered. Forests as breeding habitats match the wetlands in importance, and are only surpassed by open habitats, used largely by specialists. Of special interest is the importance of rocky habitats, which we have already identified as a speciality of Palearctic birds, allowing them to reach areas not covered by forest. Tolerant species are largely birds of forest, wetland and rocky habitats. Clearly rocks have often proven a good substitute for trees.

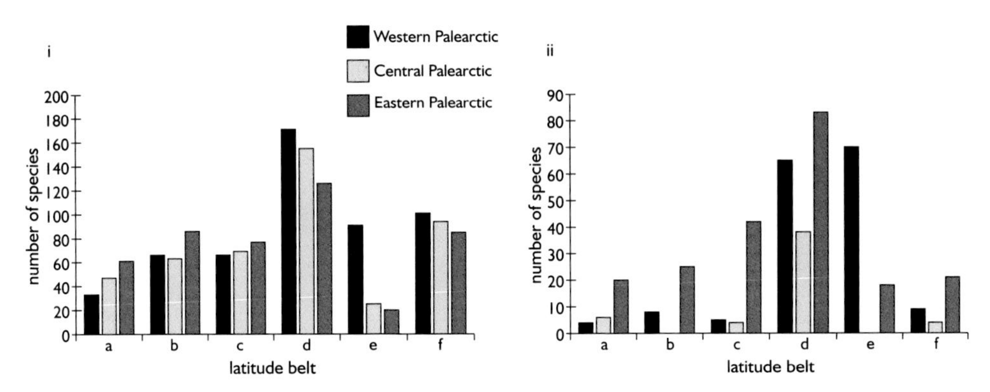

Figure 21.8 *Distribution of species across the Western, Central and Eastern Palearctic in relation to latitudinal position of geographical range (i); in (ii) only species exclusive to one of the three regions are included.*

In terms of diet, it is clear that omnivores, mixed-strategy carnivores and insectivores have been the dominant groups among Palearctic birds. The insectivores have depended on migration. The bioclimatically tolerant species are dominated by mixed-strategy carnivores and omnivores in particular. The pattern of diet categories is not dissimilar between different bioclimatic categories, suggesting that these strategies are tied to lineages that depend on other strategies, like migration or tolerance, as complements.

Can we detect any differences between the three sub-regions of the Palearctic? The Western Palearctic species predominate in the mid-latitude belt (like all other subregions, but they are most numerous here) (Figure 21.8i). There is also a large component of multi-latitude species and of the subtropical belt. Species of the Temperate to Arctic belts are more frequent in the Eastern Palearctic. The Central Palearctic occupies an intermediate position. Species exclusive to the Western Palearctic are largely those that occupy the mid-latitude and subtropical belts (Figure 21.8ii). There are even more Eastern Palearctic species in the mid-latitude belt, but few in the subtropical belt. Multi-latitude species are scarce in all sub-regions, indicating that they tend to occupy more than one sub-region. The Temperate to Arctic belts are dominated by Eastern Palearctic species. These results indicate a significant difference in distribution between sub-regions, the Western Palearctic being characterised by pan-palearctic multi-latitude species and others from the mid-latitude and subtropical belts, many of which are exclusive to the Western Palearctic. Birds of higher latitudes are relatively scarce in contrast.

These observations are borne out by the high frequency of warm species in the Western Palearctic compared to the other sub-regions, and the dominance of Eastern Palearctic species among those at the cold end

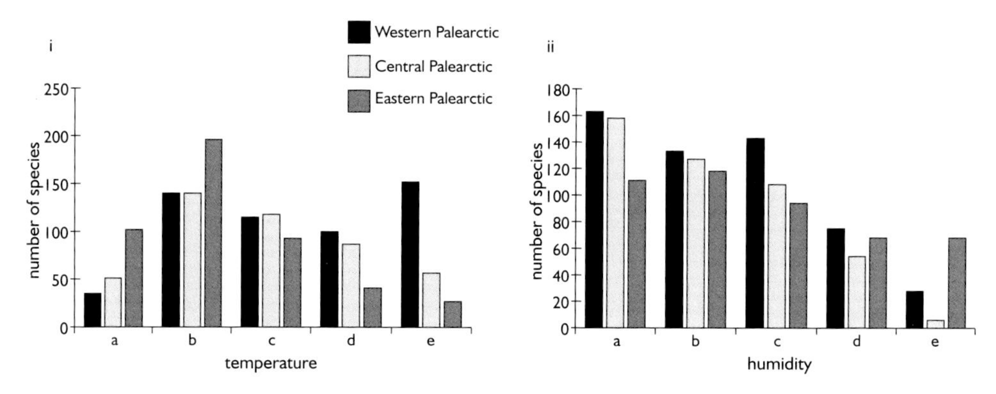

Figure 21.9 *Distribution of species across the Western, Central and Eastern Palearctic in relation to species position on the temperature (i) and humidity (ii) gradients.*

of the spectrum (Figure 21.9i). The Western Palearctic species are mainly at the dry end of the humidity gradient which probably reflects the high incidence of deserts in the region. The Central Palearctic species resemble the Western Palearctic species while the Eastern Palearctic species are mainly among the humid categories (Figure 21.9ii).

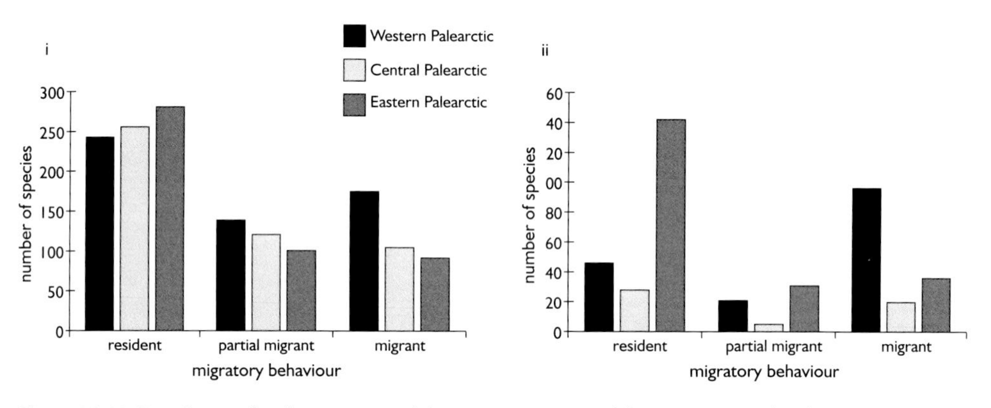

Figure 21.10 *Distribution of residency-migratory behaviour among species of the Western, Central and Eastern Palearctic (i); in (ii) only species exclusive to one of the three regions are included.*

This tendency towards low latitudes and warm climates, along with the warm, oceanic conditions and reduced seasonality of the Western Palearctic, is reflected by a higher proportion of sedentary species among Western Palearctic birds, especially when those unique to each sub-region are considered (Figures 23.14–23.15). Eastern Palearctic species include many more migratory species, with Central Palearctic species intermediate in character.

The Western Palearctic therefore has diagnostic properties that distinguish it from the other sub-regions. These seem to reflect the relative mildness of its climate, but also the presence of deserts in southerly latitudes. Together, these features have favoured birds of warm and dry climates that have occupied low latitudes and have remained fairly sedentary. This also applies at population level, as we have seen throughout this book, with Western Palearctic populations of widespread Palearctic species often tending to be sedentary while more continental ones are highly migratory. In addition, a second group of pan-palearctic species with multi-latitude ranges have complemented those largely unique to the sub-region.

The future

Many scenarios are postulated for the future of the birds of the Palearctic. I am reluctant to become either one more sage or a prophet of doom and gloom. The birds of the Palearctic are experts at surviving sharp climatic pulses. It is difficult enough to predict what the future will hold climatically, let alone predict a bird's responses. One thing seems certain. Few, if any, Palearctic birds will go extinct as a result of climate change alone. I am sure that had we had globalised media 5,000 years ago, when the green Sahara dried up, we would have seen headlines of how species would imminently go extinct as a result. They did not.

What we are likely to observe, as has happened so often before, are range changes. These will be expressed in different ways. Bioclimatically tolerant species will change the edges of their ranges, while specialists will expand and contract depending on the nature of the signal. Those of the warm belts might spread northwards in scenarios of warming, while those of cold belts will contract further into the Arctic as they did many times before during interglacials. Birds of the arid belts will do well if the planet dries and those of the more humid conditions will benefit if it becomes wetter. If the changes occur regionally then the patterns will be even more difficult to predict. If the history of the birds of the Palearctic speaks to us of one thing above all others, it is of how small changes, where a population was when a change happened, made a difference. And it also speaks to us of the almost impossible task of predicting these changes, with so many variables and contingencies involved. But those species with large geographical ranges, spanning many latitudes, would seem best buffered against change.

There is one key factor affecting the future of the Palearctic birds. It is human population pressure. This factor has caused more bird species losses than any other and will, in all likelihood, continue to do so. Imagine if the Sahara, by some climatic miracle, became wet and green tomorrow. Would it make any sense to predict how birds would move in and regain a lost wilderness? I think not. If the Sahara greened again we would fill it with people – some commensal birds might do well but the trans-Saharan migrants would have to negotiate deserts of sterilised olive groves, plastic-covered mega-orchards, water-sucking golf courses, forests of windmills, concrete jungles and vast holiday resorts. A green Sahara might, sadly, be worse (certainly not better) for birds than a dry one. This is the reality of the future and one which the red herring of climate change has successfully managed to deflect.

APPENDIX 1
Species covered in this book: bioclimatic and ecological features

Appendix 1. *The Palearctic species covered in this book and their main bioclimatic and ecological features. TOL = bioclimatic tolerance; LAT = central position of range; TEM = temperature mid-point; HUM = humidity mid-point (key for TOL, LAT, TEM and HUM is on pp. 32–33); MON = presence on high mountains (M); WP = presence Western Palearctic (in parenthesis = broken or marginal range); CP = presence Central Palearctic (in parenthesis = broken or marginal range); EP = presence Eastern Palearctic (in parenthesis = broken or marginal range); HF = foraging habitat (A aerial, Ma marine, W wetlands, O open, R rocky, M mixed, F forest); HN = nesting habitat (key as for foraging habitat); DT = diet (O omnivore, M mixed-strategy carnivore, I insectivore, H herbivore, F fish-eater, E carnivore of endothermic prey, N necrophyte (scavenger); MIG = migratory status (S sedentary, P part-migratory, M fully migratory).*

	TAXON		TOL	LAT	TEM	HUM	MON	WP	CP	EP	HF	HN	DT	MIG
Corvoidea 1a (i)	**Laniidae**													
	Red-backed Shrike	*Lanius collurio*	C	F	C	B		(+)	+	-	M	M	M	M
	Lesser Grey Shrike	*Lanius minor*	C	D	D	C		(+)	+	-	M	M	M	M
	Great Grey Shrike	*Lanius excubitor*	C	B	B	D		+	+	+	O	O	M	P
	Woodchat Shrike	*Lanius senator*	B	D	D	C		+	-	-	M	M	M	M
	Southern Grey Shrike	*Lanius meridionalis*	B	D	D	C		(+)	-	-	M	M	M	S
	Tiger Shrike	*Lanius tigrinus*	B	D	B	E		-	-	(+)	O	O	M	M
	Bull-headed Shrike	*Lanius bucephalus*	B	D	B	E		-	-	(+)	M	M	M	M
	Chinese Grey Shrike	*Lanius sphenocercus*	B	D	B	E		-	-	(+)	M	M	M	M
	Masked Shrike	*Lanius excubitor*	A	D	E	A		(+)	(+)	-	M	M	M	M
	Levantine Grey Shrike	*Lanius theresae*	A	D	E	A		(+)	-	-	O	O	M	S
	Algerian Grey Shrike	*Lanius algeriensis*	A	D	E	A		(+)	-	-	O	O	M	S
	Isabelline Shrike	*Lanius isabellinus*	A	D	E	A		-	+	(+)	O	O	M	M
	Canary Islands Grey Shrike	*Lanius koenigi*	A	E	E	E		(+)	-	-	O	O	M	S
	Black-crowned Tchagra	*Tchagra senegallus*	A	E	E	C		(+)	-	-	M	M	M	S
	Corvidae													
	Eurasian Jay	*Garrulus glandarius*	D	F	C	D		+	+	+	F	F	O	S
	Siberian Jay	*Perisoreus infaustus*	B	B	B	A		(+)	+	+	F	F	O	S
	Pander's Ground Jay	*Podoces panderi*	B	D	D	A		-	(+)	-	O	O	O	S
	Henderson's Ground Jay	*Podoces hendersoni*	B	D	D	A		-	-	(+)	O	O	O	S

	TAXON		TOL	LAT	TEM	HUM	MON	WP	CP	EP	HF	HN	DT	MIG
Corvoidea 1a (i) cont.	Biddulph's Ground Jay	*Podoces biddulphi*	A	D	C	A		-	(+)	-	O	O	O	S
	Pleske's Ground Jay	*Podoces pleskei*	A	D	E	A		-	(+)	-	O	O	O	S
	Asian Azure-winged Magpie	*Cyanopica cyanus*	C	D	C	B		-	-	(+)	M	M	O	S
	Iberian Azure-winged Magpie	*Cyanopica cooki*	B	D	D	C		(+)	-	-	M	M	O	S
	Red-billed Blue Magpie	*Urocissa erythrorhyncha*	A	D	E	E		-	-	+	F	F	O	S
	Black-billed Magpie	*Pica pica*	E	C	B	B		+	+	+	O	O	M	S
	Spotted Nutcracker	*Nucifraga caryocatactes*	E	C	B	D		(+)	+	+	F	F	O	S
	Red-billed Chough	*Pyrrhocorax pyrrhocorax*	E	D	C	D	M	(+)	(+)	+	O	R	O	S
	Alpine Chough	*Pyrrhocorax graculus*	B	D	C	E	M	(+)	(+)	-	O	R	O	S
	Common Raven	*Corvus corax*	E	F	B	B		(+)	+	+	M	R	M	S
	Carrion Crow	*Corvus corone*	D	F	B	C		+	+	+	O	O	M	P
	Rook	*Corvus frugilegus*	D	C	C	B		+	+	+	O	O	M	P
	Large-billed Crow	*Corvus macrorhynchus*	C	F	C	E		-	-	+	M	M	O	S
	Eurasian Jackdaw	*Corvus monedula*	C	F	C	B		+	+	-	O	R	O	P
	Daurian Jackdaw	*Corvus dauuricus*	C	D	C	C		-	-	+	M	M	O	M
	Collared Crow	*Corvus torquatus*	B	E	D	E		-	-	(+)	M	M	M	S
	Brown-necked Raven	*Corvus ruficollis*	A	E	E	A		+	(+)	-	O	O	O	S
	Oriolidae													
	Golden Oriole	*Oriolus oriolus*	D	F	C	B		+	+	-	F	F	O	M
	Black-naped Oriole	*Oriolus chinensis*	C	F	C	E		-	-	+	F	F	O	M
Sylvoidea 1a (ii)	**Hirundinidae**													
	Barn Swallow	*Hirundo rustica*	E	F	B	B		+	+	+	A	R	I	M
	Asian House Martin	*Delichon dasypus*	C	D	B	E	M	-	-	+	A	R	I	M
	House Martin	*Delichon urbica*	E	F	B	D		+	+	+	A	R	I	M
	Sand Martin	*Riparia riparia*	E	F	B	C		(+)	+	+	A	R	I	M
	Pale Sand Martin	*Riparia diluta*	A	C	C	A		-	(+)	-	A	R	I	M
	Red-rumped Swallow	*Cecropis daurica*	E	D	D	D		(+)	+	+	A	R	I	M
	Crag Martin	*Ptyonoprogne rupestris*	C	D	D	B	M	+	+	(+)	A	R	I	P
	Rock Martin	*Ptyonoprogne fuligula*	B	E	E	C		(+)	(+)	-	A	R	I	S
	Pycnonotidae													
	Common Bulbul	*Pycnonotus barbatus*	B	E	E	C	M	(+)	-	-	M	M	O	S
	Chinese Bulbul	*Pycnonotus sinensis*	A	E	E	E		-	-	(+)	M	M	O	S
	White-cheeked Bulbul	*Pycnonotus leucogenys*	A	E			M	(+)	-	-	M	M	O	S
	White-spectacled Bulbul	*Pycnonotus xanthopygus*	A	E	E	A		(+)	-	-	M	M	O	S
	Hypocolidae													
	Grey Hypocolius	*Hypocolius ampelinus*	B	E	E	C		(+)	(+)	-	M	M	O	M
	Cisticolidae													

	TAXON		TOL	LAT	TEM	HUM	MON	WP	CP	EP	HF	HN	DT	MIG
Sylvoidea 1a (ii) cont.	Zitting Cisticola	*Cisticola juncidis*	C	D	D	D		(+)	-	-	O	O	I	P
	Chinese Hill Warbler	*Rhopophilus pekinensis*	A	D	C	A		-	-	(+)	M	M	O	S
	Streaked Scrub Warbler	*Scotocerca inquieta*	A	E	E	A		(+)	(+)	-	M	M	O	S
	Graceful Prinia	*Prinia gracilis*	B	E	E	C	M	(+)	-	-	M	M	O	S
	Sylviidae													
	Asian Stubtail	*Urosphena squamiceps*	B	D	B	E		-	-	(+)	F	F	O	M
	Cetti's Warbler	*Cettia cetti*	C	D	D	B		+	+	-	W	W	M	P
	Manchurian Bush Warbler	*Cettia canturians*	C	C	C	E		-	-	(+)	W	W	O	M
	Japanese Bush Warbler	*Cettia diphone*	B	D	B	C	M	-	-	(+)	M	M	O	M
	Chinese Bush Warbler	*Bradypterus tacsanowskius*	C	D	B	C		-	-	(+)	F	F	O	M
	Spotted Bush Warbler	*Bradypterus thoracicus*	B	F	E	E	M	-	-	+	M	M	O	M
	Large-billed Bush Warbler	*Bradypterus major*	A	D			M	-	(+)	-	F	F	I	S
	Lanceolated Warbler	*Locustella lanceolata*	D	B	B	C		-	(+)	+	W	W	O	M
	Grasshopper Warbler	*Locustella naevia*	C	C	C	C		+	+	-	W	W	O	M
	Savi's Warbler	*Locustella luscinioides*	C	C	C	C		(+)	(+)	-	W	W	I	M
	Gray's Grasshopper Warbler	*Locustella fasciolata*	C	C	B	B		-	-	+	M	M	I	M
	Styan's Grasshopper Warbler	*Locustella pleskei*	C	D	C	D		-	-	(+)	M	M	I	M
	River Warbler	*Locustella fluviatilis*	B	C	B	B		+	(+)	-	W	W	I	M
	Pallas's Grasshopper Warbler	*Locustella certhiola*	B	C	B	A		-	(+)	+	W	W	I	M
	Middendorff's Grasshopper	*Locustella ochotensis*	B	B	A	C		-	-	(+)	W	W	O	M
	Sedge Warbler	*Acrocephalus schoenobaenus*	D	C	C	B		+	+	-	W	W	O	M
	Eurasian Reed Warbler	*Acrocephalus scirpaceus*	C	C	C	C		+	+	-	W	W	O	M
	Marsh Warbler	*Acrocephalus palustris*	C	C	C	C		+	+	-	W	W	O	M
	Aquatic Warbler	*Acrocephalus paludicola*	C	C	B	C		+	(+)	-	W	W	I	M
	Moustached Warbler	*Acrocephalus melanopogon*	C	D	D	B		(+)	+	-	W	W	O	P
	Oriental Reed Warbler	*Acrocephalus orientalis*	C	D	D	C		-	-	+	W	W	O	M
	Black-browed Reed Warbler	*Acrocephalus bistrigiceps*	C	D	B	C		-	-	(+)	W	W	O	M
	Blyth's Reed Warbler	*Acrocephalus dumetorum*	B	C	D	A		(+)	+	-	F	F	O	M
	Great Reed Warbler	*Acrocephalus arundinaceus*	B	C	D	A		(+)	+	-	W	W	O	M
	Thick-billed Warbler	*Acrocephalus aedon*	C	C	B	C		-	-	+	W	W	O	M
	Paddyfield Warbler	*Acrocephalus agricola*	B	D	D	A		(+)	(+)	(+)	W	W	M	M
	Clamorous Reed Warbler	*Acrocephalus stentoreus*	B	D	E	B		-	(+)	-	W	W	I	M
	Streaked Reed Warbler	*Acrocephalus sorghophilus*	B	D	B	E		-	(+)	-	W	W	O	M
	Cape Verde Swamp Warbler	*Acrocephalus brevipennis*	A	E	E	E		(+)	-	-	M	M	I	S
	Blunt-winged Warbler	*Acrocephalus concinens*	A	E	E	A		-	-	(+)	W	W	I	M
	Icterine Warbler	*Hippolais icterina*	C	C	C	C		+	(+)	-	M	M	O	M
	Booted Warbler	*Hippolais caligata*	B	F	D	A		(+)	+	-	M	M	I	M

	TAXON		TOL	LAT	TEM	HUM	MON	WP	CP	EP	HF	HN	DT	MIG
Sylvoidea 1a (ii) cont.	Eastern Olivaceous Warbler	*Hippolais pallida*	B	D	D	A		+	+	-	M	M	O	P
	Western Olivaceous Warbler	*Hippolais opaca*	B	D	D	A		+	-	-	M	M	O	M
	Melodious Warbler	*Hippolais polyglotta*	B	D	D	C		+	-	-	M	M	O	M
	Upcher's Warbler	*Hippolais languida*	A	D	E	A	M	(+)	+	-	M	M	O	M
	Olive-tree Warbler	*Hippolais olivetorum*	A	D	E	A		(+)	-	-	M	M	O	M
	Sykes's Warbler	*Hippolais rama*	A	D	E	A		-	(+)	-	M	M	O	M
	Common Chiffchaff	*Phylloscopus collybita*	C	F	C	B		+	+	(+)	F	F	O	P
	Dusky Warbler	*Phylloscopus fuscatus*	C	F	B	D		-	-	+	M	M	O	M
	Wood Warbler	*Phylloscopus sibilatrix*	C	C	B	C		+	(+)	-	F	F	O	M
	Radde's Warbler	*Phylloscopus schwarzi*	C	C	B	B		-	(+)	+	M	M	I	M
	Pallas's Warbler	*Phylloscopus proregulus*	C	C	C	C		-	-	+	F	F	I	M
	Pale-legged Leaf Warbler	*Phylloscopus tenellipes*	C	C	B	B		-	-	(+)	F	F	I	M
	Eastern Crowned Warbler	*Phylloscopus coronatus*	C	D	B	C		-	-	(+)	F	F	I	M
	Willow Warbler	*Phylloscopus trochilus*	C	B	B	C		+	+	+	M	M	O	M
	Yellow-browed Warbler	*Phylloscopus inornatus*	C	B	C	B		-	(+)	+	F	F	I	M
	Greenish Warbler	*Phylloscopus trochiloides*	C	A/D	C	C		(+)/(+)	+/+	+/+	M	M	I	M
	Western Bonelli's Warbler	*Phylloscopus bonelli*	B	D	D	C		+	-	-	F	F	O	M
	Eastern Bonelli's Warbler	*Phylloscopus orientalis*	B	D	D	C		+	-	-	F	F	O	M
	Hume's Leaf Warbler	*Phylloscopus humei*	B	D	E	E	M	-	(+)	-	F	F	I	M
	Arctic Warbler	*Phylloscopus borealis*	B	B	B	A	M	(+)	+	+	F	F	I	M
	Mountain Chiffchaff	*Phylloscopus sindianus*	A	D			M	(+)	(+)	-	M	M	I	M
	Iberian Chiffchaff	*Phylloscopus ibericus*	A	D	E	A		(+)	-	-	F	F	I	M
	Plain Leaf Warbler	*Phylloscopus neglectus*	A	D			M	-	(+)	-	F	F	I	M
	Sulphur-bellied Warbler	*Phylloscopus griseolus*	A	D			M	-	(+)	-	M	M	I	M
	Brooks's Leaf Warbler	*Phylloscopus subviridis*	A	D			M	-	(+)	-	F	F	I	M
	Tickell's Leaf Warbler	*Phylloscopus affinis*	A	D			M	-	(+)	-	O	O	I	M
	Yellow-streaked Warbler	*Phylloscopus armandii*	A	D			M	-	-	(+)	M	M	I	M
	Sakhalin Leaf Warbler	*Phylloscopus borealoides*	A	D	A	A	M	-	-	(+)	F	F	I	M
	Canary Islands Chiffchaff	*Phylloscopus canariensis*	A	E	E	E		(+)	-	-	F	F	I	S
	Blackcap	*Sylvia atricapilla*	D	F	C	C		+	+	-	F	F	O	P
	Greater Whitethroat	*Sylvia communis*	C	F	C	B		+	+	-	M	M	O	M
	Garden Warbler	*Sylvia borin*	C	C	B	C		+	+	-	F	F	O	M
	Sardinian Warbler	*Sylvia melanocephala*	C	D	D	D		+	-	-	M	M	O	P
	Lesser Whitethroat	*Sylvia curruca*	C	B	B	D		(+)	+	(+)	M	M	O	M
	Eastern Orphean Warbler	*Sylvia crassirostris*	B	D	D	A		+	(+)	-	M	M	O	M
	Subalpine Warbler	*Sylvian cantillans*	B	D	D	C		+	-	-	M	M	O	M
	Dartford Warbler	*Sylvia undata*	B	D	D	C		+	-	-	M	M	O	P

	TAXON		TOL	LAT	TEM	HUM	MON	WP	CP	EP	HF	HN	DT	MIG
Sylvoidea 1a (ii) cont.	Spectacled Warbler	*Sylvia conspicillata*	B	D	E	C		(+)	-	-	O	O	O	P
	Asian Desert Warbler	*Sylvia nana*	A	D	C	A		(+)	+	-	O	O	O	M
	Ménétries's Warbler	*Sylvia mystacea*	A	D	E	A		(+)	(+)	-	M	M	O	M
	Western Orphean Warbler	*Sylvian hortensis*	A	D	E	A		+	-	-	M	M	O	M
	Rüppell's Warbler	*Sylvia ruepelli*	A	D	E	A		(+)	-	-	M	M	O	M
	Cyprus Warbler	*Sylvia melanothorax*	A	D	E	A		(+)	-	-	M	M	O	P
	Tristram's Warbler	*Sylvia deserticola*	A	D	E	A	M	(+)	-	-	M	M	I	S
	Marmora's Warbler	*Sylvia sarda*	A	D	E	A	M	(+)	-	-	O	O	I	S
	Balearic Warbler	*Sylvia balearica*	A	D	E	A		(+)	-	-	O	O	I	S
	African Desert Warbler	*Sylvia deserti*	A	E	E	A		+	-	-	O	O	O	S
	Small Whitethroat	*Sylvia minula*	A	D	C	A		-	+	-	M	M	O	M
	Hume's Whitethroat	*Sylvia althaea*	A	D	E	A		-	+	-	M	M	O	M
	Margelanic Whitethroat	*Sylvia margelanica*	A	D	C	A		-	+	-	M	M	O	M
	Barred Warbler	*Sylvia nisoria*	A	C	C	A		(+)	+	-	M	M	O	M
	Aegithalidae													
	Long-tailed Tit	*Aegithalos caudatus*	B	F	C	C		+	+	+	F	F	O	S
	Alaudidae													
	Kordofan Bush Lark	*Mirafra cordofanica*	A	E	E	A		(+)	-	-	O	O	O	S
	Greater Hoopoe Lark	*Alaemon alaudipes*	B	E	E	C	M	(+)	(+)	-	O	O	O	S
	Dupont's Lark	*Chersophilus duponti*	A	D	E	A		(+)	-	-	O	O	O	S
	Black-crowned Sparrow Lark	*Eremopterix nigriceps*	B	E	E	C		(+)	-	-	O	O	O	S
	Bar-tailed Lark	*Ammomanes cincturus*	A	E	E	A		(+)	(+)	-	O	O	O	S
	Desert Lark	*Ammomanes deserti*	A	E	E	A		(+)	(+)	-	O	O	O	S
	Thick-billed Lark	*Ramphocoris clotbey*	A	D	E	A		(+)	-	-	O	O	O	S
	Bimaculated Lark	*Melanocorypha bimaculata*	B	D	D	A		(+)	+	-	O	O	O	P
	White-winged Lark	*Melanocorypha leucoptera*	B	D	D	A		(+)	+	-	O	O	O	P
	Calandra Lark	*Melanocorypha calandra*	A	D	E	A		(+)	+	-	O	O	O	S
	Tibetan Lark	*Melanocorypha maxima*	A	D			M	-	+	-	O	O	O	S
	Mongolian Lark	*Melanocorypha mongolica*	A	D	C	A		-	-	(+)	O	O	O	S
	Black Lark	*Melanocorypha yeltoniensis*	A	C	C	A		(+)	+	-	O	O	O	P
	Greater Short-toed Lark	*Calandrella brachydactyla*	C	D	D	B		(+)	+	-	O	O	O	M
	Lesser Short-toed Lark	*Calandrella rufescens*	C	D	D	B		(+)	+	-	O	O	O	P
	Hume's Short-toed Lark	*Calandrella acutirostris*	A	D			M	-	(+)	-	O	O	O	M
	Asian Short-toed Lark	*Calandrella cheleensis*	A	D	C	A		-	-	(+)	O	O	O	S
	Shore Lark	*Eremophila alpestris*	E	A/D	B	B	-/M	+/(+)	+/+	+/(+)	O	O	O	M/S
	Temminck's Lark	*Eremophila bilopha*	A	D	E	A		(+)	-	-	O	O	O	S
	Dunn's Lark	*Eremalauda dunni*	A	E	E	A		(+)	-	-	O	O	O	S

	TAXON		TOL	LAT	TEM	HUM	MON	WP	CP	EP	HF	HN	DT	MIG
	Crested Lark	*Galerida cristata*	C	F	D	B		(+)	(+)	(+)	O	O	O	P
	Thekla Lark	*Galerida theklae*	B	D	E	C		(+)	-	-	O	O	O	S
	Eurasian Skylark	*Alauda arvensis*	D	F	C	D	M	+	+	+	O	O	O	P
	Razo Lark	*Alauda razae*	A	E	E	E		(+)	-	-	O	O	O	S
	Woodlark	*Lullula arborea*	C	F	C	B		(+)	-	-	M	M	O	P
Paroidea 1a (iii)	**Remizidae**													
	Penduline Tit	*Remiz pendulinus*	C	C	C	C		(+)	(+)	(+)	W	W	O	P
	White-crowned Penduline Tit	*Remiz coronatus*	B	D	D	A		-	(+)	-	W	W	O	S
	Chinese Penduline Tit	*Remiz consobrinus*	A	C	C	A		-	-	(+)	W	W	O	S
	Paridae													
	Marsh Tit	*Poecile palustris*	B	C	C	D		+	-	+	F	F	O	S
	Willow Tit	*Poecile montana*	B	B	B	C		(+)	+	+	F	F	O	S
	Siberian Tit	*Poecile cincta*	B	B	B	A		(+)	+	+	F	F	O	S
	Sombre Tit	*Poecile lugubris*	A	D	E	A	M	+	+	-	F	F	O	S
	Coal Tit	*Periparus ater*	E	F	B	D		+	+	+	F	F	O	S
	Yellow-bellied Tit	*Periparus venustulus*	B	E	D	E		-	-	(+)	F	F	O	S
	Rufous-naped Tit	*Periparus rufonuchalis*	A	D	E	A	M	-	(+)	-	F	F	O	S
	Crested Tit	*Lophophanes cristatus*	C	F	C	B		+	(+)	-	F	F	O	S
	Great Tit	*Parus major*	E	F	C	D		+	+	+	F	F	O	S
	Varied Tit	*Parus varius*	B	D	B	E		-	-	(+)	F	F	O	S
	Turkestan Tit	*Parus bokharensis*	A	D			M	-	(+)	-	M	M	O	S
	Yellow-breasted Tit	*Cyanistes flavipectus*	B	D	C	A		-	(+)	-	F	F	O	S
	Blue Tit	*Cyanistes caeruleus*	D	F	C	C		+	(+)	-	F	F	O	S
	Azure Tit	*Cyanistes cyanus*	B	C	B	A		(+)	+	+	F	F	O	S
	African Blue Tit	*Cyanistes teneriffae*	A	E	E	E		(+)	-	-	F	F	O	S
Muscicapoidea 1a (iv)	**Muscicapidae**													
	Rufous Bush Robin	*Cercotrichas galactotes*	A	D	E	A		+	+	-	M	M	O	M
	European Robin	*Erithacus rubecula*	D	F	C	C		+	(+)	-	F	F	O	P
	Bluethroat	*Luscinia svecica*	C	B/D	B	B	M	(+)/(+)	+/+	+/+	M	M	O	M
	Common Nightingale	*Luscinia megarhynchos*	C	D	C	C		+	(+)	-	M	M	O	M
	Siberian Rubythroat	*Luscinia calliope*	C	B/D	B	B		-	(+)/(+)	+/+	M	M	O	M
	Siberian Blue Robin	*Luscinia cyane*	C	C	B	B		-	-	+	F	F	O	M
	Japanese Robin	*Luscinia akahige*	C	D	B	D		-	-	(+)	F	F	O	S
	Rufous-tailed Robin	*Luscinia sibilans*	B	B	A	C		-	-	+	F	F	O	M
	Thrush Nightingale	*Luscinia luscinia*	A	C	C	A		(+)	+	-	M	M	O	M
	White-tailed Rubythroat	*Luscinia pectoralis*	A	D			M	-	(+)	-	R	R	O	S
	White-throated Robin	*Irania gutturalis*	A	D	E	A	M	(+)	(+)	-	M	M	O	M

	TAXON		TOL	LAT	TEM	HUM	MON	WP	CP	EP	HF	HN	DT	MIG
Muscicapoidea 1a (iv) cont.	Red-flanked Bluetail	*Tarsiger cyanurus*	C	B/D	B	B		(+)/(+)	(+)/(+)	+/+	F	F	O	M
	Red-breasted Flycatcher	*Ficedula parva*	C	F	C	B		+	(+)	-	F	F	I	M
	Collared Flycatcher	*Ficedula albicollis*	C	C	B	C		(+)	+	-	F	F	O	M
	Mugimaki Flycatcher	*Ficedula mugimaki*	C	C	B	B		-	-	+	A	F	I	M
	Yellow-rumped Flycatcher	*Ficedula zanthopygia*	C	D	B	B		-	-	+	F	F	I	M
	Narcissus Flycatcher	*Ficedula narcissina*	C	D	B	D		-	-	(+)	F	F	I	M
	Pied Flycatcher	*Ficedula hypoleuca*	C	B/D	B	C		+/+	(+)/(+)	-	F	F	O	M
	Taiga Flycatcher	*Ficedula albicilla*	B	B	A	C		-	+	+	F	F	I	M
	Semi-collared Flycatcher	*Ficedula semitorquata*	A	D	E	A	M	(+)	+	-	A	F	I	M
	Atlas Flycatcher	*Ficedula speculigera*	A	D	E	A		(+)	-	-	F	F	I	M
	Brown Flycatcher	*Muscicapa daurica*	E	C	B	C	M	-	-	+	F	F	I	M
	Dark-sided Flycatcher	*Muscicapa sibirica*	D	C	B	B	M	-	-	+	A	F	I	M
	Spotted Flycatcher	*Muscicapa striata*	C	F	B	C		+	+	(+)	F	F	O	M
	Grey-streaked Flycatcher	*Muscicapa griseisticta*	B	C	A	C		-	-	(+)	F	F	I	M
	Rusty-tailed Flycatcher	*Muscicapa ruficauda*	A	D			M	-	(+)	-	F	F	I	M
	Blue-and-white Flycatcher	*Niltavia cyanomelana*	C	D	B	D		-	-	(+)	F	F	I	M
	Japanese Paradise Flycatcher	*Terpsiphone atrocaudata*	A	D	C	E		-	-	(+)	F	F	I	M
	Common Redstart	*Phoenicurus phoenicurus*	D	F	C	B		+	+	-	F	F	O	M
	Daurian Redstart	*Phoenicurus auroreus*	D	F	B	B	M	-	-	+	F	F	O	M
	Black Redstart	*Phoenicurus ochruros*	D	F	C	C	M	+	(+)	-	R	R	O	P
	Blue-capped Redstart	*Phoenicurus coeruleocephalus*	B	D	C	A	M	-	(+)	-	M	M	O	P
	Eversmann's Redstart	*Phoenicurus erythronota*	B	D	C	A	M	-	(+)	-	F	F	O	M
	Güldenstädt's Redstart	*Phoenicurus erythrogaster*	A	D			M	-	(+)	-	R	R	O	S
	Moussier's Redstart	*Phoenicurus moussieri*	A	D	E	A		(+)	-	-	F	F	O	S
	Blackstart	*Cercomela melanura*	A	E	E	A		(+)	-	-	O	O	O	S
	Siberian Stonechat	*Saxicola maurus*	E	F	C	C		(+)	+	+	O	O	O	M
	Common Stonechat	*Saxicola torquata*	C	F	C	D		+	-	-	O	O	O	P
	Whinchat	*Saxicola rubetra*	C	B/D	B	B		+/+	(+)/(+)	-	O	O	O	M
	Hodgson's Bush Chat	*Saxicola insignis*	B	C	C	A	M	-	(+)	-	O	O	O	M
	Canary Islands Chat	*Saxicola dacotiae*	A	E	E	E		(+)	-	-	O	O	O	S
	Northern Wheatear	*Oenanthe oenanthe*	D	F	B	B	M	+	+	+	O	O	O	M
	Desert Wheatear	*Oenanthe deserti*	B	D	D	A	M	(+)	+	(+)	O	O	O	P
	Isabelline Wheatear	*Oenanthe isabellina*	B	D	D	A	M	(+)	+	(+)	O	O	O	M
	Pied Wheatear	*Oenanthe pleschanka*	B	D	D	A		(+)	+	-	O	O	O	M
	Black-eared Wheatear	*Oenanthe hispanica*	A	D	E	A		+	(+)	-	O	O	O	M
	Finsch's Wheatear	*Oenanthe finschii*	A	D	E	A		(+)	+	-	O	O	O	P
	Mourning Wheatear	*Oenanthe lugens*	A	D	E	A	M	(+)	(+)	-	O	O	O	P

	TAXON		TOL	LAT	TEM	HUM	MON	WP	CP	EP	HF	HN	DT	MIG
Muscicapoidea 1a (iv) cont	Rufous-tailed Wheatear	*Oenanthe xanthprymna*	A	D	E	A	M	(+)	(+)	-	O	O	O	P
	Red-rumped Wheatear	*Oenanthe moesta*	A	D	E	A		(+)	-	-	O	O	O	S
	Black Wheatear	*Oenanthe leucura*	A	D	E	A		(+)	-	-	O	O	O	S
	Cyprus Wheatear	*Oenanthe cypriaca*	A	D	E	A		(+)	-	-	O	O	O	M
	Eastern Pied Wheatear	*Oenanthe picata*	A	D	E	A		-	(+)	-	R	R	O	M
	White-crowned Black Wheatear	*Oenanthe leucopyga*	A	E	E	A	M	(+)	-	-	O	O	O	S
	Hume's Wheatear	*Oenanthe alboniger*	A	E	E	A	M	(+)	+	-	O	O	O	S
	Hooded Wheatear	*Oenanthe monacha*	A	E	E	A		(+)	+	-	O	O	O	S
	Blue Rock Thrush	*Monticola solitarius*	D	D	C	C		+	+	+	R	R	O	P
	Rufous-tailed Rock Thrush	*Monticola saxatilis*	B	D	E	C	M	+	+	+	R	R	O	M
	White-throated Rock Thrush	*Monticola gularis*	B	D	C	A	M	-	-	(+)	F	F	O	M
	Turdidae													
	Scaly Thrush	*Zoothera dauma*	C	C	C	B	M	-	(+)	+	F	F	O	M
	Siberian Thrush	*Zoothera sibirica*	C	C	C	B		-	-	+	F	F	O	M
	Eurasian Blackbird	*Turdus merula*	E	F	C	C		+	(+)	-	F	F	O	P
	Mistle Thrush	*Turdus viscivorus*	D	F	C	C		+	(+)	-	F	F	O	P
	Pale Thrush	*Turdus pallidus*	C	C	B	B		-	-	(+)	F	F	O	M
	Grey-backed Thrush	*Turdus hortulorum*	C	C	B	B		-	-	(+)	F	F	O	M
	Fieldfare	*Turdus pilaris*	C	B	B	C		+	+	(+)	M	M	O	P
	Redwing	*Turdus iliacus*	C	B	B	B		(+)	+	(+)	F	F	O	M
	Eyebrowed Thrush	*Turdus obscurus*	C	B	B	B	M	-	-	+	F	F	O	M
	Naumann's Thrush	*Turdus naumanni*	C	B	B	B		-	-	+	F	F	O	M
	Song Thrush	*Turdus philomelos*	C	B/D	C	C		+/+	+/+	-	F	F	O	P
	Ring Ouzel	*Turdus torquatus*	C	B/D	B	C	M	+/(+)	-	-	O	O	O	M
	Red-throated Thrush	*Turdus ruficollis*	B	F	B	A		-	+	-	F	F	O	M
	Brown-headed Thrush	*Turdus chrysolaus*	B	D	A	C		-	-	(+)	F	F	O	M
	Chinese Thrush	*Turdus mupinensis*	A	F			M	-	-	(+)	F	F	O	S
	Grey-sided Thrush	*Turdus feae*	A	D			M	-	-	(+)	F	F	O	M
	Japanese Thrush	*Turdus cardis*	A	D	A	E		-	-	(+)	F	F	O	M
	Blue Whistling Thrush	*Myophonus caeruleus*	B	D	E	E	M	-	(+)	(+)	F	F	O	M
	Grey-cheeked Thrush	*Catharus minimus*	B	A	A	B		-	-	(+)	F	F	O	M
	Sturnidae													
	Eurasian Starling	*Sturnus vulgaris*	C	F	B	C		+	+	(+)	O	O	O	P
	Daurian Starling	*Sturnus sturninus*	C	D	B	B		-	-	(+)	M	M	O	M
	White-cheeked Starling	*Sturnus cineraceus*	C	D	B	B		-	-	(+)	M	M	O	P
	Rose-coloured Starling	*Sturnus roseus*	B	D	D	A		(+)	+	-	O	O	O	M
	Chestnut-cheeked Starling	*Sturnus philippensis*	B	D	A	C		-	-	(+)	M	M	O	M

	TAXON		TOL	LAT	TEM	HUM	MON	WP	CP	EP	HF	HN	DT	MIG
Muscicapoidea 1a (iv) cont	Spotless Starling	*Sturnus unicolor*	A	D	E	A		(+)	-	-	M	M	0	S
	Tristram's Starling	*Onychognathus tristramii*	A	E	E	C		(+)	-	-	O	O	O	S
	Common Myna	*Acridotheres tristis*	B	E	E	C		-	(+)	-	M	M	O	S
	Cinclidae													
	White-throated Dipper	*Cinclus cinclus*	B	A/D	C	B	M	+/+	(+)/(+)	(+)/(+)	W	W	M	S
	Brown Dipper	*Cinclus pallasii*	B	F	C	D	M	-	-	+	W	W	M	S
	Sittidae													
	Eurasian Nuthatch	*Sitta europaea*	E	F	C	D		+	+	+	F	F	O	S
	Western Rock Nuthatch	*Sitta neumayer*	B	D	D	C		+	-	-	R	R	O	S
	Chinese Nuthatch	*Sitta villosa*	B	D	C	E	M	-	-	(+)	F	F	O	S
	Eastern Rock Nuthatch	*Sitta tephronota*	A	D	E	A	M	(+)	+	-	R	R	O	S
	Corsican Nuthatch	*Sitta whiteheadi*	A	D	E	A	M	(+)	-	-	F	F	O	S
	Krüper's Nuthatch	*Sitta krueperi*	A	D	E	A	M	(+)	-	-	F	F	O	S
	Algerian Nuthatch	*Sitta ledanti*	A	D	E	A		(+)	-	-	F	F	O	S
	Tichodromidae													
	Wallcreeper	*Tichodroma muraria*	C	D			M	(+)	(+)	+	R	R	M	P
	Certhiidae													
	Eurasian Treecreeper	*Certhia familiaris*	D	C	B	D		+	+	+	F	F	O	S
	Short-toed Treecreeper	*Certhia brachydactyla*	C	F	C	D		+	-	-	F	F	O	S
	Bar-tailed Treecreeper	*Certhia himalayana*	B	D	E	E	M	-	(+)	-	F	F	O	S
	Troglodytidae													
	Wren	*Troglodytes troglodytes*	E	F	C	C		+	(+)	+	F	R	O	S
	Regulidae													
	Goldcrest	*Regulus regulus*	E	C	B	C		+	+	(+)	F	F	O	S
	Firecrest	*Regulus ignicapillus*	D	F	C	C		+	-	-	F	F	I	P
	Madeira Firecrest	*Regulus madeirensis*	A	E	E	E		(+)	-	-	F	F	I	S
	Bombycilidae													
	Bohemian Waxwing	*Bombycilla garrulus*	C	B	B	B		+	+	+	F	F	O	M
	Japanese Waxwing	*Bombycilla japonica*	B	C	A	C		-	-	(+)	F	F	O	P
Passeroidea 1a (v)	**Prunellidae**													
	Dunnock	*Prunella modularis*	C	B/D	B	B	M	+/+	-	-	M	M	O	P
	Alpine Accentor	*Prunella collaris*	C	D			M	+	(+)	+	R	R	O	P
	Brown Accentor	*Prunella fulvescens*	B	D	C	A	M	-	+	(+)	O	O	O	S
	Japanese Accentor	*Prunella rubida*	B	D	B	E	M	-	-	(+)	M	M	O	S
	Siberian Accentor	*Prunella montanella*	B	A/C	B	A		-	+	+	M	M	O	M
	Black-throated Accentor	*Prunella atrogularis*	B	A/D	A	A	M	-	(+)/(+)	-	M	M	O	P
	Radde's Accentor	*Prunella ocularis*	A	D			M	(+)	(+)	-	O	O	O	P

	TAXON		TOL	LAT	TEM	HUM	MON	WP	CP	EP	HF	HN	DT	MIG
Passeroidea 1a (v) cont.	Altai Accentor	*Prunella himalayana*	A	D			M	-	(+)	-	O	O	O	P
	Robin Accentor	*Prunella rubeculoides*	A	D			M	-	+	(+)	M	M	O	S
	Koslov's Accentor	*Prunella koslowi*	A	D			M	-	-	(+)	O	O	O	S
	Passeridae													
	Eurasian Tree Sparrow	*Passer montanus*	E	F	B	D		+	+	+	M	M	O	S
	House Sparrow	*Passer domesticus*	D	F	C	C		+	+	(+)	M	M	O	P
	Saxaul Sparrow	*Passer ammodendri*	B	D	D	A		-	+	(+)	W	W	O	S
	Russet Sparrow	*Passer rutilans*	B	E	E	E	M	-	-	(+)	F	F	O	M
	Spanish Sparrow	*Passer hispaniolensis*	A	D	E	A	M	+	(+)	-	M	M	O	P
	Dead Sea Sparrow	*Passer moabiticus*	A	D	E	A		(+)	-	-	M	M	O	M
	Cape Verde Sparrow	*Passer iagoensis*	A	E	E	E		(+)	-	-	M	M	O	S
	Desert Sparrow	*Passer simplex*	A	E	E	A		(+)	(+)	-	O	O	O	S
	Rock Sparrow	*Petronia petronia*	B	D	D	A	M	(+)	(+)	(+)	O	O	O	S
	Chestnut-sided Petronia	*Petronia xanthocollis*	B	E	E	C		-	(+)	-	M	M	O	P
	Pale Rockfinch	*Carpospiza brachydactyla*	A	D	E	A	M	(+)	(+)	-	O	O	O	M
	White-winged Snowfinch	*Montifringilla nivalis*	B	D			M	(+)	+	(+)	R	R	O	S
	Père David's Snowfinch	*Montifringilla davidiana*	A	D			M	-	(+)	-	O	O	O	S
	Fringillidae													
	Common Chaffinch	*Fringilla coelebs*	C	F	C	C		+	(+)	-	F	F	O	P
	Brambling	*Fringilla montifringilla*	B	B	B	A		+	+	+	M	M	O	P
	Blue Chaffinch	*Fringilla teydea*	A	E	E	E		(+)	-	-	F	F	O	S
	Hawfinch	*Coccothraustes coccothraustes*	E	C	B	C		+	+	+	F	F	O	P
	Yellow-billed Grosbeak	*Eophona migratoria*	C	D	B	C		-	-	(+)	F	F	O	M
	Japanese Grosbeak	*Eophona personata*	C	C	B	C		-	-	(+)	F	F	O	M
	White-winged Grosbeak	*Mycerobas carnipes*	A	D			M	-	(+)	(+)	F	F	O	S
	Pine Grosbeak	*Pinicola enucleator*	C	B	A	C		+	+	+	F	F	O	P
	Eurasian Bullfinch	*Pyrrhula pyrrhula*	D	C	B	D		+	+	+	F	F	O	P
	Asian Rosy Finch	*Rhodopechys arctoa*	C	B	B	A	M	-	-	(+)	O	O	O	M
	Mongolian Finch	*Rhodopechys mongolica*	B	D	C	A	M	-	+	-	O	O	O	S
	Trumpeter Finch	*Rhodopechys githaginea*	B	E	E	C		(+)	(+)	-	O	O	O	S
	Crimson-winged Finch	*Rhodopechys sanguinea*	B	D			M	(+)	(+)	-	O	O	O	S
	Common Rosefinch	*Carpodacus erythrinus*	D	F	B	B	M	(+)	+	+	M	M	O	M
	Long-tailed Rosefinch	*Carpodacus sibiricus*	B	C	A	D	M	-	-	(+)	M	M	O	M
	Pallas's Rosefinch	*Carpodacus roseus*	B	B	B	A	M	-	-	(+)	F	F	O	P
	Sinai Rosefinch	*Carpodacus synoicus*	B	E	E	A	M	(+)	-	-	R	R	H	S
	Great Rosefinch	*Carpodacus rubicilla*	A	D			M	+	(+)	-	O	O	O	S
	Red-breasted Rosefinch	*Carpodacus puniceus*	A	D			M	-	(+)	-	R	R	O	S

	TAXON		TOL	LAT	TEM	HUM	MON	WP	CP	EP	HF	HN	DT	MIG
Passeroidea 1a (v) cont.	Red-mantled Rosefinch	*Carpodacus rhodochlamys*	A	D			M	-	(+)	-	F	F	O	P
	Beautiful Rosefinch	*Carpodacus pulcherrimus*	A	D			M	-	-	(+)	M	M	O	S
	Common Crossbill	*Loxia curvirostra*	E	F	B	D		(+)	+	+	F	F	O	S
	Two-barred Crossbill	*Loxia leucoptera*	C	B	B	B		(+)	+	+	F	F	O	S
	Parrot Crossbill	*Loxia pytyopsittacus*	B	B	A	C		+	+	-	F	F	O	S
	Scottish Crossbill	*Loxia scotica*	A	C	A	D		(+)	-	-	F	F	O	S
	Common Redpoll	*Acanthis flammea*	B	B	B	A		(+)	+	+	O	O	O	P
	Arctic Redpoll	*Acanthis hornemanni*	A	A	A	A		(+)	+	+	O	O	O	S
	Eurasian Siskin	*Spinus spinus*	C	B/D	B	B		+/+	+/+	(+)/(+)	F	F	O	P
	European Serin	*Serinus serinus*	C	F	C	D		+	-	-	M	M	O	P
	Red-fronted Serin	*Serinus pusillus*	A	D			M	(+)	(+)	-	M	M	O	S
	Syrian Serin	*Serinus syriacus*	A	D			M	(+)	-	-	M	M	H	S
	Atlantic Canary	*Serinus canaria*	A	E	E	E		(+)	-	-	F	F	O	S
	Citril Finch	*Chloroptila citrinella*	A	D			M	(+)	-	-	M	M	O	P
	Corsican Finch	*Chloroptila corsicana*	A	D			M	(+)	-	-	M	M	O	S
	European Goldfinch	*Carduelis carduelis*	D	F	C	C		+	+	-	O	O	O	P
	Eurasian Linnet	*Linaria cannabina*	D	F	C	C		+	(+)	-	O	O	O	P
	Twite	*Linaria flavirostris*	C	B/D	B	B	M	+/(+)	+/(+)	-	O	O	O	P
	Desert Finch	*Chloris obsoleta*	A	D	D	A		(+)	+	-	M	M	O	P
	European Greenfinch	*Chloris chloris*	C	F	C	C		+	(+)	-	F	F	O	P
	Oriental Greenfinch	*Chloris sinica*	C	F	B	D		-	-	+	F	F	O	P
	Emberizidae													
	Reed Bunting	*Emberiza schoeniclus*	E	F	B	C		+	+	+	W	W	O	P
	Black-faced Bunting	*Emberiza spodocephala*	C	F	B	B		-	-	+	F	F	O	P
	Yellowhammer	*Emberiza citrinella*	C	C	C	C		+	+	-	O	O	O	P
	Rock Bunting	*Emberiza cia*	C	D	D	C	M	(+)	(+)	-	O	R	O	P
	Meadow Bunting	*Emberiza cioides*	C	D	B	C	M	-	(+)	+	O	O	O	P
	Chestnut-eared Bunting	*Emberiza fucata*	C	D	B	C	M	-	-	(+)	M	M	O	P
	Yellow-throated Bunting	*Emberiza elegans*	C	D	C	E		-	-	(+)	F	F	O	M
	Yellow-breasted Bunting	*Emberiza aureola*	C	B	B	B		(+)	+	+	M	M	O	M
	Pine Bunting	*Emberiza leucocephalos*	C	B	B	B		-	(+)	+	F	F	O	M
	Ortolan Bunting	*Emberiza hortulana*	C	B/D	C	B	M	+/+	(+)/(+)	-	O	O	O	M
	Chestnut Bunting	*Emberiza rutila*	B	C	B	A		-	-	(+)	F	F	O	M
	Tristram's Bunting	*Emberiza tristrami*	B	C	B	B		-	-	(+)	F	F	O	M
	Red-headed Bunting	*Emberiza bruniceps*	B	D	A	A	M	(+)	+	-	O	O	O	M
	Black-headed Bunting	*Emberiza melanocephala*	B	D	D	A	M	(+)	+	-	M	M	O	M
	Cirl Bunting	*Emberiza cirlus*	B	D	D	C		+	-	-	M	M	O	S

	TAXON		TOL	LAT	TEM	HUM	MON	WP	CP	EP	HF	HN	DT	MIG
Passeroidea 1a (v) cont.	Grey-necked Bunting	*Emberiza buchanani*	B	D	D	A		-	(+)	-	O	O	O	M
	Jankowski's Bunting	*Emberiza jankowskii*	B	D	A	C		-	-	(+)	O	O	O	M
	Grey Bunting	*Emberiza variabilis*	B	D	A	C	M	-	-	(+)	F	F	O	M
	Rustic Bunting	*Emberiza rustica*	B	B	B	A		(+)	+	+	M	M	O	M
	Pallas's Reed Bunting	*Emberiza pallasi*	B	B	B	A		-	-	+	W	W	O	M
	Corn Bunting	*Emberiza calandra*	B	F	D	C		+	(+)	-	O	O	O	P
	Japanese Reed Bunting	*Emberiza yessoensis*	A	C	A	E		-	-	(+)	W	W	O	M
	Cretzschmar's Bunting	*Emberiza caesia*	A	D	E	A		(+)	-	-	R	R	O	M
	House Bunting	*Emberiza striolata*	A	E	E	A		(+)	(+)	-	R	R	O	S
	Yellow-browed Bunting	*Emberiza chrysophrys*	A	B	A	A		-	-	(+)	F	F	O	M
	Cinerous Bunting	*Emberiza cineracea*	A	D	E	A		(+)	-	-	O	O	O	M
	White-capped Bunting	*Emberiza stewarti*	A	D			M	-	(+)	-	O	O	O	P
	Godlewski's Bunting	*Emberiza godlewskii*	A	D			M	-	(+)	+	M	M	O	S
	Japanese Yellow Bunting	*Emberiza sulphurata*	A	D	E	E		-	-	(+)	M	M	O	M
	Little Bunting	*Emberiza pusilla*	A	A	A	A		(+)	+	+	M	M	O	M
	Lapland Longspur	*Calcarius lapponicus*	B	A	A	B		+	+	+	M	M	O	M
	Snow Bunting	*Plectrophenax nivalis*	A	A	A	A		+	+	+	O	O	O	M
	Motacillidae													
	White Wagtail	*Motacilla alba*	E	F	B	C		+	+	+	W	W	I	P
	Grey Wagtail	*Motacilla cinerea*	E	F	B	C		+	(+)	+	W	W	I	P
	Western Yellow Wagtail	*Motacilla flava*	D	F	C	B		(+)	(+)	-	W	W	I	M
	Western Citrine Wagtail	*Motacilla werae*	C	F	C	A		(+)	+	-	O	O	M	M
	Eastern Yellow Wagtail	*Motacilla tschutensis*	B	B	B	A		-	+	+	O	O	I	M
	Japanese Wagtail	*Motacilla grandis*	B	D	C	A		-	-	(+)	W	W	I	S
	Asian Yellow Wagtail	*Motacilla taivana*	A	D	B	E		-	-	+	O	O	I	M
	Eastern Citrine Wagtail	*Motacilla citreola*	A	D	C	A		-	(+)	(+)	O	O	M	M
	Forest Wagtail	*Dendronanthus indicus*	C	D	C	E		-	-	+	F	F	I	M
	Olive-backed Pipit	*Anthus hodgsoni*	D	F	B	B	M	(+)	+	+	F	F	O	M
	Richard's Pipit	*Anthus richardi*	D	C	B	C		-	-	+	O	O	O	M
	Tawny Pipit	*Anthus campestris*	C	D	C	C		(+)	+	-	O	O	O	M
	Meadow Pipit	*Anthus pratensis*	C	B	B	B		+	-	-	O	O	O	P
	Rock Pipit	*Anthus petrosus*	C	B	B	B		+	-	-	R	R	M	P
	Buff-bellied Pipit	*Anthus rubescens*	C	B	B	B		-	-	+	O	O	O	M
	Water Pipit	*Anthus spinoletta*	C	B/D	B	B	M	+/+	(+)/(+)	+/+	O	O	O	M
	Tree Pipit	*Anthus trivialis*	C	B/D	B	B		+/+	+/+	(+)/(+)	M	M	O	M
	Blyth's Pipit	*Anthus godlewskii*	A	D	C	A		-	-	(+)	O	O	I	M
	Berthelot's Pipit	*Anthus berthelotti*	A	E	E	E		(+)	-	-	O	O	I	S

	TAXON		TOL	LAT	TEM	HUM	MON	WP	CP	EP	HF	HN	DT	MIG
	Red-throated Pipit	*Anthus cervinus*	A	A	A	A		(+)	+	+	O	O	O	M
	Pechora Pipit	*Anthus gustavi*	A	A	A	A		(+)	+	+	O	O	M	M
Falcons 1c	**Falconidae**													
	Peregrine Falcon	*Falco peregrinus*	E	F	D	C		+	+	+	A	R	E	P
	Common Kestrel	*Falco tinnunculus*	E	F	C	D		+	+	+	M	R	M	P
	Eurasian Hobby	*Falco subbuteo*	E	F	B	D		(+)	+	+	A	M	M	M
	Lesser Kestrel	*Falco naumanni*	C	D	D	B		(+)	+	(+)	O	R	M	M
	Amur Falcon	*Falco amurensis*	C	D	B	B		-	-	+	M	M	M	M
	Merlin	*Falco columbarius*	C	B	B	B		+	+	+	O	O	M	M
	Eleonora's Falcon	*Falco eleonorae*	B	D	E	C		(+)	-	-	A	R	M	M
	Gyr Falcon	*Falco rusticolus*	B	A	A	B		+	+	+	O	O	E	M
	Saker	*Falco cherrug*	B	D	D	A		(+)	+	-	O	O	M	P
	Red-footed Falcon	*Falco vespertinus*	A	C	C	A		(+)	+	-	O	O	M	M
	Sooty Falcon	*Falco concolor*	A	E	E	A		(+)	-	-	A	R	M	S
	Lanner Falcon	*Falco biarmicus*	A	E	E	A		(+)	-	-	O	O	M	S
	Barbary Falcon	*Falco pelegrinoides*	A	E	E	A		(+)	(+)	-	A	R	E	S
	Upupidae													
	Eurasian Hoopoe	*Upupa epops*	D	F	C	D		+	+	+	M	M	M	M
Terrestrial non-passerines 1d	**Picidae**													
	Eurasian Wryneck	*Jynx torquilla*	C	F	D	D		+	+	+	F	F	I	M
	Grey-headed Woodpecker	*Picus canus*	E	C	B	D		(+)	+	+	F	F	O	S
	Eurasian Green Woodpecker	*Picus viridis*	C	F	C	B		+	(+)	-	F	F	O	S
	Japanese Woodpecker	*Picus awokera*	A	E	A	E		-	-	(+)	F	F	O	S
	Levaillant's Woodpecker	*Picus vaillantii*	A	E	E	A		(+)	-	-	F	F	O	S
	Black Woodpecker	*Dryocopus martius*	C	C	B	B		(+)	+	+	F	F	O	S
	Great Spotted Woodpecker	*Dendrocopos major*	D	F	C	B		+	+	+	F	F	O	S
	White-backed Woodpecker	*Dendrocopos leucotos*	D	C	B	D		(+)	+	+	F	F	O	S
	Middle Spotted Woodpecker	*Dendrocopos medius*	C	F	C	B		(+)	-	-	F	F	O	S
	Lesser Spotted Woodpecker	*Dendrocopos minor*	C	C	B	B		(+)	+	+	F	F	O	S
	Syrian Woodpecker	*Dendrocopos syriacus*	B	D	D	A		(+)	(+)	-	F	F	O	S
	Rufous-bellied Woodpecker	*Dendrocopos hyperythrus*	A	F	E	E		-	-	+	F	F	O	S
	Grey-capped Woodpecker	*Dendrocopus canicapillus*	A	F	E	E		-	-	+	F	F	O	S
	Pygmy Woodpecker	*Dendrocopos kizuki*	A	D	E	E		-	-	(+)	F	F	O	S
	Three-toed Woodpecker	*Picoides tridactylus*	C	B/D	B	B		(+)/(+)	+/+	+/+	F	F	O	S
	Meropidae													
	European Bee-eater	*Merops apiaster*	B	D	D	A		+	+	-	A	M	I	M
	Little Green Bee-eater	*Merops orientalis*	B	E	E	D		+	+	-	A	O	I	S

	TAXON		TOL	LAT	TEM	HUM	MON	WP	CP	EP	HF	HN	DT	MIG
Terrestrial non-passerines 1d cont.	Blue-cheeked Bee-eater	*Merops persicus*	A	D	E	A		(+)	(+)	-	A	M	I	M
	Coraciidae													
	European Roller	*Coracias garrulus*	C	F	C	B		(+)	(+)	-	O	O	M	M
	Dollarbird	*Eurystomus orientalis*	D	F	C	D		-	-	+	M	M	M	M
	Alcedinidae													
	Common Kingfisher	*Alcedo atthis*	E	F	C	D		(+)	(+)	(+)	W	W	F	P
	Ruddy Kingfisher	*Halcyon coromanda*	C	E	C	E		-	-	(+)	F	F	M	M
	White-throated Kingfisher	*Halcyon smyrnensis*	B	E	E	C		+	-	-	W	W	M	S
	Black-capped Kingfisher	*Halcyon pileata*	A	E	E	E		-	-	+	W	F	M	M
	Pied Kingfisher	*Ceryle rudis*	B	E	E	D		(+)	-	-	W	W	M	S
Owls 1e	**Tytonidae**													
	Barn Owl	*Tyto alba*	D	F	C	C		+	-	-	O	R	M	S
	Strigidae													
	European Scops Owl	*Otus scops*	C	D	D	B		(+)	+	-	F	F	M	M
	Oriental Scops Owl	*Otus sunia*	C	F	C	D		-	-	+	F	F	M	M
	Collared Scops Owl	*Otus lettia*	A	F			M	-	-	+	F	F	M	S
	Pallid Scops Owl	*Otus brucei*	A	D	E	A		(+)	+	-	M	M	M	M
	Japanese Scops Owl	*Otus semitorques*	A	D	A	E		-	-	(+)	R	R	M	S
	Tawny Owl	*Strix aluco*	E	F	C	D		+	(+)	(+)	F	F	M	S
	Great Grey Owl	*Strix nebulosa*	C	B	B	A		(+)	+	+	F	F	M	S
	Ural Owl	*Strix uralensis*	C	B/D	B	D		(+)/(+)	+/-	+/-	F	F	M	S
	Hume's Owl	*Strix butleri*	A	E	E	A		(+)	-	-	R	R	M	S
	Eagle Owl	*Bubo bubo*	E	F	B	D		(+)	+	+	M	R	E	S
	Snowy Owl	*Bubo scandiacus*	B	A	A	B		(+)	+	-	O	O	E	M
	Blakiston's Fish Owl	*Bubo blakistoni*	B	C	A	C		-	-	(+)	W	F	M	S
	Brown Fish Owl	*Bubo zeylonensis*	B	E	E	C		(+)	-	-	W	M	M	S
	Long-eared Owl	*Asio otus*	E	F	B	C		(+)	+	+	F	F	M	P
	Short-eared Owl	*Asio flammeus*	E	B	C	C		(+)	+	+	O	O	M	M
	Marsh Owl	*Asio capensis*	B	E	E	C		(+)	-	-	W	W	M	S
	Brown Hawk Owl	*Ninox scutulata*	C	F	C	E		-	-	+	F	F	I	M
	Tengmalm's Owl	*Aegolius funereus*	D	B/D	B	B		(+)/(+)	+/+	+/+	F	F	M	S
	Little Owl	*Athene noctua*	D	D	C	C		+	+	+	M	R	M	S
	Lilith Owlet	*Athene lilith*	A	D	E	A		(+)	-	-	M	R	M	S
	Eurasian Pygmy Owl	*Glaucidium passerinum*	C	B	B	C		(+)	+	+	F	F	M	S
	Northern Hawk Owl	*Surnia ulula*	C	B/D	B	B		+/+	+/+	+/+	F	F	M	S
	Pandionidae													
	Eurasian Osprey	*Pandion haliaetus*	E	F	B	D		(+)	+	+	W	W	F	M

	TAXON		TOL	LAT	TEM	HUM	MON	WP	CP	EP	HF	HN	DT	MIG
Acciptriformes 1 (f)	**Acciptridae**													
	Black-winged Kite	*Elanus caeruleus*	B	E	E	C		(+)	-	-	M	M	M	S
	European Honey-buzzard	*Pernis apivorus*	C	C	C	B		+	+	-	F	F	M	M
	Oriental Honey-buzzard	*Pernis ptilorhynchus*	C	C	C	B		-	-	+	F	F	M	M
	Bearded Vulture	*Gypaetus barbatus*	C	D	D	B	M	(+)	+	(+)	O	R	N	S
	Egyptian Vulture	*Neophron percnopterus*	D	D	D	B	M	(+)	+	-	O	R	N	M
	Short-toed Snake-eagle	*Circaetus gallicus*	C	F	D	C		(+)	+	-	M	M	M	M
	Eurasian Griffon Vulture	*Gyps fulvus*	C	D	D	B	M	(+)	+	-	O	R	N	P
	Eurasian Black Vulture	*Aegypius monachus*	C	D	D	B		(+)	+	-	O	M	N	S
	Lappet-faced Vulture	*Torgos tracheliotus*	B	E	E	B		(+)	-	-	O	O	N	S
	Mountain Hawk Eagle	*Nisaetus nipalensis*	C	D	C	E	M	-	-	(+)	F	F	E	S
	Booted Eagle	*Hieraaetus pennatus*	C	D	D	B		(+)	+	(+)	M	M	M	M
	Golden Eagle	*Aquila chrysaetos*	E	F	C	B		+	+	+	M	R	E	P
	Bonelli's Eagle	*Aquila fasciata*	D	D	D	C		(+)	(+)	(+)	M	R	E	S
	Verreaux's Eagle	*Aquila verreauxi*	C	E	E	C		+	-	-	R	R	E	S
	Imperial Eagle	*Aquila heliaca*	B	D	D	A		(+)	+	(+)	M	M	M	P
	Steppe Eagle	*Aquila nipalensis*	A	D	C	A		(+)	+	(+)	O	O	M	M
	Spanish Imperial Eagle	*Aquila adalberti*	A	D	E	C		(+)	-	-	M	M	M	S
	Tawny Eagle	*Aquila rapax*	A	E	E	A		+	-	-	O	O	M	S
	Lesser Spotted Eagle	*Lophaetus pomarinus*	B	C	D	A		+	-	-	M	M	M	M
	Greater Spotted Eagle	*Lophaetus clanga*	A	C	C	A		(+)	+	+	M	M	M	M
	Dark Chanting Goshawk	*Melierax metabates*	B	E	E	B		(+)	-	-	M	M	M	S
	Eurasian Sparrowhawk	*Accipiter nisus*	E	F	C	D		+	+	+	F	F	E	P
	Northern Goshawk	*Accipiter gentilis*	E	C	B	B		+	+	+	F	F	E	S
	Shikra	*Accipiter badius*	C	D	D	B		-	+	-	M	M	M	M
	Japanese Sparrowhawk	*Accipiter gularis*	C	D	B	D		-	-	(+)	F	F	E	P
	Chinese Sparrowhawk	*Accipiter soloensis*	B	E	D	E		-	-	(+)	F	F	E	M
	Levant Sparrowhawk	*Accipiter brevipes*	A	C	E	A		+	+	-	M	M	M	M
	Hen Harrier	*Circus cyaneus*	E	B	B	B		(+)	+	+	O	O	M	M
	Montagu's Harrier	*Circus pygargus*	C	F	C	B		(+)	+	-	O	O	M	M
	Western Marsh Harrier	*Circus aeruginosus*	C	C	C	C		(+)	+	-	W	W	M	P
	Eastern Marsh Harrier	*Circus spilonotus*	C	C	B	B		-	-	+	W	W	M	M
	Pied Harrier	*Circus melanoleucos*	C	C	B	B		-	-	(+)	W	W	M	M
	Pallid Harrier	*Circus macrourus*	A	C	C	A		(+)	+	-	O	O	M	M
	Black Kite	*Milvus migrans*	E	F	C	C		+	+	+	M	M	M	M
	Red Kite	*Milvus milvus*	C	C	C	D		+	-	-	M	M	M	P
	Yellow-billed Kite	*Milvus aegyptius*	B	E	E	C		+	-	-	M	M	M	S

	TAXON		TOL	LAT	TEM	HUM	MON	WP	CP	EP	HF	HN	DT	MIG
Accipitriformes 1 (f) cont.	White-tailed Eagle	*Haliaeetus albicilla*	D	F	B	B		(+)	+	+	W	W	M	M
	Pallas's Fish Eagle	*Haliaeetus leucoryphus*	B	D	D	B		-	+	-	W	W	M	M
	Steller's Sea Eagle	*Haliaeetus pelagicus*	B	B	A	C		-	-	(+)	W	R	M	P
	Grey-faced Buzzard	*Butastur indicus*	B	C	A	C		-	-	(+)	M	M	M	M
	Common Buzzard	*Buteo buteo*	E	F	B	C		+	+	+	M	F	M	P
	Rough-legged Buzzard	*Buteo lagopus*	C	A	B	B		+	+	+	O	O	E	M
	Long-legged Buzzard	*Buteo rufinus*	B	D	D	A		+	+	-	O	O	M	P
Charadriiformes 2	**Laridae**													
	Herring Gull	*Larus argentatus*	C	B	B	D		+	+	+	W	R	M	P
	Common Gull	*Larus canus*	C	B	B	B		+	+	+	W	W	O	P
	Lesser Black-backed Gull	*Larus fuscus*	C	B	B	B		+	+	-	W	W	O	M
	Greater Black-backed Gull	*Larus marinus*	C	B	B	B		+	-	-	W	R	O	P
	Black-tailed Gull	*Larus crassirostris*	C	D	B	D		-	-	(+)	W	R	M	P
	Yellow-legged Gull	*Larus michahellis*	B	D	D	B		(+)	-	-	W	R	O	P
	Glaucous-winged Gull	*Larus glaucescens*	B	B	B	A		-	-	(+)	W	R	M	M
	Glaucous Gull	*Larus hyperboreus*	B	A	A	B		+	+	+	W	R	O	P
	Iceland Gull	*Larus glaucoides*	B	A	A	D		+	-	-	W	R	M	M
	Caspian Gull	*Larus cachinnans*	B	D	D	A		(+)	+	(+)	W	R	O	M
	Slaty-backed Gull	*Larus schistisagus*	B	F	A	C		-	-	+	W	R	O	P
	Armenian Gull	*Larus armenicus*	A	D	E	A		(+)	-	-	W	W	M	M
	Mediterranean Gull	*Ichthyaetus melanocephalus*	C	C/D	D	B		(+)	-	-	W	W	M	M
	Pallas's Gull	*Ichthyaetus ichthyaetus*	B	D	D	A		(+)	+	-	W	W	O	P
	Relict Gull	*Ichthyaetus relictus*	A	C	C	A	M	-	-	(+)	W	W	M	M
	Audouin's Gull	*Ichthyaetus audouinii*	A	D	E	A		(+)	-	-	Ma	R	M	M
	Black-headed Gull	*Chroicocephalus ridibundus*	E	F	B	C		+	+	+	W	W	M	P
	Grey-headed Gull	*Chroicocephalus cirrocephalus*	B	E	E	B		(+)	-	-	W	W	M	S
	Slender-billed Gull	*Chroicocephalus genei*	A	D	D	A		(+)	(+)	-	W	W	M	M
	Saunders's Gull	*Saundersilarus saundersi*	A	D	A	E		-	-	+	W	W	M	M
	Little Gull	*Hydrocoloeus minutus*	B	B	B	A		(+)	+	(+)	W	W	M	M
	Ross's Gull	*Hydrocoloeus rosea*	B	A	A	B		-	(+)	(+)	W	W	M	M
	Ivory Gull	*Pagophila eburnea*	B	A	A	B		-	+	-	Ma	R	M	M
	Sabine's Gull	*Xema sabini*	B	A	A	B		-	-	+	W	W	M	M
	Black-legged Kittiwake	*Rissa tridactyla*	C	B	B	C		+	+	+	Ma	R	M	P
	Red-legged Kittiwake	*Rissa brevirostris*	A	A	A	A		-	-	(+)	Ma	R	M	P
	Sternidae													
	Aleutian Tern	*Onychoprion aleutica*	A	A	A	A		-	-	(+)	W	W	M	P
	Little Tern	*Sternula albifrons*	D	F	D	D		(+)	+	(+)	W	W	M	M

	TAXON		TOL	LAT	TEM	HUM	MON	WP	CP	EP	HF	HN	DT	MIG
Charadriiformes 2 cont.	Gull-billed Tern	*Gelochelidon nilotica*	C	C/D	C	C		(+)/(+)	(+)/(+)	(+)/(+)	W	W	M	M
	Caspian Tern	*Hydroprogne caspia*	D	B/D	C	B		+/(+)	-/(+)	-/(+)	W	W	M	M
	White-winged Black Tern	*Chlidonias leucopterus*	C	C	C	B		(+)	(+)	(+)	W	W	M	M
	Black Tern	*Chlidonias niger*	C	C	D	B		(+)	+	-	W	W	M	M
	Whiskered Tern	*Chlidonias hybridus*	C	D	D	C		(+)	(+)	(+)	W	W	M	M
	Sandwich Tern	*Thalasseus sandvicensis*	E	C/D	C	C		+/(+)	+/(+)	-	W	W	M	P
	Lesser Crested Tern	*Thalasseus bengalensis*	B	E	E	D		+	-	-	Ma	W	M	M
	Royal Tern	*Thalasseus maximus*	B	E	E	C		(+)	-	-	W	W	M	M
	Common Tern	*Sterna hirundo*	E	F	B	D		(+)	+	+	W	W	M	M
	Roseate Tern	*Sterna dougallii*	C	C	C	D		+	-	-	Ma	R	M	M
	Arctic Tern	*Sterna paradisea*	B	A	A	B		+	+	+	W	O	M	M
	Stercoraridae													
	Arctic Skua	*Stercorarius parasiticus*	B	A	A	B		+	+	+	O	O	O	M
	Long-tailed Skua	*Stercorarius longicaudus*	B	A	B	A		(+)	+	+	O	O	M	M
	Pomarine Skua	*Stercorarius pomarinus*	A	A	A	A		-	+	+	O	O	M	M
	Great Skua	*Catharacta skua*	A	B	A	E		(+)	-	-	Ma	R	M	P
	Dromadidae													
	Crab Plover	*Dromas ardeola*	A	E	E	A		(+)	-	-	W	W	M	P
	Alcidae													
	Razorbill	*Alca torda*	C	B	B	C		+	-	-	Ma	R	M	P
	Great Auk	*Pinguinus impennis*	C	B	B	D		+	-	-	Ma	R	M	P
	Little Auk	*Alle alle*	B	A	A	B		-	(+)	-	Ma	R	M	M
	Common Guillemot	*Uria aalge*	C	B	B	D		+	-	+	Ma	R	M	P
	Brünnich's Guillemot	*Uria lomvia*	B	A	A	B		+	-	+	Ma	R	M	M
	Ancient Murrelet	*Synthliboramphus antiquus*	B	D	B	B		-	-	(+)	Ma	R	M	P
	Japanese Murrelet	*Synthliboramphus wumizusume*	A	D	A	E		-	-	(+)	Ma	R	M	P
	Pigeon Guillemot	*Cepphus columba*	C	B	B	A		-	-	(+)	Ma	R	M	P
	Black Guillemot	*Cepphus grylle*	C	A	B	B		+	(+)	+	Ma	R	M	P
	Spectacled Guillemot	*Cepphus carbo*	B	C	A	C		-	-	(+)	Ma	R	M	P
	Marbled Murrelet	*Brachyramphus marmoratus*	C	C	B	B		-	-	(+)	Ma	F	F	M
	Long-billed Murrelet	*Brachyramphus perdix*	B	B	B	B		-	-	(+)	Ma	F	F	P
	Kittlitz's Murrelet	*Brachyramphus brevirostris*	A	B	A	A		-	-	(+)	Ma	R	M	P
	Crested Auklet	*Aethia cristalleta*	A	B	A	A		-	-	(+)	Ma	R	M	P
	Whiskered Auklet	*Aethia pygmaea*	A	B	A	A		-	-	(+)	Ma	R	M	P
	Least Auklet	*Aethia pusilla*	A	B	A	A		-	-	(+)	Ma	R	M	P
	Parakeet Auklet	*Cyclorrhynchus psittacula*	A	B	A	A		-	-	(+)	Ma	R	M	P
	Rhinoceros Auklet	*Cerorhinca monocerata*	A	D	A	A		-	-	(+)	Ma	R	M	P

	TAXON		TOL	LAT	TEM	HUM	MON	WP	CP	EP	HF	HN	DT	MIG
Charadriiformes 2 cont.	Atlantic Puffin	*Fratercula arctica*	C	B	B	D		+	-	-	Ma	R	M	P
	Tufted Puffin	*Fratercula cirrhata*	C	B	B	A		-	-	(+)	Ma	R	M	M
	Horned Puffin	*Fratercula corniculata*	A	B	A	A		-	-	(+)	Ma	R	M	P
	Glareolidae													
	Cream-coloured Courser	*Cursorius cursor*	A	E	E	A		(+)	(+)	-	O	O	I	P
	Collared Pratincole	*Glareola pratincola*	C	D	D	B		(+)	(+)	-	A	W	I	M
	Oriental Pratincole	*Glareola maldivarum*	C	D	D	C		-	-	+	A	W	I	M
	Black-winged Pratincole	*Glareola nordmanni*	A	C	C	A		(+)	+	-	A	W	I	M
	Turnicidae													
	Small Buttonquail	*Turnix sylvatica*	B	D	E	C		(+)	-	-	O	O	O	S
	Yellow-legged Buttonquail	*Turnix tanki*	C	D	C	E		-	-	+	O	O	O	M
	Rostratulidae													
	Painted Snipe	*Rostratula benghalensis*	A	D	E	E		-	-	+	W	W	M	M
	Scolopacidae													
	Whimbrel	*Numenius phaeopus*	C	B	B	B		+	(+)	(+)	W	W	O	M
	Eurasian Curlew	*Numenius arquata*	C	B	B	B		(+)	+	(+)	O	O	M	P
	Far-eastern Curlew	*Numenius madagascariensis*	B	B	A	C		-	-	+	W	W	M	M
	Slender-billed Curlew	*Numenius tenuirostris*	A	C	C	A		-	(+)	-	W	W	M	M
	Little Curlew	*Numenius minutus*	A	A	A	A		-	-	(+)	M	M	M	M
	Black-tailed Godwit	*Limosa limosa*	C	C	B	C		(+)	+	(+)	W	W	O	M
	Bar-tailed Godwit	*Limosa lapponica*	A	A	A	A		(+)	+	+	O	O	O	M
	Asian Dowitcher	*Limnodromus semipalmatus*	A	C	C	A		-	(+)	(+)	W	W	M	M
	Long-billed Dowitcher	*Limnodromus scolopaceus*	A	A	A	A		-	-	+	W	W	M	M
	Jack Snipe	*Lymnocryptes minimus*	B	B	B	A		(+)	+	+	W	W	O	M
	Eurasian Woodcock	*Scolopax rusticola*	E	C	B	D		(+)	+	+	F	F	M	M
	Common Snipe	*Gallinago gallinago*	E	B	B	B		(+)	+	+	W	W	M	P
	Solitary Snipe	*Gallinago solitaria*	C	F	A	C	M	-	(+)	(+)	W	W	O	S
	Latham's Snipe	*Gallinago hardwickii*	B	D	B	E		-	-	(+)	M	M	M	M
	Great Snipe	*Gallinago media*	B	B	B	A		+	+	-	W	W	O	M
	Pin-tailed Snipe	*Gallinago stenura*	B	B	B	A		(+)	+	+	W	W	M	M
	Swinhoe's Snipe	*Gallinago megala*	A	C	C	A		-	(+)	(+)	W	W	M	M
	Red-necked Phalarope	*Phalaropus lobatus*	B	A	A	C		+	+	+	W	O	M	M
	Grey Phalarope	*Phalaropus fulicarius*	B	A	A	D		(+)	(+)	+	W	O	M	M
	Terek Sandpiper	*Xenus cinereus*	B	B	B	A		(+)	+	+	W	W	O	M
	Common Sandpiper	*Actitis hypoleucos*	D	F	B	C		(+)	+	+	W	W	M	M
	Common Redshank	*Tringa totanus*	D	F	B	C		(+)	+	(+)	W	W	M	M
	Common Greenshank	*Tringa nebularia*	C	B	B	B		+	+	+	W	W	M	M

	TAXON		TOL	LAT	TEM	HUM	MON	WP	CP	EP	HF	HN	DT	MIG
Charadriiformes 2 cont.	Green Sandpiper	*Tringa ochropus*	B	B	B	A		(+)	+	+	W	W	M	M
	Wood Sandpiper	*Tringa glareola*	B	B	B	A		(+)	+	+	W	W	M	M
	Marsh Sandpiper	*Tringa stagnatilis*	A	C	C	A		+	+	-	W	W	M	M
	Spotted Redshank	*Tringa erythropus*	A	A	A	A		(+)	+	+	M	M	M	M
	Grey-tailed Tattler	*Tringa brevipes*	A	B	A	A		-	(+)	(+)	W	W	M	M
	Ruddy Turnstone	*Arenaria interpres*	A	A	A	A		+	+	+	O	O	O	M
	Buff-breasted Sandpiper	*Tryngites subruficollis*	A	A	A	A		-	-	(+)	O	O	M	M
	Ruff	*Philomachus pugnax*	B	B	B	A		(+)	+	(+)	W	W	O	M
	Broad-billed Sandpiper	*Limicola falcinellus*	A	A	A	A	M	(+)	(+)	(+)	W	W	O	M
	Spoonbill Sandpiper	*Eurynorhynchus pygmeus*	A	A	A	A		-	-	(+)	O	O	M	M
	Long-toed Stint	*Calidris subminuta*	B	B	B	A		-	+	+	W	W	M	M
	Dunlin	*Calidris alpina*	B	A	A	C		+	+	+	O	O	M	M
	Purple Sandpiper	*Calidris maritima*	B	A	A	B		+	+	-	O	O	O	M
	Temminck's Stint	*Calidris temmincki*	A	A	A	A		+	+	+	O	O	O	M
	Little Stint	*Calidris minuta*	A	A	A	A		-	+	-	O	O	O	M
	Pectoral Sandpiper	*Calidris melanotos*	A	A	A	A		-	+	+	O	O	M	M
	Curlew Sandpiper	*Calidris ferruginea*	A	A	A	A		-	+	+	O	O	O	M
	Red Knot	*Caldris canutus*	A	A	A	A		-	(+)	(+)	O	O	O	M
	Sanderling	*Caldris alba*	A	A	A	A		-	(+)	(+)	O	O	O	M
	Red-necked Stint	*Calidris ruficollis*	A	A	A	A		-	-	+	O	O	M	M
	Sharp-tailed Sandpiper	*Calidris acuminate*	A	A	A	A		-	-	+	O	O	M	M
	Great Knot	*Calidris tenuirostris*	A	A	A	A		-	-	(+)	O	O	M	M
	Western Sandpiper	*Calidris mauri*	A	A	A	A		-	-	(+)	O	O	M	M
	Baird's Sandpiper	*Calidris bairdii*	A	A	A	A		-	-	(+)	O	O	M	M
	Rock Sandpiper	*Calidris ptilocnemis*	A	A	A	A		-	-	(+)	O	O	M	M
	Burhinidae													
	Eurasian Stone Curlew	*Burhinus oedicnemus*	C	D	D	C		(+)	(+)	-	O	O	M	P
	Senegal Thick-knee	*Burhinus senegalensis*	B	E	E	C		(+)	-	-	O	O	M	S
	Recurvirostridae													
	Black-winged Stilt	*Himantopus himantopus*	C	D	D	C		(+)	(+)	-	W	W	M	M
	Avocet	*Recurvirostra avosetta*	C	D	D	B		(+)	(+)	(+)	W	W	M	M
	Haematopodidae													
	Eurasian Oystercatcher	*Haematopus ostralegus*	D	F	B	C		(+)	(+)	(+)	W	W	M	P
	Canary Islands Oystercatcher	*Haematopus meadewaldowi*	A	E	E	E		(+)	-	-	W	W	M	S
	Pluvialidae													
	Eurasian Golden Plover	*Pluvialis apricaria*	C	B	B	B		+	+	-	O	O	O	M
	Pacific Golden Plover	*Pluvialis fulva*	B	A	A	B		-	+	+	O	O	M	M

	TAXON		TOL	LAT	TEM	HUM	MON	WP	CP	EP	HF	HN	DT	MIG
Charadriiformes 2 cont.	Grey Plover	*Pluvialis squatarola*	A	A	A	A		-	+	+	O	O	M	M
	Charadriidae													
	Northern Lapwing	*Vanellus vanellus*	C	F	B	C		(+)	+	+	W	W	O	P
	Grey-headed Lapwing	*Vanellus cinereus*	C	D	B	C		-	-	(+)	W	W	M	P
	White-tailed Lapwing	*Vanellus leucurus*	B	D	D	A		(+)	+	-	W	W	M	P
	Spur-winged Lapwing	*Vanellus spinosus*	B	E	E	C		(+)	(+)	-	W	W	O	M
	Red-wattled Lapwing	*Vanellus indicus*	B	E	E	C		-	+	-	O	O	M	S
	Sociable Lapwing	*Vanellus gregarius*	A	C	C	A		(+)	+	-	O	O	O	M
	Little Ringed Plover	*Charadrius dubius*	E	F	B	D		(+)	+	+	W	W	M	M
	Kentish Plover	*Charadrius alexandrinus*	D	D	D	C		(+)	(+)	(+)	W	W	M	P
	Lesser Sand Plover	*Charadrius mongolus*	C	F	B	A	M	-	-	(+)	O	O	M	M
	Long-billed Plover	*Charadrius placidus*	C	D	B	D		-	-	+	W	W	I	P
	Common Ringed Plover	*Charadrius hiaticula*	C	A	B	B		+	+	+	W	W	M	M
	Oriental Plover	*Charadrius veredus*	B	C	B	A		-	-	(+)	W	W	I	M
	Greater Sand Plover	*Charadrius leschenaultii*	B	D	D	A		(+)	+	-	O	O	M	M
	Caspian Plover	*Charadrius asiaticus*	B	D	D	A		(+)	+	-	O	O	M	M
	Kittlitz's Plover	*Charadrius pecuarius*	B	E	E	B		(+)	-	-	W	W	M	S
	Gaviidae													
	Black-throated Diver	*Gavia arctica*	C	B	B	B		+	+	+	W	W	M	M
	Red-throated Diver	*Gavia stellata*	B	A	A	C		+	+	+	W	W	M	M
	Great Northern Diver	*Gavia immer*	B	A	A	D		+	-	-	W	W	M	M
	Pacific Diver	*Gavia pacifica*	B	A	A	B		-	-	(+)	W	W	M	M
	White-billed Diver	*Gavia adamsii*	A	A	A	A		-	+	+	W	W	M	M
	Procellaridae													
	Northern Fulmar	*Fulmarus glacialis*	B	B	A	C		+	+	-	Ma	R	M	M
	Cory's Shearwater	*Calonectris diomedea*	B	D	E	C		(+)	-	-	Ma	R	M	M
	Cape Verde Shearwater	*Calonectris edwardsii*	A	E	E	E		(+)	-	-	Ma	R	M	M
	Streaked Shearwater	*Calonectris leucomelas*	A	E	E	E		-	-	(+)	Ma	R	M	M
	Manx Shearwater	*Puffinus puffinus*	C	C	B	E		(+)	-	-	Ma	R	M	M
	Little Shearwater	*Puffinus assimilis*	C	D	C	C		+	-	-	Ma	R	M	M
	Balearic Shearwater	*Puffinus mauretanicus*	A	D	E	A		(+)	-	-	Ma	R	M	M
	Yelkouan Shearwater	*Puffinus yelkouan*	A	D	E	A		(+)	-	-	Ma	R	M	M
	Zino's Petrel	*Pterodroma madeira*	A	E	E	E		(+)	-	-	Ma	R	M	M
	Fea's Petrel	*Pterodroma feae*	A	E	E	E		(+)	-	-	Ma	R	M	M
	Bulwer's Petrel	*Bulweria bulwerii*	A	E	E	E		(+)	-	-	Ma	R	M	M
	Hydrobatidae													
	Leach's Storm-petrel	*Oceanodroma leucorhoa*	B	B	B	B		(+)	-	(+)	Ma	R	M	M

	TAXON		TOL	LAT	TEM	HUM	MON	WP	CP	EP	HF	HN	DT	MIG
Charadriiformes 2 cont.	Swinhoe's Storm-petrel	*Oceanodroma monorhis*	C	E	C	E		-	-	+	Ma	R	M	M
	Madeiran Storm-petrel	*Oceanodroma castro*	A	E	E	E		(+)	-	(+)	Ma	R	M	M
	European Storm-petrel	*Hydrobates pelagicus*	C	F	C	D		+	-	-	Ma	R	M	M
	White-faced Storm-petrel	*Pelagodroma marina*	C	D	D	D		+	-	-	Ma	R	M	M
	Ciconiidae													
	Black Stork	*Ciconia nigra*	D	C	C	B		(+)	+	+	W	W	M	M
	White Stork	*Ciconia ciconia*	C	F	C	B		(+)	(+)	-	W	M	M	M
	Oriental Stork	*Ciconia boyciana*	A	D	A	E		-	-	(+)	W	W	M	M
	Fregatidae													
	Magnificant Frigatebird	*Fregata magnificens*	A	E	E	E		(+)	-	-	Ma	M	M	S
	Phalacrocoracidae													
	European Shag	*Phalacrocorax aristotelis*	D	F	B	B		+	-	-	Ma	R	F	S
	Pelagic Cormorant	*Phalacrocorax pelagicus*	D	B	B	E		-	-	+	Ma	R	F	S
	Great Cormorant	*Phalacrocorax carbo*	D	B/D	C	D		+/(+)	-/(+)	-/(+)	W	W	F	P
	Japanese Cormorant	*Phalacrocorax capillatus*	B	D	B	E		-	-	(+)	Ma	R	M	P
	Red-faced Cormorant	*Phalacrocorax urile*	B	D	A	C		-	-	(+)	Ma	R	M	P
	Long-tailed Cormorant	*Phalacrocorax africanus*	B	E	E	B		(+)	-	-	W	W	M	S
	Pygmy Cormorant	*Phalacrocorax pygmaeus*	A	D	E	A		(+)	(+)	-	W	W	M	P
	Sulidae													
	Northern Gannet	*Morus bassanus*	A	B	A	E		+	-	-	Ma	R	M	M
	Threskiornithidae													
	Eurasian Spoonbill	*Platalea leucorodia*	C	D	D	C		(+)	(+)	+	W	W	M	M
	Black-faced Spoonbill	*Platalea minor*	B	B	A	C		-	-	(+)	W	W	M	M
	Glossy Ibis	*Plegadis falcinellus*	C	D	D	C		(+)	(+)	-	W	W	M	M
	Northern Bald Ibis	*Geronticus eremita*	B	D	D	C		(+)	-	-	O	R	O	M
	Black-headed Ibis	*Threskiornis melanocephalus*	A	C	E	E		-	-	+	W	W	M	M
	Crested Ibis	*Nipponia nippon*	A	D	C	E		-	-	(+)	W	W	M	S
	Ardeidae													
	Great Bittern	*Botaurus stellaris*	E	C	C	D		(+)	+	+	W	W	M	P
	Little Bittern	*Ixobrychus minutus*	D	D	D	D		(+)	+	-	W	W	M	M
	Schrenck's Bittern	*Ixobrychus eurythmus*	C	D	C	E		-	-	+	W	W	M	M
	Yellow Bittern	*Ixobrychus sinensis*	B	E	D	E		-	-	+	W	W	M	M
	Cinnamon Bittern	*Ixobrychus cinnamomeus*	A	E	E	E		-	-	+	W	W	M	M
	Japanese Night Heron	*Gorsachius goisagi*	A	E	E	E		-	-	(+)	W	W	M	M
	Black-crowned Night Heron	*Nycticorax nycticorax*	E	D	C	D		(+)	(+)	(+)	W	W	M	M
	Striated Heron	*Butorides striatus*	C	D	D	D		-	-	+	W	W	M	M
	Squacco Heron	*Ardeola ralloides*	C	D	D	B		(+)	(+)	-	W	W	M	M

	TAXON		TOL	LAT	TEM	HUM	MON	WP	CP	EP	HF	HN	DT	MIG
Charadriiformes 2 cont.	Chinese Pond Heron	*Ardeola bacchus*	A	D	E	E		-	-	+	W	W	M	M
	Cattle Egret	*Bubulcus ibis*	D	D	D	D		(+)	(+)	(+)	O	O	M	P
	Great White Egret	*Casmerodius albus*	E	D	D	C		(+)	(+)	+	W	W	M	M
	Grey Heron	*Ardea cinerea*	D	F	D	D		(+)	+	+	W	W	M	P
	Purple Heron	*Ardea purpurea*	D	D	D	D		(+)	(+)	(+)	W	W	M	M
	Little Egret	*Egretta garzetta*	C	D	D	C		(+)	(+)	-	W	W	M	M
	Pacific Reef Egret	*Egretta sacra*	C	E	D	D		-	-	(+)	W	W	M	S
	Western Reef Egret	*Egretta gularis*	B	E	E	D		+	-	-	W	W	M	S
	Intermediate Egret	*Egretta intermedia*	B	E	E	C		-	-	+	W	W	M	M
	Chinese Egret	*Egretta eulophotes*	A	D	E	E		-	-	(+)	W	W	M	M
	Pelecanidae													
	Great White Pelican	*Pelecanus onocrotalus*	C	D	D	B		(+)	(+)	-	W	W	M	M
	Dalmatian Pelican	*Pelecanus crispus*	B	D	D	A		(+)	(+)	-	W	W	F	P
Gruiformes and Cuculiformes 4	**Otididae**													
	Little Bustard	*Tetrax tetrax*	C	D	D	B		(+)	(+)	-	O	O	O	P
	Great Bustard	*Otis tarda*	B	D	D	A		(+)	(+)	(+)	O	O	O	P
	MacQueen's Bustard	*Chlamydotis macqueenii*	B	D	D	A		-	(+)	-	O	O	O	M
	Houbara Bustard	*Chlamydotis undulata*	A	E	E	A		(+)	-	-	O	O	O	S
	Gruidae													
	Siberian Crane	*Leucogeranus leucogeranus*	A	A	A	A		-	(+)	(+)	W	W	O	M
	Sandhill Crane	*Grus canadensis*	C	A	C	B		-	-	+	W	W	O	M
	Common Crane	*Grus grus*	C	B/D	C	B		(+)/(+)	+/(+)	-	W	W	O	M
	Hooded Crane	*Grus monacha*	A	C	C	A		-	-	(+)	W	W	O	M
	White-naped Crane	*Grus vipio*	A	C	C	A		-	-	(+)	W	W	O	M
	Red-crowned Crane	*Grus japonensis*	A	C	A	E		-	-	(+)	W	W	O	M
	Demoiselle Crane	*Anthropoides virgo*	A	C	C	A	M	(+)	+	-	O	O	O	M
	Rallidae													
	Swinhoe's Rail	*Coturnicops exquisitus*	B	C	A	C		-	-	(+)	W	W	O	M
	Water Rail	*Rallus aquaticus*	C	C	C	D		(+)	+	+	W	W	O	P
	Corncrake	*Crex crex*	C	C	B	C		(+)	+	-	O	O	O	M
	Baillon's Crake	*Porzana pusilla*	D	D	C	D		(+)	+	+	W	W	O	M
	Little Crake	*Porzana parva*	C	C	C	C		(+)	(+)	-	W	W	O	M
	Spotted Crake	*Porzana porzana*	C	C	C	C		(+)	(+)	-	W	W	O	M
	Band-bellied Crake	*Porzana paykullii*	B	D	B	E		-	-	(+)	W	W	O	M
	Ruddy-breasted Crake	*Porzana fusca*	A	D	E	E		-	-	+	W	W	O	M
	Common Coot	*Fulica atra*	D	F	C	D		(+)	+	+	W	W	O	P
	Red-knobbed Coot	*Fulica cristata*	B	E	E	C		(+)	-	-	W	W	O	S

	TAXON		TOL	LAT	TEM	HUM	MON	WP	CP	EP	HF	HN	DT	MIG
Gruiformes and Cuculiformes 4 cont.	Common Moorhen	*Gallinula chloropus*	D	F	C	D		+	+	+	W	W	O	P
	Purple Swamphen	*Porphyrio porphyrio*	C	D	D	D		(+)	-	-	W	W	O	S
	Cuculidae													
	Common Cuckoo	*Cuculus canorus*	E	F	C	D		+	+	+	F	F	I	M
	Oriental Cuckoo	*Cuculus saturatus*	D	E	B	C		-	+	+	F	F	I	M
	Hodgson's Hawk Cuckoo	*Cuculus fugax*	C	D	C	E		-	-	+	F	F	I	M
	Indian Cuckoo	*Cuculus micropterus*	C	D	C	E		-	-	+	F	F	I	M
	Lesser Cuckoo	*Cuculus poliocephalus*	C	D	C	E	M	-	-	+	F	F	I	M
	Great Spotted Cuckoo	*Clamator glandarius*	B	D	E	B		(+)	(+)	-	M	M	M	M
Caprimulgiformes and Apodiformes 5	**Caprimulgidae**													
	Grey Nightjar	*Caprimulgus indicus*	C	F	C	D		-	-	+	A	M	I	M
	Eurasian Nightjar	*Caprimulgus europaeus*	C	F	C	B		(+)	(+)	(+)	A	F	I	M
	Egyptian Nightjar	*Caprimulgus aegypticus*	B	D	D	A		(+)	(+)	-	A	O	I	M
	Nubian Nightjar	*Caprimulgus nubicus*	A	E	E	A		(+)	-	-	A	O	I	M
	Red-necked Nightjar	*Caprimulgus ruficollis*	A	D	E	A		(+)	-	-	A	M	I	M
	Apodidae													
	White-throated Needletail	*Hirundapus caudacutus*	C	F	C	D		-	-	+	A	R	I	M
	Alpine Swift	*Tachymarptis melba*	C	D	D	D	M	+	+	-	A	R	I	M
	Fork-tailed Swift	*Apus pacificus*	D	F	B	C		-	-	+	A	R	I	M
	Common Swift	*Apus apus*	C	F	C	B		+	(+)	(+)	A	R	I	M
	Little Swift	*Apus affinis*	B	E	E	C		(+)	(+)	-	A	R	I	P
	White-rumped Swift	*Apus caffer*	B	E	E	C		(+)	-	-	A	R	I	M
	Pallid Swift	*Apus pallidus*	A	D	E	A		+	-	-	A	R	I	M
	Plain Swift	*Apus unicolor*	A	E	E	E		(+)	-	-	A	R	I	M
Columbiformes and allies 6	**Phoenicopteridae**													
	Greater Flamingo	*Phoenicopterus ruber*	C	D	D	C		(+)	(+)	-	W	W	O	P
	Podicipedidae													
	Great Crested Grebe	*Podiceps cristatus*	D	F	D	D		+	+	+	W	W	M	P
	Black-necked Grebe	*Podiceps nigricollis*	C	C	D	B		(+)	+	(+)	W	W	M	P
	Red-necked Grebe	*Podiceps grisegena*	C	C	B	B		(+)	(+)	(+)	W	W	M	M
	Slavonian Grebe	*Podiceps auritus*	C	B	B	B		+	+	+	W	W	M	M
	Little Grebe	*Tachybaptus ruficollis*	C	F	D	D		+	-	+	W	W	M	P
	Phaethontidae													
	Red-billed Tropicbird	*Phaethon aethereus*	B	E	E	D		(+)	-	-	Ma	R	M	M
	Pteroclididae													
	Pallas's Sandgrouse	*Syrrhaptes paradoxus*	B	D	D	A		-	+	(+)	O	O	H	S
	Pin-tailed Sandgrouse	*Pterocles alchata*	B	D	D	A		(+)	+	-	O	O	H	P

	TAXON		TOL	LAT	TEM	HUM	MON	WP	CP	EP	HF	HN	DT	MIG
Columbiformes and allies 6 cont.	Black-bellied Sandgrouse	*Pterocles orientalis*	B	D	D	A		(+)	+	-	O	O	H	P
	Chestnut-bellied Sandgrouse	*Pterocles exustus*	B	E	E	C		+	+	-	O	O	H	S
	Spotted Sandgrouse	*Pterocles senegallus*	A	E	E	A		(+)	-	-	O	O	O	S
	Liechtenstein's Sandgrouse	*Pterocles lichtensteinii*	A	E	E	A		(+)	-	-	O	O	H	S
	Crowned Sandgrouse	*Pterocles coronatus*	A	E	E	A		(+)	(+)	-	O	O	H	S
	Columbidae													
	White-bellied Green Pigeon	*Treron sieboldii*	A	D	E	E		-	-	+	F	F	O	P
	Namaqua Dove	*Oena capensis*	B	E	E	C		(+)	-	-	O	M	O	S
	Rock Dove	*Columba livia*	D	F	D	D		+	+	(+)	M	R	O	S
	Wood Pigeon	*Columba palumbus*	D	F	C	C		+	(+)	-	F	F	O	P
	Stock Dove	*Columba oenas*	C	F	C	B		(+)	(+)	-	M	M	O	P
	Hill Pigeon	*Columba rupestris*	C	D	B	C	M	-	-	+	O	R	O	P
	Trocaz Pigeon	*Columba trocaz*	A	E	E	E	M	(+)	-	-	F	F	H	S
	Bolle's Pigeon	*Columba bollii*	A	E	E	E	M	(+)	-	-	F	F	H	S
	Laurel Pigeon	*Columba junoniae*	A	E	E	E	M	(+)	-	-	F	F	H	S
	Japanese Wood Pigeon	*Columba janthina*	A	E	E	E		-	-	(+)	F	F	O	S
	Oriental Turtle Dove	*Streptopelia orientalis*	E	F	B	D		-	(+)	+	M	M	O	M
	Eurasian Collared Dove	*Streptopelia decaocto*	D	F	C	D		+	(+)	(+)	M	M	O	S
	Eurasian Turtle Dove	*Streptopelia turtur*	D	F	C	C		+	(+)	-	M	M	O	M
	Laughing Dove	*Streptopelia senegalensis*	B	E	E	D		(+)	+	-	O	O	O	S
	Red Turtle Dove	*Streptopelia tranquebarica*	A	E	E	E		-	-	+	O	O	O	M
Anseriformes and Galliformes 7	**Anatidae**													
	White-headed Duck	*Oxyura leucocephala*	B	D	D	A		(+)	(+)	-	W	W	O	P
	Mute Swan	*Cygnus olor*	B	C	C	B		(+)	(+)	(+)	W	W	O	P
	Whooper Swan	*Cygnus cygnus*	B	B	B	A		+	+	+	W	W	O	M
	Bewick's Swan	*Cygnus columbianus*	A	A	A	A		-	+	+	O	O	H	M
	Greylag Goose	*Anser anser*	C	C	B	B		(+)	+	+	W	W	H	P
	Swan Goose	*Anser cygnoides*	B	C	B	B		-	-	(+)	W	W	H	M
	Bean Goose	*Anser fabalis*	B	B	B	A		-	+	+	O	O	O	M
	Bar-headed Goose	*Anser indicus*	A	D			M	-	-	(+)	W	W	H	M
	Pink-footed Goose	*Anser brachyrhynchus*	A	A	A	A		+	-	-	O	O	H	M
	Lesser White-fronted Goose	*Anser erythropus*	A	A	A	A		-	(+)	-	O	O	H	M
	White-fronted Goose	*Anser albifrons*	A	A	A	A		-	+	+	O	O	H	M
	Emperor Goose	*Anser canagicus*	A	A	A	A		-	-	(+)	W	W	H	M
	Snow Goose	*Anser caerulescens*	A	A	A	A		-	-	(+)	O	O	O	M
	Barnacle Goose	*Branta leucopsis*	A	A	A	A		-	+	-	O	O	O	M
	Red-breasted Goose	*Branta ruficollis*	A	A	A	A		-	(+)	-	O	O	H	M

	TAXON		TOL	LAT	TEM	HUM	MON	WP	CP	EP	HF	HN	DT	MIG
Anseriformes and Galliformes 7 cont.	Brent Goose	*Branta bernicla*	A	A	A	A		-	+	+	O	O	O	M
	Egyptian Goose	*Alopochen aegyptiacus*	B	E	E	C		(+)	-	-	W	W	H	S
	Common Shelduck	*Tadorna tadorna*	C	B/D	C	C		(+)/(+)	+/(+)	-	W	W	O	P
	Ruddy Shelduck	*Tadorna ferruginea*	C	D	C	E		(+)	+	+	W	W	O	P
	Crested Shelduck	*Tadorna cristata*	A	D	A	E	M	-	-	(+)	F	F	O	M
	Long-tailed Duck	*Clangula hyemalis*	B	A	A	C		+	+	+	W	W	O	M
	Common Eider	*Somateria mollissima*	B	A	A	B		+	-	+	W	W	O	P
	King Eider	*Somateria spectabilis*	B	A	A	B		-	+	+	W	W	O	M
	Spectacled Eider	*Somateria fischeri*	A	A	A	A		-	-	+	W	O	O	M
	Steller's Eider	*Polysticta stelleri*	A	A	A	A		-	-	+	W	O	O	M
	Harlequin Duck	*Histrionicus histrionicus*	C	B	B	B		+	-	+	W	W	M	P
	Common Scoter	*Melanitta nigra*	B	A	A	C		+	+	+	W	W	O	M
	Velvet Scoter	*Melanitta fusca*	B	B	B	A		+	+	+	W	W	O	M
	Goosander	*Mergus merganser*	D	B	B	B		+	+	+	W	W	M	P
	Red-breasted Merganser	*Mergus serrator*	B	B	B	B		+	+	+	W	W	M	P
	Scaly-sided Merganser	*Mergus squamatus*	B	C	A	C		-	-	(+)	W	W	M	M
	Smew	*Mergellus albellus*	B	B	B	A		(+)	+	+	W	W	M	M
	Common Goldeneye	*Bucephala clangula*	C	B	B	B		+	+	+	W	W	O	M
	Barrow's Goldeneye	*Bucephala islandica*	C	A	B	B		(+)	-	-	W	W	M	M
	Mandarin Duck	*Aix galericulata*	A	D	A	E		-	-	(+)	W	W	O	P
	Marbled Duck	*Marmaronetta angustirostris*	A	D	E	A		(+)	(+)	-	W	W	O	M
	Red-crested Pochard	*Netta rufina*	C	D	C	D		(+)	+	-	W	W	O	P
	Common Pochard	*Aythya ferina*	C	C	C	C		(+)	+	(+)	W	W	O	P
	Ferruginous Duck	*Aythya nyroca*	C	D	D	B		(+)	+	(+)	W	W	O	P
	Tufted Duck	*Aythya fuligula*	C	B	B	B		+	+	+	W	W	O	P
	Greater Scaup	*Aythya marila*	C	B	B	B		+	+	+	W	W	O	M
	Baer's Pochard	*Aythya baeri*	B	C	A	C		-	-	(+)	W	W	O	M
	Mallard	*Anas platyrhynchos*	E	F	B	C	+	+	+	W	W	M	P	
	Shoveler	*Anas clypeata*	D	F	B	B		(+)	+	+	W	W	O	P
	Gadwall	*Anas strepera*	C	C	B	B		(+)	+	(+)	W	W	O	P
	Garganey	*Anas querquedula*	C	C	B	B		(+)	+	+	W	W	O	M
	Spot-billed Duck	*Anas poecilorhyncha*	C	D	C	C		-	-	+	W	W	O	S
	Eurasian Wigeon	*Anas penelope*	C	B	B	B		+	+	+	W	W	O	M
	Common Teal	*Anas crecca*	C	B	B	B		+	+	+	W	W	O	P
	Pintail	*Anas acuta*	C	B	B	B		+	+	+	W	W	O	M
	Falcated Duck	*Anas falcata*	B	C	A	C		-	-	+	W	W	O	M
	Baikal Teal	*Anas formosa*	B	B	B	A		-	-	+	W	W	O	M

	TAXON		TOL	LAT	TEM	HUM	MON	WP	CP	EP	HF	HN	DT	MIG
Anseriformes and Galliformes 7 cont.	**Numidae**													
	Helmeted Guineafowl	*Numida meleagris*	B	E	E	C		(+)	-	-	O	O	O	S
	Tetraonidae													
	Siberian Grouse	*Dendragapus falcipennis*	A	C			M	-	-	(+)	F	F	O	S
	Willow Grouse	*Lagopus lagopus*	C	B	B	B		+	+	+	O	O	O	S
	Rock Ptarmigan	*Lagopus mutus*	C	B/D	B	B	M	(+)/(+)	+/+	+/+	O	O	O	S
	Western Capercaillie	*Tetrao urogallus*	C	B/D	B	B		(+)/(+)	+/+	-	F	F	O	S
	Black Grouse	*Tetrao tetrix*	C	C	B	C		(+)	+	+	M	M	O	S
	Black-billed Capercaillie	*Tetrao parvirostris*	B	B	A	C		-	-	+	F	F	O	S
	Caucasian Black Grouse	*Tetrao mlokosiewiczi*	A	D			M	(+)	-	-	F	F	O	S
	Hazel Grouse	*Bonasa bonasia*	C	B	B	B		(+)	+	+	F	F	O	S
	Phasianidae													
	Caspian Snowcock	*Tetraogallus caspius*	A	D			M	(+)	-	-	M	M	H	S
	Caucasian Snowcock	*Tetraogallus caucasicus*	A	D			M	(+)	-	-	O	O	H	S
	Altai Snowcock	*Tetraogallus altaicus*	A	C			M	-	(+)	-	O	O	O	S
	Red-legged Partridge	*Alectoris rufa*	C	D	D	C		+	-	-	O	O	O	S
	Barbary Partridge	*Alectoris barbara*	B	D	E	A	M	(+)	-	-	M	M	O	S
	Chukar	*Alectoris chukar*	B	D	E	A		(+)	+	+	M	M	O	S
	Rock Partridge	*Alectoris graeca*	B	D	C	E	M	+	-	-	M	M	O	S
	Sand Partridge	*Ammoperdix heyi*	A	E	E	A		+	-	-	M	M	O	S
	See-see Partridge	*Ammoperdix griseogularis*	A	E	E	A		-	(+)	-	O	O	O	S
	Double-spurred Francolin	*Francolinus bicalcaratus*	B	E	E	C		(+)	-	-	M	M	O	S
	Black Francolin	*Francolinus francolinus*	A	E	E	A		+	-	-	M	M	O	S
	Grey Partridge	*Perdix perdix*	B	C	C	C		+	+	-	O	O	O	S
	Daurian Partridge	*Perdix dauricae*	A	D	C	A		-	-	+	M	M	O	S
	Common Quail	*Coturnix coturnix*	D	F	D	D		+	+	-	O	O	O	M
	Japanese Quail	*Coturnix japonica*	B	F	B	B		-	-	+	O	O	O	M
	Struthionidae													
	Ostrich	*Struthio camelus*	A	E	E	C		(+)	-	-	O	O	O	S

APPENDIX 2

European Pleistocene fossil birds

Appendix 2. *Categorisation of European Pleistocene fossil birds by period and frequency. Early Pleistocene (1.8 mya–0.78 mya); Middle Pleistocene (0.78 mya–0.125 mya); Late Pleistocene (0.125 mya–0.001 mya). Rankings represent no sites (-), 5 or fewer sites (A), 6–20 sites (B), 21–100 sites (C), 101 or more sites (D).*

Species		Early	Middle	Late
Red-backed Shrike	*Lanius collurio*	A	-	C
Lesser Grey Shrike	*Lanius minor*	A	-	B
Great Grey Shrike	*Lanius excubitor*	A	A	C
Woodchat Shrike	*Lanius senator*	-	A	B
Masked Shrike	*Lanius nubicus*	-	A	A
Eurasian Jay	*Garrulus glandarius*	A	C	D
Siberian Jay	*Perisoreus infaustus*	-	A	B
Iberian Azure-winged Magpie	*Cyanopica cooki*	-	-	A
Black-billed Magpie	*Pica pica*	B	C	D
Spotted Nutcracker	*Nucifraga caryocatactes*	A	B	C
Alpine Chough	*Pyrrhocorax graculus*	B	C	D
Red-billed Chough	*Pyrrhocorax pyrrhocorax*	B	C	D
Eurasian Jackdaw	*Corvus monedula*	B	C	D
Rook	*Corvus frugilegus*	A	A	C
Carrion Crow	*Corvus corone*	A	C	D
Collared Crow	*Corvus torquatus*	-	A	A
Brown-necked Raven	*Corvus ruficollis*	-	A	A
Common Raven	*Corvus corax*	B	C	D
Fan-tailed Raven	*Corvus rhipidurus*	-	-	A
Golden Oriole	*Oriolus oriolus*	A	-	C
Sand Martin	*Riparia riparia*	-	A	B
Crag Martin	*Ptyonoprogne rupestris*	-	B	C
Barn Swallow	*Hirundo rustica*	B	B	D
Red-rumped Swallow	*Cecropis daurica*	-	A	C
House Martin	*Delichon urbica*	A	A	C
Graceful Prinia	*Prinia gracilis*	-	-	A

Species		Early	Middle	Late
Grasshopper Warbler	*Locustella naevia*	-	-	A
Aquatic Warbler	*Acrocephalus paludicola*	-	-	A
Sedge Warbler	*Acrocephalus schoenobaenus*	-	-	A
Marsh Warbler	*Acrocephalus palustris*	A	-	A
Eurasian Reed Warbler	*Acrocephalus scirpaceus*	-	-	A
Great Reed Warbler	*Acrocephalus arundinaceus*	A	-	B
Melodious Warbler	*Hippolais polyglotta*	-	A	-
Icterine Warbler	*Hippolais icterina*	-	-	A
Sardinian Warbler	*Sylvia melanocephala*	-	A	A
Orphean Warbler	*Sylvia hortensis*	A	A	A
Barred Warbler	*Sylvia nisoria*	-	A	-
Lesser Whitethroat	*Sylvia curruca*	-	-	A
Greater Whitethroat	*Sylvia communis*	A	A	A
Garden Warbler	*Sylvia borin*	-	A	A
Blackcap	*Sylvia atricapilla*	B	A	A
Bonelli's Warbler	*Phylloscopus bonelli*	-	-	A
Wood Warbler	*Phylloscopus sibilatrix*	A	-	A
Common Chiffchaff	*Phylloscopus collybita*	-	-	A
Willow Warbler	*Phylloscopus trochilus*	-	-	A
Calandra Lark	*Melanocorypha calandra*	A	B	C
Bimaculated Lark	*Melanocorypha bimaculata*	-	A	-
White-winged Lark	*Melanocorypha leucoptera*	-	-	A
Black Lark	*Melanocorypha yeltoniensis*	-	A	A
Greater Short-toed Lark	*Calandrella brachydactyla*	A	A	B
Lesser Short-toed Lark	*Calandrella rufescens*	A	A	A
Crested Lark	*Galerida cristata*	A	A	C
Thekla Lark	*Galerida theklae*	-	A	A
Woodlark	*Lullula arborea*	A	B	C
Eurasian Skylark	*Alauda arvensis*	B	A	D
Shore Lark	*Eremophila alpestris*	A	A	C
Penduline Tit	*Remiz pendulinus*	-	A	A
Marsh Tit	*Poecile palustris*	-	A	A
Sombre Tit	*Poecile lugubris*	A	-	-
Willow Tit	*Poecile montana*	-	-	A
Crested Tit	*Lophophanes cristatus*	A	A	A
Coal Tit	*Periparus ater*	A	A	A
Blue Tit	*Cyanistes caeruleus*	-	A	A
Great Tit	*Parus major*	A	A	C

Species		Early	Middle	Late
Rufous Bush Robin	*Cercotrichas galactotes*	-	A	A
European Robin	*Erithacus rubecula*	A	A	C
Thrush Nightingale	*Luscinia luscinia*	-	A	B
Common Nightingale	*Luscinia megarhynchos*	-	-	A
Bluethroat	*Luscinia svecica*	-	A	A
Black Redstart	*Phoenicurus ochruros*	-	-	B
Common Redstart	*Phoenicurus phoenicurus*	-	A	B
Blackstart	*Cercomela melanura*	A	-	-
Whinchat	*Saxicola rubetra*	-	A	B
Common Stonechat	*Saxicola torquata*	A	A	A
Northern Wheatear	*Oenanthe oenanthe*	-	A	C
Black-eared Wheatear	*Oenanthe hispanica*	-	A	A
Finsch's Wheatear	*Oenanthe finschii*	-	-	A
Black Wheatear	*Oenanthe leucura*	-	-	A
Rufous-tailed Rock Thrush	*Monticola saxatilis*	-	A	B
Blue Rock Thrush	*Monticola solitarius*	-	A	B
Scaly Thrush	*Zoothera dauma*	-	A	B
Ring Ouzel	*Turdus torquatus*	-	B	C
Eurasian Blackbird	*Turdus merula*	A	C	D
Mistle Thrush	*Turdus viscivorus*	A	B	D
Fieldfare	*Turdus pilaris*	-	B	D
Song Thrush	*Turdus philomelos*	A	A	C
Redwing	*Turdus iliacus*	A	B	C
Dunnock	*Prunella modularis*	A	A	B
Alpine Accentor	*Prunella collaris*	-	B	C
House Sparrow	*Passer domesticus*	-	A	C
Spanish Sparrow	*Passer hispaniolensis*	-	A	A
Dead Sea Sparrow	*Passer moabiticus*	-	A	-
Eurasian Tree Sparrow	*Passer montanus*	-	A	B
Rock Sparrow	*Petronia petronia*	A	B	C
White-winged Snowfinch	*Montifringilla nivalis*	-	A	C
Common Chaffinch	*Fringilla coelebs*	A	B	C
Brambling	*Fringilla montifringilla*	-	-	C
Red-fronted Serin	*Serinus pusillus*	-	-	A
European Serin	*Serinus serinus*	A	A	A
Citril Finch	*Chloroptila citrinella*	-	A	A
European Greenfinch	*Chloris chloris*	A	B	C
European Goldfinch	*Carduelis carduelis*	A	A	C

Species		Early	Middle	Late
Eurasian Siskin	*Spinus spinus*	A	A	B
Eurasian Linnet	*Linaria cannabina*	-	A	C
Common Redpoll	*Acanthis flammea*	-	A	B
Two-barred Crossbill	*Loxia leucoptera*	-	-	A
Common Crossbill	*Loxia curvirostra*	-	A	C
Parrot Crossbill	*Loxia pytyopsittacus*	-	A	A
Common Rosefinch	*Carpodacus erythrinus*	-	A	A
Great Rosefinch	*Carpodacus rubicilla*	-	-	A
Pine Grosbeak	*Pinicola enucleator*	A	A	B
Eurasian Bullfinch	*Pyrrhula pyrrhula*	-	B	C
Hawfinch	*Pinicola enucleator*	A	B	C
Lapland Longspur	*Pyrrhula pyrrhula*	-	-	A
Snow Bunting	*Plectrophenax nivalis*	-	B	C
Yellowhammer	*Emberiza citrinella*	-	B	C
Cirl Bunting	*Emberiza cirlus*	-	-	A
Rock Bunting	*Emberiza cia*	-	-	A
Ortolan Bunting	*Emberiza hortulana*	-	A	A
Cretzschmar's Bunting	*Emberiza caesia*	-	-	A
Reed Bunting	*Emberiza schoeniclus*	-	-	B
Black-headed Bunting	*Emberiza melanocephala*	-	A	A
Tawny Pipit	*Anthus campestris*	A	A	B
Tree Pipit	*Anthus trivialis*	A	A	B
Meadow Pipit	*Anthus pratensis*	A	A	B
Red-throated Pipit	*Anthus cervinus*	-	A	-
Water Pipit	*Anthus spinoletta*	-	B	B
Yellow Wagtail sp.	*Motacilla flava*	-	B	B
Grey Wagtail	*Motacilla cinerea*	A	A	A
White Wagtail	*Motacilla alba*	A	B	C
Lesser Kestrel	*Falco naumanni*	A	B	C
Common Kestrel	*Falco tinnunculus*	B	C	D
Red-footed Falcon	*Falco vespertinus*	A	B	C
Merlin	*Falco columbarius*	-	B	C
Eurasian Hobby	*Falco subbuteo*	A	B	C
Eleonora's Falcon	*Falco eleonorae*	-	A	B
Saker	*Falco cherrug*	A	A	B
Gyr Falcon	*Falco rusticolus*	-	A	C
Peregrine Falcon	*Falco peregrinus*	A	B	C
Eurasian Hoopoe	*Upupa epops*	-	B	B

Species		Early	Middle	Late
Eurasian Wryneck	*Jynx torquilla*	A	A	B
Great Spotted Woodpecker	*Dendrocopos major*	B	B	C
Syrian Woodpecker	*Dendrocopos syriacus*	-	A	-
Middle Spotted Woodpecker	*Dendrocopos medius*	A	A	B
White-backed Woodpecker	*Dendrocopos leucotos*	-	A	B
Lesser Spotted Woodpecker	*Dendrocopos minor*	A	A	B
Three-toed Woodpecker	*Picoides tridactylus*	-	A	A
Black Woodpecker	*Dryocopus martius*	A	-	B
Eurasian Green Woodpecker	*Picus viridis*	A	B	C
Grey-headed Woodpecker	*Picus canus*	-	A	C
European Bee-eater	*Merops apiaster*	A	A	B
European Roller	*Coracias garrulus*	A	A	C
Common Kingfisher	*Alcedo atthis*	-	-	A
White-throated Kingfisher	*Halcyon smyrnensis*	-	-	A
Pied Kingfisher	*Ceryle rudis*	-	-	A
Barn Owl	*Tyto alba*	B	B	C
Eagle Owl	*Bubo bubo*	B	C	D
Brown Fish Owl	*Bubo zeylonensis*	A	-	A
Snowy Owl	*Bubo scandiacus*	A	B	D
Tawny Owl	*Strix aluco*	B	B	C
Hume's Owl	*Strix butleri*	A	-	-
Ural Owl	*Strix uralensis*	A	B	B
Great Grey Owl	*Strix nebulosa*	A	A	B
European Scops Owl	*Otus scops*	B	A	C
Long-eared Owl	*Asio otus*	A	B	C
Short-eared Owl	*Asio flammeus*	B	C	D
Marsh Owl	*Asio capensis*	A	-	-
Northern Hawk Owl	*Surnia ulula*	A	A	C
Eurasian Pygmy Owl	*Glaucidium passerinum*	B	A	B
Little Owl	*Athene noctua*	B	C	C
Tengmalm's Owl	*Aegolius funereus*	A	B	C
Eurasian Osprey	*Pandion haliaetus*	A	A	B
European Honey-buzzard	*Pernis apivorus*	A	A	B
Bearded Vulture	*Gypaetus barbatus*	A	B	C
Egyptian Vulture	*Neophron percnopterus*	-	A	B
Short-toed Snake-eagle	*Circaetus gallicus*	A	-	B
Eurasian Griffon Vulture	*Gyps fulvus*	-	B	C
Eurasian Black Vulture	*Aegypius monachus*	A	B	C

Species		Early	Middle	Late
Lesser Spotted Eagle	*Lophaetus pomarinus*	A	A	B
Greater Spotted Eagle	*Lophaetus clanga*	-	A	B
Booted Eagle	*Hieraaetus pennatus*	-	A	A
Steppe Eagle	*Aquila nipalensis*	A	A	B
Spanish Imperial Eagle	*Aquila adalberti*	-	-	A
Imperial Eagle	*Aquila heliaca*	A	A	B
Golden Eagle	*Aquila chrysaetos*	A	B	D
Bonelli's Eagle	*Aquila fasciata*	-	A	B
Eurasian Sparrowhawk	*Accipiter nisus*	A	B	C
Northern Goshawk	*Accipiter gentilis*	A	B	C
Western Marsh Harrier	*Circus aeruginosus*	-	A	C
Hen Harrier	*Circus cyaneus*	A	B	C
Montagu's Harrier	*Circus pygargus*	-	A	B
Pallid Harrier	*Circus macrourus*	-	A	B
Black Kite	*Milvus migrans*	A	A	B
Red Kite	*Milvus milvus*	-	A	B
White-tailed Eagle	*Haliaeetus albicilla*	A	B	C
Common Buzzard	*Buteo buteo*	A	B	C
Long-legged Buzzard	*Buteo rufinus*	-	B	C
Rough-legged Buzzard	*Buteo lagopus*	-	-	C
Eurasian Oystercatcher	*Haematopus ostralegus*	-	-	A
Black-winged Stilt	*Himantopus himantopus*	-	A	B
Avocet	*Recurvirostra avosetta*	-	A	A
Eurasian Stone Curlew	*Burhinus oedicnemus*	-	A	A
Collared Pratincole	*Glareola pratincola*	-	A	A
Little Ringed Plover	*Charadrius dubius*	-	A	A
Common Ringed Plover	*Charadrius hiaticula*	-	-	B
Kentish Plover	*Charadrius alexandrinus*	-	-	A
Eurasian Dotterel	*Charadrius morinellus*	-	B	C
Pacific Golden Plover	*Pluvialis fulva*	-	A	-
Eurasian Golden Plover	*Pluvialis apricaria*	-	A	C
Grey Plover	*Pluvialis squatarola*	A	B	C
Spur-winged Lapwing	*Vanellus spinosus*	-	-	A
Northern Lapwing	*Vanellus vanellus*	A	B	C
Red Knot	*Caldris canutus*	-	A	B
Sanderling	*Caldris alba*	-	-	A
Little Stint	*Calidris minuta*	-	-	A
Temminck's Stint	*Calidris temmincki*	-	-	A

Species		Early	Middle	Late
Curlew Sandpiper	*Calidris ferruginea*	-	-	B
Purple Sandpiper	*Calidris maritima*	-	-	A
Dunlin	*Calidris alpina*	-	A	C
Broad-billed Sandpiper	*Limicola falcinellus*	-	-	A
Ruff	*Philomachus pugnax*	A	A	C
Jack Snipe	*Lymnocryptes minimus*	A	A	C
Common Snipe	*Gallinago gallinago*	A	B	C
Great Snipe	*Gallinago media*	A	B	C
Solitary Snipe	*Gallinago solitaria*	-	-	A
Eurasian Woodcock	*Scolopax rusticola*	A	B	C
Black-tailed Godwit	*Limosa limosa*	A	A	C
Bar-tailed Godwit	*Limosa lapponica*	-	A	A
Whimbrel	*Numenius phaeopus*	A	A	B
Slender-billed Curlew	*Numenius tenuirostris*	-	A	A
Eurasian Curlew	*Numenius arquata*	-	A	C
Spotted Redshank	*Tringa erythropus*	A	A	B
Common Redshank	*Tringa totanus*	A	A	C
Marsh Sandpiper	*Tringa stagnatilis*	-	A	A
Common Greenshank	*Tringa nebularia*	-	A	B
Green Sandpiper	*Tringa ochropus*	A	A	B
Wood Sandpiper	*Tringa glareola*	A	A	B
Terek Sandpiper	*Xenus cinereus*	-	A	-
Common Sandpiper	*Actitis hypoleucos*	A	A	B
Ruddy Turnstone	*Arenaria interpres*	-	A	B
Grey Phalarope	*Phalaropus fulicarius*	A	A	-
Pomarine Skua	*Stercorarius pomarinus*	-	A	A
Arctic Skua	*Stercorarius parasiticus*	-	-	B
Long-tailed Skua	*Stercorarius longicaudus*	-	A	A
Great Skua	*Catharacta skua*	-	-	A
Pallas's Gull	*Ichthyaetus ichthyaetus*	-	-	A
Mediterranean Gull	*Ichthyaetus melanocephalus*	-	-	A
Little Gull	*Hydrocoloeus minutus*	-	A	B
Black-headed Gull	*Chroicocephalus ridibundus*	A	A	B
Audouin's Gull	*Ichthyaetus audouinii*	-	-	A
Common Gull	*Larus canus*	A	A	B
Lesser Black-backed Gull	*Larus fuscus*	-	A	B
Herring Gull	*Larus argentatus*	-	A	C
Glaucous Gull	*Larus hyperboreus*	-	A	A

Species		Early	Middle	Late
Lesser Black-backed Gull	*Larus fuscus*	-	-	A
Black-legged Kittiwake	*Rissa tridactyla*	A	-	B
Ivory Gull	*Pagophila eburnea*	-	-	A
Sandwich Tern	*Thalasseus sandvicensis*	-	-	A
Common Tern	*Sterna hirundo*	A	A	B
Arctic Tern	*Sterna paradisea*	-	-	A
Little Tern	*Sternula albifrons*	-	A	A
Whiskered Tern	*Chlidonias hybridus*	-	A	-
Black Tern	*Chlidonias niger*	-	A	B
White-winged Tern	*Chlidonias leucopterus*	-	-	A
Common Guillemot	*Alle alle*	-	A	B
Razorbill	*Alca torda*	A	A	B
Great Auk	*Pinguinus impennis*	A	-	B
Little Auk	*Alle alle*	-	A	B
Atlantic Puffin	*Fratercula arctica*	-	A	B
Northern Fulmar	*Fulmarus glacialis*	-	-	A
Cory's Shearwater	*Calonectris diomedea*	-	-	B
Manx Shearwater	*Puffinus puffinus*	-	-	B
Yelkouan Shearwater	*Puffinus yelkouan*	-	-	A
Balearic Shearwater	*Puffinus mauretanicus*	-	-	A
European Storm-petrel	*Hydrobates pelagicus*	-	A	A
Leach's Storm-petrel	*Oceanodroma leucorhoa*	-	A	-
Northern Gannet	*Morus bassanus*	-	-	A
Great Cormorant	*Phalacrocorax carbo*	A	A	B
European Shag	*Phalacrocorax aristotelis*	-	-	C
Pygmy Cormorant	*Phalacrocorax pygmaeus*	-	-	A
Long-tailed Cormorant	*Phalacrocorax africanus*	A	-	A
Great White Pelican	*Pelecanus onocrotalus*	-	-	A
Dalmatian Pelican	*Pelecanus crispus*	-	-	A
Great Bittern	*Botaurus stellaris*	-	A	B
Little Bittern	*Ixobrychus minutus*	-	A	A
Black-crowned Night Heron	*Nycticorax nycticorax*	-	A	A
Striated Heron	*Butorides striatus*	A	-	-
Cattle Egret	*Bubulcus ibis*	-	A	A
Little Egret	*Egretta garzetta*	-	A	A
Great White Egret	*Casmerodius albus*	-	A	A
Grey Heron	*Ardea cinerea*	A	A	C
Purple Heron	*Ardea purpurea*	A	A	A

Species		Early	Middle	Late
Black Stork	*Ciconia nigra*	-	A	A
White Stork	*Ciconia ciconia*	-	A	B
Glossy Ibis	*Plegadis falcinellus*	-	A	A
Northern Bald Ibis	*Geronticus eremita*	A	A	A
Eurasian Spoonbill	*Platalea leucorodia*	-	A	A
Red-throated Diver	*Gavia stellata*	-	A	B
Black-throated Diver	*Gavia arctica*	-	-	A
Great Northern Diver	*Gavia immer*	A	-	A
Great Spotted Cuckoo	*Clamator glandarius*	-	A	-
Common Cuckoo	*Cuculus canorus*	-	B	B
Water Rail	*Rallus aquaticus*	A	B	C
Spotted Crake	*Porzana porzana*	A	B	C
Little Crake	*Porzana parva*	A	A	A
Baillon's Crake	*Porzana pusilla*	-	A	A
Corncrake	*Crex crex*	A	B	C
Common Moorhen	*Gallinula chloropus*	-	A	C
Purple Swamphen	*Porphyrio porphyrio*	A	A	A
Common Coot	*Fulica atra*	A	B	C
Red-knobbed Coot	*Fulica cristata*	-	-	A
Common Crane	*Grus grus*	-	A	C
Siberian Crane	*Leucogeranus leucogeranus*	-	A	-
Demoiselle Crane	*Anthropoides virgo*	-	A	A
Little Bustard	*Tetrax tetrax*	-	A	B
Houbara Bustard	*Chlamydotis undulata*	-	-	A
Great Bustard	*Otis tarda*	-	B	C
Common Swift	*Apus apus*	-	B	C
Pallid Swift	*Apus pallidus*	-	A	-
Alpine Swift	*Tachymarptis melba*	-	B	C
Little Swift	*Apus affinis*	-	A	A
Eurasian Nightjar	*Caprimulgus europaeus*	-	A	A
Red-necked Nightjar	*Caprimulgus ruficollis*	-	A	A
Rock Dove	*Columba livia*	A	B	D
Stock Dove	*Columba oenas*	A	B	C
Wood Pigeon	*Columba palumbus*	A	B	D
Trocaz Pigeon	*Columba trocaz*	-	-	A
Laurel Pigeon	*Columba junoniae*	-	-	A
Eurasian Collared Dove	*Streptopelia decaocto*	-	-	A
European Turtle Dove	*Streptopelia turtur*	-	A	B

Species		Early	Middle	Late
Laughing Dove	*Streptopelia senegalensis*	-	-	A
Namaqua Dove	*Oena capensis*	-	-	A
Little Grebe	*Tachybaptus ruficollis*	A	-	B
Great Crested Grebe	*Podiceps cristatus*	A	A	B
Red-necked Grebe	*Podiceps grisegena*	-	A	A
Slavonian Grebe	*Podiceps auritus*	A	A	B
Black-necked Grebe	*Podiceps nigricollis*	A	A	B
Greater Flamingo	*Phoenicopterus ruber*	-	-	A
Mute Swan	*Cygnus olor*	-	A	B
Bewick's Swan	*Cygnus columbianus*	-	B	B
Whooper Swan	*Cygnus cygnus*	A	B	C
Bean Goose	*Anser fabalis*	A	A	C
Pink-footed Goose	*Anser brachyrhynchus*	-	A	B
(Greater) White-fronted Goose	*Anser albifrons*	A	B	C
Lesser White-fronted Goose	*Anser erythropus*	A	A	B
Greylag Goose	*Anser anser*	-	B	C
Barnacle Goose	*Branta leucopsis*	-	A	B
Brent Goose	*Branta bernicla*	-	A	B
Red-breasted Goose	*Branta ruficollis*	-	A	A
Ruddy Shelduck	*Tadorna ferruginea*	-	A	B
Common Shelduck	*Tadorna tadorna*	A	A	C
Eurasian Wigeon	*Anas penelope*	A	B	C
Gadwall	*Anas strepera*	A	B	C
Eurasian Teal	*Anas crecca*	A	B	D
Mallard	*Anas platyrhynchos*	A	C	D
Northern Pintail	*Anas acuta*	A	C	C
Garganey	*Anas querquedula*	A	B	C
Shoveler	*Anas clypeata*	A	B	C
Marbled Duck	*Marmaronetta angustirostris*	-	A	A
Red-crested Pochard	*Netta rufina*	A	A	B
Common Pochard	*Aythya ferina*	A	B	C
Ferruginous Duck	*Aythya nyroca*	A	B	C
Tufted Duck	*Aythya fuligula*	A	B	C
Greater Scaup	*Aythya marila*	-	A	B
Common Eider	*Somateria mollissima*	-	-	B
King Eider	*Somateria spectabilis*	-	-	A
Long-tailed Duck	*Clangula hyemalis*	A	A	C
Common Scoter	*Melanitta nigra*	-	A	C

Species		Early	Middle	Late
Velvet Scoter	*Melanitta fusca*	-	A	B
Common Goldeneye	*Bucephala clangula*	A	B	C
Smew	*Mergellus albellus*	-	A	C
Red-breasted Merganser	*Mergus serrator*	-	B	C
Goosander	*Mergus merganser*	-	A	C
White-headed Duck	*Oxyura leucocephala*	-	-	A
Hazel Grouse	*Bonasa bonasia*	-	A	C
Willow Grouse	*Lagopus lagopus*	A	B	D
Rock Ptarmigan	*Lagopus mutus*	-	B	D
Black Grouse	*Tetrao tetrix*	A	C	D
Caucasian Black Grouse	*Tetrao mlokosiewiczi*	-	A	B
Black-billed Capercaillie	*Tetrao parvirostris*	A	B	D
Caucasian Snowcock	*Tetraogallus caucasicus*	-	A	A
Altai Snowcock	*Tetraogallus altaicus*	-	-	A
Chukar	*Alectoris chukar*	A	A	B
Rock Partridge	*Alectoris graeca*	-	B	C
Red-legged Partridge	*Alectoris rufa*	-	A	C
Sand Partridge	*Ammoperdix heyi*	-	A	A
Black Francolin	*Francolinus francolinus*	-	-	A
Grey Partridge	*Perdix perdix*	A	B	D
Common Quail	*Coturnix coturnix*	B	C	D

References

Alerstam, T. and Enckell, P. H. (1979). Unpredictable Habitats and Evolution of Bird Migration. *Oikos* **33**: 228-232.

Allen, F. G. H. and Brudenell-Bruce, P. G. C. (1967). The White-rumped Swift *Apus affinis* in southern Spain. *Ibis* **109**: 113-115.

Allen, J. R. M., Brandt, U., Brauer, A. *et al.* (1999). Rapid environmental changes in southern Europe during the last glacial period. *Nature* **400**: 740-743.

Allende, L. M., Rubio, I., Ruíz-del-Valle, V., Guillén, J., Martínez-Laso, J., Lowy, E., Varela, P., Zamora, J. and Arnaiz-Villena, A. (2001). The Old World Sparrows (Genus Passer) Phylogeography and Their Relative Abundance of Nuclear mtDNA Pseudogenes. *J. Mol. Evol.* **53**: 144-154.

Alley, R. B., Lynch-Stieglitz, J. and Severinghaus, J. P. (1999). Global climate change. *Proc. Natl. Acad. Sci. USA* **96**: 1331-1334.

Alström, P., Davidson, P., Duckworth, J. W. *et al.* (2010). Description of a new species of *Phylloscopus* warbler from Vietnam and Laos. *Ibis* **152**: 145-168.

Alström, P., Ericson, P. G. P., Olsson, U. and Sundberg, P. (2006). Phylogeny and classification of the avian superfamily Sylvioidea. *Mol. Phylog. Evol.* **38**: 381-397.

Alström, P. and Olsson, U. (1995). A new species of *Phylloscopus* warbler from Sichuan Province, China. *Ibis*: **137**: 459-468.

Alström, P., Olsson, U. and Colston, P. R. (1992). A new species of *Phylloscopus* warbler from central China. *Ibis*: **134**: 329-334.

Alström, P., Olsson, U., Lei, F., Wang, H-t., Gao, W. and Sundberg, P. (2008). Phylogeny and classification of the Old World Emberizini (Aves, Passeriformes). *Mol. Phylog. Evol.* **47**: 960-973.

Amaral, F. B., Sheldon, F. H., Gamauf, A., Haring, E., Riesing, M., Silveira, L. F. and Wajntal, A. (2009). Patterns and processes of diversification in a widespread and ecologically diverse avian group, the buteonine hawks (Aves, Accipitridae). *Mol. Phylogen. Evol.* Doi:10.1016/j.ympev.2009.07.020.

Andreev, A. A., Siegert, C., Klimanov, V. A., Derevyagin, A. Y., Shilova, G. N. & Melles, M. (2002). Late Pleistocene and Holocene Vegetation and Climate on the Taymyr Lowland, Northern Siberia. *Quaternary Research*, 57, 138-150.

Antonov, A., Stokke, B. G., Moksnes, A. and Røskaft, E. (2007). Aspects of breeding ecology of the eastern olivaceous warbler (*Hippolais pallida*). *J. Orn.* **148**: 443-451.

Arnaiz-Villena, A., Álvarez-Tejado, M., Ruíz-del-Valle, V., García-de-la-Torre, C., Varela, P., Recio, M. J., Ferre, S. and Martínez-Laso, J. (1998). Phylogeny and rapid Northern and Southern Hemisphere speciation of goldfinches during the Miocene and Pliocene Epochs. *Cell. Mol. Life. Sci.* **54**: 1031-1041.

Arnaiz-Villena, A., Álvarez-Tejado, M., Ruíz-del-Valle, V., García-de-la-Torre, C., Varela, P., Recio, M. J., Ferre, S. and Martínez-Laso, J. (1999). Rapid Radiation of Canaries (Genus *Serinus*). *Mol. Biol. Evol.* **16**: 2-11.

Arnaiz-Villena, A., Moscoso, J., Ruíz-del-Valle, V., Gonzalez, J., Reguera, R., Ferri, A., Wink, M. and Serrano_vela, J. I. (2008). Mitochondrial DNA Phylogenetic Definition of a Group of 'Arid-Zone' Carduelini Finches. *Open Ornithol. J.* **1**: 1-7.

Arnaiz-Villena, A., Moscoso, J., Ruíz-del-Valle, V., Gonzalez, J., Reguera, R., Wink, M. and Serrano-Vela, J. I. (2007). Bayesian phylogeny of Fringillinae birds: status of the singular African oriole finch *Linurgus olivaceus* and evolution and heterogeneity of the genus *Carpodacus. Acta Zool. Sinica* **53**: 826-834.

Austin, J. J. (1996). Molecular Phylogenetics of *Puffinus* Shearwaters: Preliminary Evidence from Mitochondrial Cytochrome *b* Gene Sequences. *Mol. Phylog. Evol.* **6**: 77-88.

Baker, A. J., Pereira, S. L. and Paton, T. A. (2007). Phylogenetic relationships and divergence times of Charadriiformes genera: multigene evidence for the Cretaceous origin of at least 14 clades of shorebirds. *Biol. Lett.* **3**: 205-209.

Balbontín, J., Negro, J. J., Hernán Sarasola, J., Ferrero, J. J. and Rivera, D. (2008). Land-use changes may explain the recent range expansion of the Black-shouldered Kite *Elanus caeruleus* in southern Europe. *Ibis* **150**: 707-716.

Barker, F. K. (2004). Monophyly and relationships of wrens (Aves: Troglodytidae) a congruence analysis of heterogeneous mitochondrial and nuclear DNA sequence data. *Mol. Phylog. Ecol.* **31**: 486-504.

Barnes, K., Bloomer, P. and Ryan, P. (2006). Phylogeny and speciation in African larks (Alaudidae). *J. Orn.* **147** (suppl.): 134.

Barron, E., van Andel, T. H. and Pollard, D. (2004). Glacial Environments II: Reconstructing the climate of Europe in the last glaciation. In *Neanderthals and Modern Humans in the European Landscape during the Last Glaciation* (van Andel, T. H. and Davies, W., eds.), pp 57-78. MacDonald Institute Monographs, Cambridge.

Barrowclough, G. F., Groth, J. G. and Mertz, L. A. (2006). The RAG-1 exon in the avian order Caprimulgiformes: Phylogeny, heterozygosity, and base composition. *Mol. Phylog. Evol.* **41**: 238-248.

Beard, K. C. (2008). The Oldest North American Primate and Mammalian Biogeography during the Palaeocene-Eocene Thermal Maximum. *Proc. Natl. Acad. Sci. USA* **105**: 3815-3818.

Begun, D. R. (2003). Planet of the Apes. *Sci. Amer.* **289**: 64-73.

Bennett, K. D. (1997). *Evolution and Ecology. The Pace of Life.* Cambridge University Press, Cambridge.

Bennett, K. D., Tzedakis, P. C. & Willis, K. J. (1991). Quaternary Refugia of North European Trees. *Journal of Biogeography*, **18**, 103-115.

Bensch, S., Åkesson, S. and Irwin, D. E. (2002a). The use of AFLP to find an informative SNP: genetic differences across a migratory divide in willow warblers. *Mol. Ecol.* **11**: 2359-2366.

Bensch, S., Grahn, M., Müller, N. *et al.* (2009). Genetic, morphological, and feather isotope variation of migratory willow warblers show gradual divergence in a ring. *Mol. Ecol.* **19**: 3087-3096.

Bensch, S., Helbig, A., J., Salomon, M. and Seibold, I. (2002b). Amplified fragment length polymorphism analysis identifies hybrids between two subspecies of warblers. *Mol. Ecol.* **11**: 473-481.

Bensch, S., Irwin, D. E., Irwin, J. H., Kvist, L. and Åkesson, S. (2006). Conflicting patterns of mitochondrial and nuclear DNA diversity in *Phylloscopus* warblers. *Mol. Ecol.* **15**: 161-171.

Benson, C. W., Brooke, R. K., Stuart Irwin, M. P. and Steyn, P. (1968). White-rumped Swifts in Spain. *Ibis* **110**: 106.

Bermingham, E., Rohwer, S., Freeman, S. and Wood, C. (1992). Vicariance biogeography in the Pleistocene and speciation in North American wood warblers: A test of Mengel's model. *Proc. Natl. Acad. Sci. USA* **89**: 6624-6628.

Berthold, P. and Helbig, A. J. (1992). The genetics of bird migration: stimulus, timing and direction. *Ibis* **134** *suppl. 1*: 35-40.

Berthold, P., Helbig, A. J., Mohr, G. and Querner, U. (1992). Rapid microevolution of migratory behaviour in a wild bird species. *Nature* **360**: 668-670.

Bildstein, K. L. (2006). *Migrating Raptors of the World. Their Ecology and Conservation.* Cornell University Press, Ithaca. 320pp.

Bird, M. I., Taylor, D. and Hunt, C. (2005). Palaeoenvironments of insular Southeast Asia during the Last Glacial Period: a savanna corridor in Sundaland? *Quat. Sci. Rev.* **24**: 2228-2242.

Birks, H. J. B. (1986). Late Quaternary biotic changes in terrestrial and lacustrine environments, with particular reference to north-west Europe. In *Handbook of Holocene Palaeoecology and Palaeohydrology.* Chichester, John Wiley & Sons.

Blanc, P-L. (2002). The opening of the Plio-Quaternary Gibraltar Strait: assessing the size of a cataclysm. *Geodyn. Acta* **15**: 303-317.

Blance, P-L. (2006). Improved modelling of the Messinian salinity crisis and conceptual implications. *Palaeogeog., Palaeoclimatol. & Palaeoecol.* **238**: 349-372.

Blondel, J. and Mourer-Chauviré, C. (1998). Evolution and history of the western Palaearctic avifauna. *Trends Ecol. Evol.* **13**: 488-492.

Blondel, J., Catzeflis, F. and Perret, P. (1996). Molecular phylogeny and the historical biogeography of the warblers of the genus *Sylvia* (Aves). *J. Evol. Biol.* **9**: 871-891.

Blondel, J. and Vigne, J-D. (1993). Space, Time, and Man as Determinants of Diversity of Birds and Mammals in the Mediterranean Region. In Ricklefs, R. E. and Schluter, D. (eds.). *Species Diversity in Ecological Communities. Historical and Geographical Perspectives.* Chicago University Press, Chicago.

Böhning-Gaese, K. and Oberrath, R. (2003). Macroecology of habitat choice in long-distance migratory birds. *Oecologia* **137**: 296-303.

Bond, G. C. and Lotti, R. (1995). Iceberg discharges into the North Atlantic on millennial time scales during the last glaciation. *Science* **267**: 1005-1010.

Bourdon, E., Amaghzaz, M. and Bouya, B. (2008). A new seabird (Aves: cf. Phaethontidae) from the Lower Eocene phosphates of Morocco. *Geobios* **41**: 455-459.

Brambilla, M., Vitulano, S., Spina, F. *et al.* (2008). A molecular phylogeny of the *Sylvia cantillans* complex: Cryptic species within the Mediterranean basin. *Mol. Phylog. Evol.* **48**: 461-472.

Bridge, E. S., Jones, A. W. and Baker, A. J. (2005). A phylogenetic framework for the terns (Sternini) inferred from mtDNA sequences: implications for taxonomy and plumage evolution. *Mol. Phylog. Evol.* **35**: 459-469.

Brown, J. W., Rest, J. S., García-Moreno, J., Sorenson, M. D. and Mindell, D. P. (2008). Strong mitochondrial DNA support for a Cretaceous origin of modern avian lineages. *BMC Biology* **6**: 6.

Brown, L. (1970). *African Birds of Prey.* Collins, London. 320pp.

Burroughs, W. J. (2005). *Climate Change in Prehistory: The End of the Reign of Chaos.* Cambridge University Press, Cambridge.

Burton, J. F. (1995). *Birds & Climate Change.* Christopher Helm, London.

Carrascal, L. M. and Moreno, E. (1993). Food caching versus immediate consumption in the nuthatch: the effect of social context. *Ardea* **81**: 135-141.

Calvo, E., Villanueva, J., Grimalt, J. O., *et al.* (2001). New insights into the glacial latitudinal temperature gradients in the North Atlantic. Results from $U^{K'}_{37}$ sea surface temperatures and terrigenous inputs. *Earth Sci. Plan. Lett.* **188**: 509-519.

Carrión, J. S. (2003). *Evolución Vegetal.* DM, Murcia.

Carrión, J. S. and Fernández, S. (2009). The survival of the 'natural potential vegetation' concept (or the power of tradition). *J. Biogeog.* **36**: 2202-2203.

Carrión, J. S., Munuera, M., Navarro, C. and Saez, F. (2000). Paleoclimas e historia de la vegetación cuaternaria en España a través del análisis polínico. *Complutum* **11**: 115-142.

Cerling, T. E., Harris, J. M., MacFadden, B. J. *et al.* (1997). Global vegetation change through the Miocene/Pliocene boundary. *Nature* **389**: 153-158.

Chubb, A. L. (2004). New nuclear evidence for the oldest divergence among neognath birds: the phylogenetic utility of ZENK (i). *Mol. Phylog. Evol.* **30**: 140-151.

Churcher, C. S. & Smith, P. E. L. (1972). Kom Ombo: Preliminary Report on the Fauna of Late Paleolithic Sites in Upper Egypt, *Science* 177: 259-261.

Cicero, C. and Johnson, N. K. (2006). The tempo of avian diversification: reply. *Evolution* **60**: 413-414.

Clark, M. K., House, M. A., Royden, L. H., Whipple, K. X., Burchfiel, B. C., Zhang, X. and Tang, W. (2005). Late Cenozoic uplift of southeastern Tibet. *Geology* **33**: 525-528.

Clark, P. U. and Mix, A. C. (2002). Ice sheets and sea level of the Last Glacial Maximum. *Quat. Sci. rev.* **21**: 1-7.

Clarke, J. A., Tambussi, C. P., Noriega, J. I., Erickson, G. M. and Ketchman, R. A. (2005). Definitive fossil evidence for the extant avian radiation in the Cretaceous. *Nature* **433**: 305-308.

CLIMAP Project Members. (1976). The Surface of the Ice-Age Earth. *Science,* **191**, 1131-1137.

Coope, G. R. (2002). Changes in the Thermal Climate in Northwestern Europe during Marine Oxygen Isotope Stage 3, Estimated from Fossil Insect assemblages. *Quaternary Research*, **57**, 401-408.

Cooper, A. and Penny, D. (1997). Mass Survival of Birds Across the Cretaceous-Tertiary Boundary: Molecular Evidence. *Science* **275**: 1109-1113.

Cooper, J. H. (2000). First Fossil Record of Azure-winged Magpie *Cyanopica cyanus* in Europe. *Ibis* **142**: 150-151.

Cox, G. W. (1985). The evolution of avian migration systems between temperate and tropical regions of the New World. *Am. Nat.* **126**: 451-474.

Cracraft, J. (1969). Notes on fossil hawks (Accipitridae). *Auk* **86**: 353-354.

Cracraft, J. (2000). Avian evolution, Gondwana biogeography and the Cretaceous-Tertiary mass extinction event. *Proc. Roy. Sco. Lond. B* **268**: 459-469.

Cramp, S. (ed.). (1980). *Handbook of the Birds of Europe, the Middle East and North Africa. The Birds of the Western Palearctic.* Volume 2, Hawks to Bustards. Oxford University Press, Oxford. 687pp.

Crombie, M. K., Arvidson, R. E., Sturchio, N. C., El Alfy, Z. & Abu Zeid, K. (1997). Age and isotopic constraints on Pleistocene pluvial episodes in the Western Desert, Egypt. *Palaeogeog., Palaeoclimatol., Palaeoecol.*,**130**: 337-355.

Crosby, G. T. (1972). Spread of the Cattle Egret in the Western Hemisphere. *Bird Banding* **43**: 205-212.

Crowe, T. M., Bowie, R. C. K., Bloomer, P., Mandiwana, T. G., Hedderson, T. A. J., Randi, E., Pereira, S. L. and Wakeling, J. (2006). Phylogenetics, biogeography and classification of, and character evolution in gamebirds (Aves: Galliformes): effects of character exclusion, data partitioning and missing data. *Cladistics* **22**: 495-532.

Cubo, J. and Mañosa, S. (1999). Evidence for heterochrony in the evolution of the goshawk *Accipiter gentilis* (Accipitridae, Aves). *Ann. Sci. Nat.* **2**: 67-72.

Curry-Lindahl, K. (1981a). *Bird Migration in Africa. Movement between six continents.* Volume1. Academic Press, London. 444pp.

Curry-Lindahl, K. (1981b). *Bird Migration in Africa. Movement between six continents.* Volume 2. Academic Press, London. 250pp.

Dansgaard, W., Johnsen, S. J., Clausen, H. B. *et al.* (1993). Evidence for general instability of past climate from a 250-kyr ice core record. *Nature* **364**: 218-220.

Dansgaard, W., White, J. W. C. and Johnsen, S. J. (1989). The abrupt termination of the Younger Dryas climatic event. *Nature* **339**: 532-534.

Decandido, R., Nualsri, C., Allen, D. and Bildstein, K. L. (2004). Autumn 2003 raptor migration at Chumphon, Thailand: a globally significant raptor migration watch site. *Forktail* **20**: 49-54.

deMenocal, P. (1995). Plio-Pleistocene African Climate. *Science* **270**: 53-39.

deMenocal, P. B. (2008). Africa on the edge. *Nature Geoscience* **1**: 650-651.

Dennell, R. (2009). *The Palaeolithic Settlement of Asia.* Cambridge World Archaeology. Cambridge.

Dennell, R. and Roebroeks, W. (2005). An Asian Perspective on Early Human Dispersal from Africa. *Nature* **438**: 1099-1104.

Doherty, P. F., Jr., Grubb, T. C., Jr. and Bronson, C. L. (1996). Territories and caching-related behaviour of Red-headed Woodpeckers wintering in a beech grove. *Wilson Bull.* **108**: 740-747.

Donne-Gaussé, C., Laudet, V. and Hänni, C. (2002). A molecular phylogeny of anseriformes based on mitochondrial DNA analysis. *Mol. Phylog. Evol.* **23**: 339-356.

Dor, R., Safran, R. J., Sheldon, F. H. *et al.* (2010). Phylogeny of the genus *Hirundo* and the Barn Swallow subspecies complex. *Mol. Phylog. Evol.* **56**: 409-418.

Drake, N. and Bristow, C. (2006). Shorelines in the Sahara: geomorphological evidence for an enhanced monsoon from palaeolake Megachad. *The Holocene* **16**: 901-911.

Drovetski, S. V., Zink, R. M., Fadeev, I. V. *et al.* (2004). Mitochondrial phylogeny of *Locustella* and related genera. *J. Avian Biol.* **35**: 105-110.

Dupont, L. M. (1993). Vegetation zones in NW Africa during the Brunhes Chron reconstructed from marine palynological data. *Quaternary Science Reviews* **12**: 189-202.

Ericson, P. G. P., Anderson, C. L., Britton, T., Elzanowski, A., Johansson, U. S., Källersö, M., Ohlson, J. I., Parsons, T. J., Zuccon, D. and Mayr, G. (2006). Diversification of Neoaves: integration of molecular sequence data and fossils. *Biol. Lett.* **2**: 543-547.

Ericson, P.G.P., Christidis, L., Cooper, A. *et al.* (2001). *Proc. Roy. Soc. Lond. B.* **269**: 235-241.

Ericson, P. G. P., Envall, I., Irestedt, M. and Norman, J. A. (2003). Inter-familial relationships of the shorebirds (Aves: Charadriiformes) based on nuclear DNA sequence data. *BMC Evol. Biol.* **3**: 16.

Feduccia, A. (2003). 'Big bang' for Tertiary birds? *Trends Ecol. Evol.* **18**: 172-176.

Field, J. S. and Lahr, M. M. (2005). Assessment of the Southern Dispersal: GIS-Based Analyses of Potential Routes at Oxygen Isotopic Stage 4. *J. World Prehist.* **19**: 1-45.

Filipelli, G. M. and Flores, J-A. (2009). From the warm Pliocene to the cold Pleistocene: A tale of two oceans. *Geology* **37**: 959-960.

Finlayson, C. (1992). *Birds of the Strait of Gibraltar.* T & A D Poyser, London. 534pp.

Finlayson, C. (2004). *Neanderthals and Modern Humans. An Ecological and Evolutionary Perspective.* Cambridge University Press, Cambridge.

Finlayson, C. (2007). *Al-Andalus. How nature has shaped history.* Santana Books, Málaga.

Finlayson, C. (2009). *The Humans Who Went Extinct. Why Neanderthals died out and we survived.* Oxford University Press, Oxford.

Finlayson, C. and Carrión, J. S. (2007). Rapid ecological turnover and its impact on Neanderthal and other human populations. *Trends Ecol. Evol.* **22**: 213-222.

Finlayson, C., Fa, D. A. and Finlayson, G. (2000). Biogeography of Human Colonizations and Extinctions in the Pleistocene. *Mem. GIBCEMED* **1**: 1-69.

Finlayson, C., Giles Pacheco, F., Rodríguez-Vidal, J. *et al.* (2006). Late survival of Neanderthals at the southernmost extreme of Europe. *Nature* **443**: 850-853.

Finlayson, G. (2006). *Climate, vegetation and biodiversity – a multiscale study of the south of the Iberian Peninsula.* PhD Thesis, Anglia-Ruskin University, Cambridge.

Finlayson, J. C. (1979). Movements of the Fan-tailed Warbler *Cisticola juncidis* at Gibraltar. *Ibis* **121**: 487-489.

Fok, K. W., Wade, C. M. and Parkin, D. T. (2002). Inferring the phylogeny of disjunct populations of the azure-winged magpie *Cyanopica cyanus* from mitochondrial control region sequences. *Proc. R. Soc. Lond.* B, **269**: 1671-1679.

Forschler, M. I., Senar, J. C., Perret, P. and Björklund, M. (2009). The species status of the Corsican Finch *Carduelis corsicana* assessed by three genetic markers with different

rates of evolution. *Mol. Phylog. Evol.* **52**: 234-240.
Fortelius, M., Eronen, J., Liu, L. *et al.* (2006). Late Miocene and Pliocene large land mammals and climatic changes in Eurasia. *Palaeogeog., Palaeoclimatol., Palaeoecol.* **238**: 219-227.
Friesen, V. L. and Anderson, V. J. (1997). Phylogeny and Evolution of the Sulidae (Aves: Pelecaniformes): A Test of Alternative Modes of Speciation. *Mol. Phylog. Evol.* **7**: 252-260.
Gamauf, A. and Haring, E. (2004). Molecular phylogeny and biogeography of Honey-buzzards (genera *Pernis* and *Henicopernis*). *J. Zool. Syst. Evol. Res.* **42**: 145-153.
Gause, G. F. (1934). *The Struggle for Existence.* Williams & Wilkins, Baltimore.
Gauthreaux, S. (1982). The ecology and evolution of avian migration systems. In Farner, D. S., King, J. R. & Parkes, K. C. (eds.), *Avian Biology.* Academic Press, New York, pp. 93-168.
Gibb, G. C. and Penny, D. (2010). Two aspects along the continuum of pigeon evolution: A South-Pacific radiation and the relationship of pigeons within Neoaves. *Mol. Phylog. Evol.* **56**: 698-706.
Gill, F. B., Slikas, B. and Sheldon, F. H. (2005). Phylogeny of Titmice (Paridae): II. Species relationships based on sequences of the mitochondrial cytochrome-*B* gene. *Auk* **122**: 121-143.
Gladenkov, Y. B. and Sinel'nikova, V. N. (2009). Oligocene biogeography of the North Pacific (Implications of Mollusks). *Strat. and Geol. Correl.* **17**: 98-110.
Godoy, J. A., Negro, J. J., Hiraldo, F. and Donázar, J. A. (2004). Phylogeography, genetic structure and diversity in the endangered bearded vulture (*Gypaetus barbatus,* L.) as revealed by mitochondrial DNA. *Mol. Ecol.* **13**: 371-390.
Gómez-Díaz, E., González-Sólis, J., Peinado, M. A. and Page, R. D. M. (2006). Phylogeography of the *Calonectris* shearwaters using molecular and morphometric data. *Mol. Phylog. Evol.* **41**: 322-332.
Gonzalez, J., Delgado Castro, G., Garcia-del-Rey, E. *et al.* (2009). Use of mitocondrial and nuclear genes to infer the origino f two endemic pigeons from the Canary Islands. *J. Ornithol.* **150**: 357-367.
Gould, S. J. (1989). *Wonderful Life. The Burgess Shale and the Nature of History.* Hutchinson Radius, London.
Green, R. E., Barnes, K. N. and de L. Brooke, M. (2009). How the longspur won its spurs: a study of claw and toe length in ground-dwelling passerine birds. *J. Zool.* **277**: 126-133.
Griffin, D. L. (2002). Aridity and humidity: two aspects of the late Miocene climate of North Africa and the Mediterranean. *Palaeogeog., Palaeoclimatol., Palaeoecol.* **182**: 65-91.
Griffiths, C. (1999). Phylogeny of the Falconidae inferred from molecular and morphological data. *Auk* **116**: 116-130.
Griffiths, C. S., Barrowclough, G. F., Groth, J. G. and Mertz, L. A. (2007). Phylogeny, diversity, and classification of the Accipitridae based on DNA sequences of the RAG-1 exon. *J. Avian Biol.* **38**: 587-602.
Groombridge, J. J., Jones, C. G., Bayes, M. K., van Zyl, A. J., Carrillo, J., Nichols, R. A. and Bruford, M. W. (2002). A molecular phylogeny of African kestrels with reference to divergence across the Indian Ocean. *Mol. Phylogenet. Evol.* **25**: 267-277.
Grossman, M. L. and Hamlet, J. (1964). *Birds of Prey of the World.* Cassell, London. 496pp.
Guillamet, A., Pons, J-M., Godelle, B. and Crochet, P-A. (2006). History of the Crested Lark in the Mediterranean region as revealed by mtDNA sequences and morphology. *Mol. Phylog. Evol.* **39**: 645-656.
Guo, Z. T., Ruddiman, W. F., Hao, Q. Z. *et al.* (2002). Onset of Asian desertification by 22 Myr ago inferred from loess deposits in China. *Nature* **416**: 159-163.
Hackett, S. J. *et al.* (2008). A Phylogenomic Study of Birds Reveals Their Evolutionary History. *Science* **320**, 1763–1768.
Hagemeijer, W. J. M. and Blair, M. J. (eds.) (1997). *The EBCC Atlas of European Breeding Birds. Their distribution and abundance.* T & A D Poyser, London. 903pp.
Han, K-L., Robbins, M. B. and Braun, M. J. (2010). A multi-gene estimate of phylogeny in the nightjars and nighthawks (Caprimulgidae). *Mol. Phylog. Evol.* **55**: 443-453.
Harrison, C. J. O. (1979). Small non-passerine birds of the Lower Tertiary as exploiters of ecological niches now occupied by passerines. Nature 281: 562-563.
Heidrich, P., Amengual, J. and Wink, M. (1998). Phylogenetic Relationships in Mediterranean and North Atlantic shearwaters (Aves: Procellariidae) based on nucleotide sequences of mtDNA. *Biochem. Syst. Ecol.* **26**: 145-170.
Heinrich, H. (1988). Origin and consequences of cyclic ice rafting in the northeast Atlantic during the past 130,000 years. *Quat. Res.* **28**: 142-152.
Helbig, A. J. (1996). Genetic basis, mode of inheritance and evolutionary changes of migratory directions in Palearctic warblers (Aves: Sylviidae). *J. Exp. Biol.* **199**: 49-55.
Helbig, A. J. (2003). Evolution of migration: a phylogenetic and biogeographic perspective. In Berthold, P., Gwinner, E. & Sonnenschein, E. (eds.), *Avian Migration.* Springer, Heidelberg, pp. 3-20.
Helbig, A. J., Kocum, A., Seibold, I. and Braun, M. J. (2005). A multi-gene phylogeny of aquiline eagles (Aves: Accipitriformes) reveals extensive paraphyly at the genus level. *Mol. Phylogen. Evol.* **35**: 147-164.
Helbig, A. J., Seibold, I., Bednarek, W., Gaucher, P., Ristow, D., Scharlau, W., Schmidl, D. and Wink, M. (1994). Phylogenetic relationships among Falcon species (genus *Falco*) according to DNA sequence variation of the cytochrome b gene. In Meyburg, B-U and Chancellor, R. D. (eds.), *Raptor Conservation Today,* WWGBP/The Pica Press.
Helbig, A. J., Seibold, I., Martens, J. *et al.* (1995). Genetic Differentiation and Phylogenetic R elationships of Bonelli's Warbler *Phylloscopus bonelli* and Green Warbler *P. nitidus. J. Avian Biol.* **26**: 139-153.
Hewitt, G. M. (1996). Some genetic consequences of ice ages, and their role in divergence and speciation. *Biol. J. Linn. Soc.* **58**: 247-276.
Hewitt, G. M. (2000). The genetic legacy of the Quaternary ice ages. *Nature* **405**: 907-913.
Hooghiemstra, H., Stalling, H., Agwu, C. O. C. & Dupont, L. M. (1992). Vegetational and climatic changes at the northern fringe of the Sahara 250,000-5000 years BP: evidence from 4 marine pollen records located between Portugal and the Canary Islands. *Rev. Palaeobot. and Palynol.* **74**: 1-53.
Hopkins, D. M., Matthews, J. V., Wolfe, J. A. and Silberman, M. L. (1971). A Pliocene flora and insect fauna from the Bering Strait region. *Palaeogeog., Palaeoclimatol., Palaeoecol.* **9**: 211-231.
Hourlay, F., Libois, R., D'Amico, F., Sarà, M., O'Halloran, J. and Michaux, J. R. (2008). Evidence of a highly com-

plex phylogeographic structure on a specialist river bird species, the dipper (*Cinclus cinclus*). *Mol. Phylog. Evol.* **49**: 435-444.

Houston, D. C. (1979). The Adaptations of Scavengers. In Sinclair, A. R. E. and Norton-Griffiths, M. (eds.), *Serengeti. Dynamics of an Ecosystem.* University of Chicago Pres, Chicago. pp. 263-286.

Hu, Y., Meng, J., Wang, Y. and Li, C. (2005). Large Mesozoic mammals fed on young dinosaurs. *Nature* **433**: 149-152.

Huntley, B. & Birks, H. J. B. (1983). *An Atlas of past and present Pollen maps of Europe.* Cambridge, Cambridge University Press.

Irby, L. H. (1895). *The Ornithology of the Straits of Gibraltar.* 2nd Edition, revised and enlarged. London, Taylor & Francis.

Irwin, D. E. (2002). Phylogeographic breaks without geographic barriers to gene flow. *Evolution* **56**: 2383-2394.

Irwin, D. E., Alström, P., Olsson, U. and Benowitz-Fredericks, Z. M. (2001a). Cryptic species in the genus *Phylloscopus* (Old World leaf warblers). *Ibis* **143**: 233-247.

Irwin, D. E., Bensch, S. and Price, T. D. (2001b). Speciation in a ring. *Nature* **409**: 333-337.

Irwin, D. E., Bensch, S., Irwin, J. H. and Price, T. D. (2005). Speciation by Distance in a Ring Species. *Science* **307**: 414-416.

Jahns, S., Huls, M. & Sarnthein, M. (1998). Vegetation and climate history of west equatorial Africa based on a marine pollen record off Liberia (site GIK 16776) covering the last 400,000 years. *Rev. Palaeobot. and Palynol.* **102**: 277-288.

Janis, C. (1993). Tertiary Mammal Evolution in the Context of Changing Climates, Vegetation and Tectonic Events. *Ann. Rev. Ecol. Syst.* **24**: 467-500.

Ji, Q., Luo Z-X., Yuan, C-X. and Tabrum, A. R. (2006). A Swimming Mammaliaform from the Middle Jurassic and Ecomorphological Diversification of Early Mammals. *Science* **311**: 1123-1127.

Johansson, U. S., Fjeldså, J. and Bowie, R. C. K. (2008). Phylogenetic relationships within Passerida (Aves: Passeriformes): A review and a new molecular phylogeny based on three nuclear intron markers. *Mol. Phylog. Evol.* **48**: 858-876.

Johnson, J. A., Lerner, H. R. L., Rasmussen, P. C. And Mindell, D. P. (2006). Systematics within *Gyps* vultures: a clade at risk. *BMC Evol. Biol.* **6**: 65.

Johnson, J. A., Watson, R. T. and Mindell, D. P. (2005). Prioritizing species conservation: does the Cape Verde kite exist? *Proc. R. Soc. B.* **272**: 1365-1371.

Johnson, K. P., de Kort, S., Dinwoodey, K., Mateman, A. C., ten Cate, C., Lessels, C. M. and Clayton, D. H. (2001). A molecular phylogeny of the dove genera *Streptopelia* and *Columba*. *Auk* **118**: 874-887.

Johnson, N. K. and Cicero, C. (2004). New mitochondrial DNA data affirm the importance of Pleistocene speciation in North American birds. *Evolution* **58**: 1122-1130.

Kennedy, M. and Page, R. D. M. (2002). Seabird Supertrees: Combining partial estimates of Procelariiform Phylogeny. *Auk* **119**: 88-108.

Kennedy, M., Gray, R. D. and Spencer, H. G. (2000). The Phylogenetic Relationships of the Shags and Cormorants: Can Sequence Data Resolve a Disagreement between Behaviour and Morphology? *Mol. Phylog. Evol.* **17**: 345-359.

Kennett, J. P. and Stott, L. D. (1991). Abrupt deep-sea warming, palaeoceanographic changes and benthic extinctions at the end of the Palaeocene. *Nature* **353**: 225-229.

Klicka, J. and Zink, R. M. (1997). The Importance of Recent Ice Ages in Speciation: A Failed Paradigm. *Science* **277**: 1666-1669.

Kopij, G. (2008). Range and population expansion of the Cattle Egret *Bubulcus ibis* in Lesotho. *Ostrich* **79**: 245-248.

Korpimäki, E. (1987). Prey caching of breeding Tengmalm's Owls *Aegolius funereus* as a buffer against temporary food shortage. *Ibis* **129**: 499-510.

Krajweski, C., Sipiorski, J. T. and Anderson, F. E. (2010). Complete mitochondrial genome sequences and the phylogeny of cranes (Gruiformes: Gruidae). *Auk* **127**: 440-452.

Kruckenhauser, L., Haring, E., Pinsker, W., Riesing, M. J., Winkler, H., Wink, M. and Gamauf, A. (2004). Genetic vs. morphological differentiation of Old World buzzards (genus *Buteo,* Accipitridae). *Zool. Script.* **33**: 197-211.

Kryukov, A., Iwasa, M. A., Kakizawa, R., Suzuki, H., Pinsker, W. and Haring, E. (2004). Synchronic east-west divergence in azure-winged magpies (*Cyanopica cyanus*) and magpies (*Pica pica*). *J. Zool. Syst. Evol. Res.* **42**: 342-351.

Kukla, G. J., Bender, M. L., de Beaulieu, J-L., *et al.* (2002). Last Interglacial Climates. *Quaternary Research*, **58**, 2-13.

Lack, D. (1956). *Swifts in a tower.* Methuen, London.

Lack, D. (1971). *Ecological Isolation in Birds.* Blackwell, Oxford.

Lambeck, K., Esat, T. M. and Potter, E-K. (2002a). Links between climate and sea levels for the past three million years. *Nature* **419**: 199-206.

Lambeck, K., Yokoyama, Y. and Purcell, T. (2002b). In to and out of the Last Glacial Maximum: sea-level change during Oxygen Isotope Stages 3 and 2. *Quat. Sci. Rev.* **21**: 343-360.

Lanner, R. M. (1996). *Made for each other. A Symbiosis of Birds and Pines.* Oxford University Press, New York.

Larsen, C., Speed, M., Harvey, N. and Noyes, H. A. (2007). A molecular phylogeny of the nightjars (Aves: Caprimulgidae) suggests extensive conservation of primitive morphological traits across multiple lineages. *Mol. Phylog. Evol.* **42**: 789-796.

Lei, X., Lian, Z-M., Lei, F-M, Yin, Z-H. And Zhao, H-F. (2007). Phylogeny of some Muscicapinae species based on cty *b* mitochondrial gene sequences. *Acta Zool. Sinica* **53**: 95-105.

Lerner, H. R. L. and Mindell, D. P. (2005). Phylogeny of eagles, Old World vultures, and other Accipitridae based on nuclear and mitochondrial DNA. *Mol. Phylogen. Evol.* **37**: 327-346.

Lerner, H. R. L., Klaver, M. C. and Mindell, D. P. (2008). Molecular phylogenetics of the buteonine birds of prey (Accipitridae). *Auk* **304**: 304-315.

Lezine, A-M., Turon, J-L. & Buchet, G. (1995). Pollen analyses off Senegal: evolution of the coastal palaeoenvironment during the last deglaciation. *J. Quat. Sci.* **10**: 95-105.

Liang, G., Li, T., Yin, Z-h and Lei, F-m. (2008). Molecular Phylogenetic Analysis of Some Fringillidae Species Based on Mitochondrial CoI Gene Sequences. *Zool. Res.* **29**: 465-475.

Liebers, D., de Knijff, P. and Helbig, A. J. (2004). The herring gull complex is not a ring species. *Proc. Roy. Soc. Lond. B.* **271**: 893-901.

Liebers, D. and Helbig, A. J. (2002). Phylogeography and colonization history of Lesser Black-backed Gulls (*Larus fuscus*) as revealed by mtDNA sequences. *J. Evol. Biol.* **15**: 1021-1033.

Liu, Z., Pagani, M., Zinniker, D., DeConto, R., Huber, M., Brinkhuis, M., Shah, S. R., leckie, R. M. and Pearson, A. (2009). Global Cooling During the Eocene-Oligocene Climate Transition. *Science* **323**: 1187-1190.

Livezey, B. C. (1995). Phylogeny and evolutionary ecology of modern sea ducks (Anatidae: Mergini). *Condor* **97**: 233-255.

Loget, N., van den Driessche, J. and Davy, P. (2005). How did the Messinian salinity crisis end? *Terra Nova* 17: 414-419

Loget, N. and van den Driessche, J. (2006). On the origin of the Strait of Gibraltar. *Sedim. Geol.* **188-189**: 341-356.

Lõhmus, A. and Väli, U. (2005). Habitat use by the Vulnerable greater spotted eagle *Aquila clanga* interbreeding with the lesser spotted eagle *Aquila pomarina* in Estonia. *Oryx* **39**: 170-177.

Louchart, A. (2008). Emergence of long distance bird migrations: a new model integrating global climate changes. *Naturwissenschaften* **95**: 1109-1119.

Louchart, A., Mourer-Chauviré, C., Mackaye, H. T., Likius, A., Vignaud, P. and Brunet, M. (2004). The birds of the Djurab Pliocene faunas, Chad, Central Africa. *Bull. Soc. Geo. France* **175**: 413-421.

Lovette, I. J. (2005). Glacial cycles and the tempo of avian speciation. *Trends. Ecol. Evol.* **20**: 57-59.

Lovette, I. J. and Rubenstein, D. R. (2007). A comprehensive molecular phylogeny of the starlings (Aves: Sturnidae) and mockingbirds (Aves: Mimidae): Congruent mtDNA and nuclear trees for a cosmopolitan avian radiation. *Mol. Phylog. Evol.* **44**: 1031-1056.

Maarleveld, G. C. (1976). Periglacial phenomena and the mean annual temperature during the last glacial time in the Netherlands. *Biuletyn Peryglacjalny*, **26**, 57-78.

Maclean, G. L. (1983). Water Transport by Sandgrouse. *Bioscience* **33**: 365-369.

Markova, A. K., Simakova, A. N., Puzachenko, A. Y. & Kitaev, L. M. (2002). Environments of the Russian Plain during the Middle Valdai Briansk Interstade (33,000-24,000 yr BP) Indicated by Fossil Mammals and Plants. *Quaternary Research*, **57**, 391-400.

Marshall, H. D. and Baker, A. J. (1999). Colonization History of Atlantic Island Common Chaffinches (*Fringilla coelebs*) Revealed by Mitochondrial DNA. *Mol. Phylog. Evol.* **11**: 201-212.

Matthiesen, D. G. (1990). Avian medullary bone in the fossil record, an example from the Early Pleistocene of Olduvai Gorge, Tanzania. *J. Vert. Palaeontol.* **9**: 34A.

Mayr, E. (1942). *Systematics and the Origin of Species from the Viewpoint of a Zoologist.* Columbia University Press, New York.

Mayr, E. (1963). *Animal Species and Evolution.* Belknap, Harvard.

Mayr, G. (2000). A new moosebird (Coliiformes: Coliidae) from the Oligocene of Germany. *J. Ornithol.* **141**: 85-92.

Mayr, G. (2003). A new Eocene swift-like bird with a peculiar feathering. *Ibis* **145**: 382-391.

Mayr, G. (2004). Morphological evidence for sister group relationship between flamingoes (Aves: Phoenicopteridae) and grebes (Podicipedidae). *Zool. J. Linn. Soc.* **140**: 157-169.

Mayr, G. (2005a). New trogons from the early Tertiary of Germany. *Ibis* **147**: 512-518.

Mayr, G. (2005b). A tiny barbet-like bird from the Lower Oligocene of Germany: the smallest species and earliest substantial fossil record of the Pici (Woodpeckers and Allies). *Auk* **122**: 1055-1063.

Mayr, G. (2005c). The Paleogene fossil record of birds in Europe. *Biol. Rev.* **80**: 515-542.

Mayr, G. (2006a). New specimens of the Eocene Messelirrisoridae (Aves: Bucerotes) with comments on the preservation of uropygial gland waxes in fossil birds from Mssel and the phylogenetic affinities of Bucerotes. *Paläontologische Zeit.* **80**: 390-405.

Mayr, G. (2006b). New specimens of the Early Eocene stem group galliform *Paraortygoides* (Gallinuloididae), with comments on the evolution of a crop in the stem lineage of Galliformes. *J. Orn.* **147**: 31-37.

Mayr, G. (2009). A well-preserved second trogon skeleton (Aves: Trogonidae) from the Middle Eocene of Messel, Germany. *Palaeobio., Palaeoenv.* **89**: 1-6.

Mayr, G. and Mourer-Chauviré, C. (2004). Unusual tarsometatarsus of a moosebird from the Paleogene of France and the relationships of *Selmes* Peters, 1999. *J. Vert. Paleontol.* **24**: 366-372.

Mayr, G. and Peters, D. S. (1998). The mousebirds (Aves: Coliiformes) from the Middle Eocene of Grube Messel (Hessen, Germany). *Sencken. Leth.* **78**: 179-197.

McCracken, K. G. and Sheldon, F. H. (1998). Molecular and Osteological Heron Phylogenies: Sources of Incongruence. *Auk* **115**: 127-141.

McElwain, J. C. (1998). Do fossil plants signal palaeo-atmospheric CO_2 concentration in the geologic past? *Phil. Trans. Roy. Soc. Lond. B* **353**: 1-15.

Mengel, R. M. (1964). The probable history of species formation in some northern wood warblers (Parulidae). *Living Bird* **3**: 9-43.

Menon, S., Zafar-ul, I., Soberón, J. and Townsend Peterson, A. (2008). Preliminary analysis of the ecology and geography of the Asian nuthatches (Aves: Sittidae). *Wilson J. Orn.* **120**: 692-699.

Mlíkovský, J. (2002). *Cenozoic Birds of the World. Part 1: Europe.* Ninox Press, Prague. 416 pp.

Mlíkovský, J. (2009). Evolution of the Cenozoic marine avifaunas of Europe. *Ann. Naturhist. Mus. Wien* **111A**: 357-374.

Molinero, F. C. and Ferrero Cantisán, J. J. (1985). Ecology and Status of the Black-shouldered Kite in Extremadura, Western Spain. In Newton, I. and Chancellor, R. D. (eds.).*Conservation Studies on Raptors.* ICBP Technical Publication 5, Norwich. Pp137-142.

Montuire, S. & Marcolini, F. (2002). Palaeoenvironmental significance of the mammalian faunas of Italy since the Pliocene. *Journal of Quaternary Science*, **17**, 87-96.

Moreau, R. E. (1954). The main vicissitudes of the European Avifauna since the Pliocene. *Ibis* **96**: 411-431.

Moreau, R. E. (1966). *The bird faunas of Africa and its islands.* Academic Press, London.

Moreau, R. E. (1972). *The Palaearctic-African Bird Migration Systems.* Academic Press, London. 384pp.

Moum, T., Arnason, U. and Arnason, E. (2002). Mitochondrial DNA Sequence Evolution and Phylogeny of the Atlantic Alcidae, Including the Extinct Great Auk (*Pinguinus impennis*). *Mol. Biol. Evol.* **19**: 1434-1439.

Moyle, R. G. and Marks, B. D. (2006). Phylogenetic relationships of the bulbuls (Aves: Pycnonotidae) based on mitochondrial and nuclear DAN sequence data. *Mol. Phylog. Evol.* **40**: 687-695.

Murray, A. M. (2004). Late Eocene and Early Oligocene Teleost and associated Ichthyofauna of the Jebel Qatrani Formation, Fayum, Egypt. *Palaeontology* 47: 711-724.

Narcisi, B. (2001). Palaeoenvironmental and palaeoclimatic implications of the Late-Quaternary sediment record of Vico volcanic lake (central Italy). *Journal of Quaternary Science*, **16**, 245-255.

Negro, J. J. and Hiraldo, F. (1994). Lack of allozyme variation in the Spanish Imperial Eagle *Aquila adalberti. Ibis* **136**: 87-90.

Newton, I. (2003). *The Speciation and Biogeography of Birds.* Academic Press, Amsterdam.

Newton, I. (2008). *The Migration Ecology of Birds.* Academic Press, Amsterdam.

Nguembock, B., Fjeldså, J., Couloux, A. and Pasquet, E. (2009). Molecular phylogeny of Carduelinae (Aves, Passeriformes, Fringillidae) proves polyphyletic origin of the genera *Serinus* and *Carduelis* and suggests redefined generic limits. *Mol. Phylog. Evol.* **51**: 169-181.

Nittinger, F., Gamauf, A., Pinsker, W., Wink, M. And Haring, E. (2007). Phylogeography and population structure of the saker falcon (*Falco cherrug*) and the influence of hybridization: mitochondrial and microsatellite data. *Mol. Ecol.* **16**: 1497-1517.

Nittinger, F., Haring, E., Pinsker, W., Wink, M. and Gamauf, A. (2005). Out of Africa? Phylogenetic relationships between *Falco biarmicus* and the other hierofalcons (Aves: Falconidae). *J. Zool. Syst. Evol. Res.* **43**: 321-331.

Olsson, U., Alström, P., Ericson, P. G. P. and Sundberg, P. (2005). Non-monophyletic taxa and cryptic species – Evidence from a molecular phylogeny of leaf-warblers (*Phylloscopus*, Aves). *Mol. Phylog. Evol.* **36**: 261-276.

Olsson, U., Alström, P., Gelang, M., Ericson, P. G. P. And Sundberg, P. (2006). Phylogeography of Indonesian and Sino-Himalayan region bush warblers (*Cettia*, Aves). *Mol. Phylog. Evol.* **41**: 556-565.

Ottosson, U., Bensch, S., Svensson, L. and Waldenström, J. (2005). Differentiation and phylogeny of the Olivaceous Warbler *Hippolais pallida* species complex. *J. Orn.* **146**: 127-136.

Outlaw, D. C. and Voelker, G. (2006). Systematics of *Ficedula* flycatchers (Muscicapidae): A molecular reassessment of a taxonomic enigma. *Mol. Phylog. Evol.* **41**: 118-126.

Outlaw, R. K., Voelker, G. and Outlaw, D. C. (2007). Molecular systematics and historical biogeography of the rock-thrushes (Muscicapidae: *Monticola*). *Auk* **124**: 561-577.

Päckert, M., Dietzen, C., Martens, J. e*t al.*(2006). Radiation of Atlantic goldcrests *Regulus regulus* spp.: evidence of a new taxon from the Canary Islands. *J. Avian Biol.* **37**: 364-380.

Päckert, M., Martens, J. and Sun, Y-H. (2010). Phylogeny of long-tailed tits and allies inferred from mitochondrial and nuclear markers (Aves: Passeriformes, Aegithalidae). *Mol. Phylog. Evol.* **55**: 952-967.

Pagani, M., Freeman, K. H. and Arthur, M. A. (1999). Late Miocene Atmospheric CO_2 Concentrations and the Expansion of C_4 Grasses. *Science* **285**: 876-879.

Pagani, M., Zachos, J. C., Freeman, K. H. *et al.* (2005). Marked Decline in Atmospheric Carbon Dioxide Concentrations During the Paleogene. *Science* **309**: 600-603.

Parry, S. J., Clark, W. S. and Prakash, V. (2002). On the taxonomic status of the Indian Spotted Eagle *Aquila hastata. Ibis* **144**: 665-675.

Pasquet, E. (1998). Phylogeny of the nuthatches of the *Sitta canadensis* group and its evolutionary and biogeographic implications. *Ibis* **140**: 150-156.

Paton, T. A. and Baker, A. J. (2006). Sequences from 14 mitochondrial genes provide a well-supported phylogeny of the Charadriiform birds congruent with the nuclear RAG-1 tree. *Mol. Phylog. Evol.* **39**: 657-667.

Paton, T. A., Baker, A. J., Groth, J. G. and Barrowclough, G. F. (2003). RAG-1 sequences resolve phylogenetic relationships within Charadriiform birds. *Mol. Phylog. Evol.* **29**: 268-278.

Pavlova, A., Zink, R. M., Drovetski, S. V., Red'kin, Y. and Rohwer, S. (2003). Phylogeographic patterns in *Motacilla flava* and *Motacilla citreola*: species limits and population history. *Auk* **120**: 744-758.

Pavlova, A., Zink, R. M., Rohwer, S., Koblik, E. A., Red'kin, Y. A., Fadeev, I. V. and Nesterov, E. V. (2005). Mitochondrial DNA and plumage evolution in the White Wagtail *Motacilla alba. J. Avian Biol.* **36**: 322-336.

Pereira, S. L. and Baker, A. J. (2008). DNA evidence for a Paleocene origin of the Alcidae (Aves: Charadriiformes) in the Pacific and multiple dispersals across northern oceans. *Mol. Phylog. Evol.* **46**: 430-445.

Pereira, S.L., Johnson, K. P., Clayton, D. H. and Baker, A. J. (2007). Mitochondrial and Nuclear DNA Seuquences Support a Cretaceous Origin of Columbiformes and a Dispersal-Driven Radiation in the Palaeogene. *Syst. Biol.* **54**: 656-672.

Peteet, D. (2000). Sensitivity and rapidity of vegetational response to abrupt climate change. *Proceedings of the National Academy of Sciences, USA*, **97**, 1359-1361.

Pickford, M. & Morales, J. (1994). Biostratigraphy and palaeobiogeography of East Africa and the Iberian Peninsula. *Palaeogeog., Palaeoclimatol., Palaeoecol.* **112**: 297-322.

Pitra, C., Lieckfeldt, D., Frahnert, S. and Fickel, J. (2002). Phylogenetic Relationships and Ancestral Areas of the Bustards (Gruiformes: Otididae), Inferred from Mitochondrial DNA and Nuclear Intron Sequences. *Mol. Phylog. Evol.* **23**: 63-74.

Poe, S. and Chubb, A. L. (2004). Birds in a bush: five genes indicate explosive evolution of avian orders. *Evolution* **58**: 404-415.

Pratt, R. C., Gibb, G. C., Morgan-Richards, M., Phillips, M. J., Hendy, M. D. and Penny, D. (2009). Toward Resolving Deep Neoaves Phylogeny: Data, Signal Enhancement, and Priors. *Mol. Biol. Evol.* **26**: 313-326.

Prokopenko, A. A., Karabanov, E. B., Williams, D. F. & Khursevich, G. K. (2002). The Stability and the Abrupt Ending of the Last Interglaciation in Southeastern Siberia. *Quaternary Research*, **58**, 56-59.

Pulido, F. and Berthold, P. (2010). Current selection for lower migratory activity will drive the evolution of residency in a migratory bird population. *Proc. Natl. Acad. Sci. USA* **107**: 7341-7346.

Qu, Y., Ericson, P. G. P., Lei, F., Gebauer, A., Kaiser, M. and Helbig, A. J. (2006). Molecular phylogenetic relationship of snow finch complex (genera *Montifringilla, Pyrgilauda* and *Onychostruthus*) from the Tibetan plateau. *Mol. Phylog. Evol.* **40**: 218-226.

Randi, E. (1996). A Mitochondrial Cytochrome *B* Phylogeny of the *Alectoris* Partridges. *Mol. Phylog. Evol.* **6**: 214-227.

Rappole, J. H. (1995). *Ecology of migrant birds: a neotropical perspective.* Smithsonian Inst., Washington.

Rappole, J. H. (2005). Evolution of Old and New World migration systems: a response to Bell. *Ardea* **93**: 125-131.

Rappole, J. H. and Jones, P. (2002). Evolution of Old and New World migration systems. *Ardea* **90**: 525-537.

Rasmussen, D. T., Olson, S. L. and Simons, E. L. (1987). Fossil Birds from the Oligocene Jebel Qatrani Formation, Fayum Province, Egypt. *Smiths. Contrib. Paleobiol.* **62**: 1-20.

Rea, D. K., Zachos, J. C., Owen, R. M. and Gingerich, P. D. (1990). Global Change at the Palaeocene-Eocene Boundary: climatic and evolutionary consequences of tectonic events. *Palaeogeog., Palaeoclimatol., Palaeoecol.* **79**: 117-128.

Reeves, A. B., Drovetski, S. V. and Fadeev, I. V. (2008). Mitochondrial DNA data imply a stepping-stone colonization of Beringia by arctic warbler *Phylloscopus borealis. J. Avian Biol.* **39**: 567-575.

Rhymer, J. M., McAuley, D. G. and Ziel, H. L. (2005). Phylogeography of the american woodcock (*Scolopax minor*): are management units based on band recovery data reflected in genetically based management units? *Auk* **122**: 1149-1160.

Riesing, M. J., Kruckenhauser, L., Gamauf, A. and Haring, E. (2003). Molecular phylogeny of the genus *Buteo* (Aves: Accipitridae) based on mitochondrial marker sequences. *Mol. Phylogen. Evol.* **27**: 328-342.

Ritz, M. S., Millar, C., Miller, G. D., Phillips, R. A., Ryan, P., Sternkopf, V., Liebers-Helbig, D. and Peter, H-U. (2008). Phylogeography of the southern skua complex—rapid colonization of the southern hemisphere during a glacial period and reticulate evolution. *Mol. Phylog. Evol.* **49**: 292-303.

Rivas-Martínez, S. (1981). Les étages bioclimatiques de la vegetation de la Péninsule Ibérique. *Actas III Congreso Optima. Anales Jardin Botanico Madrid*, **37**, 251-268.

Rivas-Martínez, S. (1987). Memoria del mapa de series de vegetación de España. Madrid, ICONA.

Roberts, N. (1998). *The Holocene. An Environmental History.* Blackwell, Oxford.

Rohling, E. J., Fenton, M., Jorissen, F. J., *et al.* (1998). Magnitudes of sea-level lowstands of the past 500,000 years. *Nature* **394**: 162-165.

Rolhausen, G., Segelbacher, G., Hobson, K. A. and Schaefer, H. M. (2009). Contemporary Evolution of Reproductive Isolation and Phenotypic Divergence in Sympatry along a Migratory Divide. *Curr. Biol.* **19**: 2097-2101.

Rose, J., Meng, X. & Watson, C. (1999). Palaeoclimate and palaeoenvironmental responses in the western Mediterranean over the last 140 ka: evidence from Mallorca, Spain. *Journal of the Geological Society, London*, **156**, 435-448.

Rousseau, D-D., Gerasimenko, N., Matviischina, Z. & Kukla, G. (2001). Late Pleistocene Environments of the Central Ukraine. *Quaternary Research*, **56**, 349-356.

Ruddiman, W. F. and Kutzbach, J. E. (1991). Plateau Uplift and Climatic Change. *Sci. Am.* March 1991: 66-75.

Ruddiman, W. F. and McIntyre, A. (1977). Late quaternary surface ocean kinematics and climate change in the high-latitude North Atlantic. *J. Geophys. Res.* **82**: 3877-3887.

Ruddiman, W. F., McIntyre, A. And Raymo, M. (1986). Palaeoenvironmental results from North Atlantic Sites 607 and 609. *Initial Reports of the Deep-Sea Drilling Project* **94**: 855-878.

Rudoy, A. N. (2002). Glacier-Dammed Lakes and Geological Work on Glacial Superfloods in the Late Pleistocene. *Quat. Int.* **87**: 119-140.

Safriel, U. N. (1995). The evolution of Palaearctic migration – the case for southern ancestry. *Isr. J. Zool.* **41**: 417-431.

Salewski, V. and Bruderer, B. (2007). The evolution of bird migration – a synthesis. *Naturwissenschaften* **94**: 268-279.

Salomon, M., Voisin, J-F. and Bried, J. (2007). On the taxonomic status and denomination of the Iberian Chiffchaffs. *Ibis* **145**: 87-97.

Sánchez Marco, A. (2007a). New occurrences of the extinct vulture *Gyps melitensis* (Falconiformes, Aves) and a reappraisal of the paleospecies. *J. Vert. Paleontol.* **27**: 1057-1061.

Sánchez Marco, A. (2007b). Presencia de Halcón Sacre *Falco cherrug* en el sur de Iberia durante el ultimo pleniglacial. *Ardeola* **54**: 345-347.

Scheider, J., Wink, M., Stubbe, M., Hille, S. And Wiltschko, W. (2004). Phylogeographic Relationships of the Black Kite *Milvus migrans.* In Chancellor, R. D. & Meyburg, B-U. (eds), *Raptors Worldwide,* WWGBP/MME, Budapest. 890pp.

Schreiber, A., Stubbe, M. and Stubbe, A. (2000). Red kite (*Milvus milvus*) and black kite (*M. migrans*): minute genetic interspecies distance of two raptors breeding in a mixed community (Falconiformes: Accipitridae). *Biol. J. Linn. Soc.* **69**: 351-365.

Schuster, M., Roquin, C., Duringer, P. *et al.* (2005). Holocene Lake Mega-Chad palaeoshorelines from space. *Quat. Sci. Rev.* **24**: 1821-1827.

Ségalen, L., Lee-Thorp, J. A. and Cerling, T. (2007). Timing of C_4 grass expansion across sub-Saharan Africa. *J. Hum. Evol.* **53**: 549-559.

Seibold, I. and Helbig, A. J. (1995), Evolutionary history of New and Old World vultures inferred from nucleotide sequences of the mitochondrial cytochrome *b* gene. *Phil. Trans. Roy. Soc. Lond. B.* **350**: 163-178.

Seibold, I., Helbig, A. J. and Wink, M. (1993). Molecular Systematics of Falcons (Family Falconidae). *Naturwissenschaften* **80**: 87-90.

Seiffert, E. R. (2006). Revised age estimates for the later Paleogene mammal faunas of Egypt and Oman. *Proc. Natl. Acad. Sci. USA* 103: 5000-5005.

Seki, S-I. (2006). The origin of the East Asian *Erithacus* robin, *Erithacus komadori,* inferred from cytochrome *b* sequence data. *Mol. Phylog. Evol.* **39**: 899-905.

Sepulchre, P., Schuster, M., Ramstein, G. *et al.* (2008). Evolution of Lake Chad Basin hydrology during the mid-Holocene: A preliminary approach from lake to climate modelling. *Gobal Planet. Change* **61**: 41-48.

Shackleton, N. J. and Opdyke, N. D. (1976). Oxygen isotope and palaeomagnetic stratigraphy of equatorial Pacific core V28-239, late Pliocene to latest Pleistocene. *Geol. Soc. Amer. Mem.* **145**: 449-464.

Sheldon, F. H., Jones, C. E. and McCracken, K. G. (2000). Relative Patterns and Rates of Evolution in Heron Nuclear and Mitochondrial DNA. *Mol. Biol. Evol.* **17**: 437-450.

Sheldon, F. H., Whittingham, L. A., Moyle, R. G., Slikas, B. and Winkler, D. W. (2005). Phylogeny of swallows (Aves: Hirundinidae) estimated from nuclear and mitochondrial DNA sequences. *Mol. Phylog. Evol.* **35**: 254-270.

Shen Y-Y., Liang, L., Sun, Y-B., Yang, X-J., Murphy, R. W. and Zhang, Y-P. (2010). A mitogenomic perspective on the ancient, rapid radiation in the Galliformes with an emphasis on the Phasianidae. *BMC Evol. Biol.* **10**: 132.

Sibley, C. G. and Monroe, Jr., B. L. (1990). *Distribution and Taxonomy of Birds of the World.* Yale University Press, Yale.

Simmons, R. E. and Legra, L. A. T. (2009). Is the Papuan Harrier *Circus spilothorax* a globally threatened species? Ecology, climate change threat and first population estimates from Papua New Guinea. *Bird Cons. Int.* doi:10.1017/S095927090900851X.

Simpson, G. G. (1944). *Tempo and Mode in Evolution.* Columbia University Press, New York.

Slikas, B. (1997). Phylogeny of the Avian Family Ciconiidae (Storks) Based on Cytochrome *b* Sequences and DNA-DNA Hybridixation Distances. *Mol. Phylog. Evol.* **8**: 275-300.

Smith, C. C. and Reichman, O. J. (1984). The evolution of food caching by birds and mammals. *Ann. Rev. Ecol. Syst.* **15**: 329-351.

Snow, D. W. (1978). Relationships between the European and African Avifaunas. *Bird Study* **25**: 134-148.

Soffer, O. (1985). *The Upper Paleolithic of the Central Russian Plain.* San Diego, Academic Press.

Solheim, R. (1984). Caching behaviour, prey choice and surplus killing by Pygmy Owls Glaucidium passerinum during winter, a functional response of a generalist predator. *Ann. Zool. Fenn.* **21**: 301-308.

Stewart, J. R. and Lister, A. M. (2001). Cryptic northern refugia and the origins of the modern biota. *Trends Ecol. Evol.* **16**: 608-613.

Štorchová, Z., Landová, E. and Frynta, D. (2010). Why some tits store food and others do not: evaluation of ecological factors. *J. Ethol.* **28**: 207-219.

Stuiver, M. and Grootes, P. M. (2000). GISP2 Oxygen Isotope Ratios. *Quat. Res.* **53**: 277-284.

Sturmbauer, C., Berger, B., Dallinger, R. and Föger, M. (1998). Mitochondrial Phylogeny of the Genus *Regulus* and Implications on the Evolution of Breeding Behavior in Sylvioid Songbirds. *Mol. Phylog. Evol.* **10**: 144-149.

Suárez, N. M., Betancor, E., Klassert, T. E., Almeida, T., Hernández, M. and Pestano, J. J. (2009). Phylogeography and genetic structure of the Canarian common chaffinch (*Fringilla coelebs*) inferred with mtDNA and microsatellite loci. *Mol. Phylog. Evol.* **53**: 556-564.

Suc, J. P., Drivaliari, A., Bessais, E., Guiot, J., Bertini, A., Zheng, Z., Abdelmalek, S. M., Diniz, F., Combourieu-Nebout, N., Leroy, S., Cheddadi, R., Ferrier, J. & Duzer, D. (1994). Mediterranean Pliocene Vegetation and Climate: How to Quantify The Climate Parameters? *US Geological Survey Open-File Report*, 94-23(10).

Svenning, J-C. (2002). A review of natural vegetation openness in north-western Europe. *Biol. Cons.* **104**: 133-148.

Taberlet, P. and Cheddadi, R. (2002). Quaternary Refugia and Persistence of Biodiversity. *Science* **297**: 2009-2010.

Taberlet, P., Fumagalli, L., Wust-Saucy, A. G. and Cossons, J-F. (1998). Comparative phylogeography and postglacial colonization routes in Europe. *Mol. Ecol.* **7**: 453-464.

Thévenot, M., Vernon, R. and Bergier, P. (2003). *The birds of Morocco.* BOU Checklist Series:20. 594 pp.

Thomas, G. H., Wills, M. A. and Székeley, T. (2004a). A supertree approach to shorebird phylogeny. *BMC Evol. Biol.* **4**: 28.

Thomas, G. H., Wills, M. A. and Székeley, T. (2004b). Phylogeny of shorebirds, gulls, and alcids (Aves: Charadrii) from the cytochrome-b gene: parsimony, Bayesian inference, minimum evolution, and quartet puzzling. *Mol. Phylog. Evol.* **30**: 516-526.

Tietze, D. T., Martens, J. and Sun, Y-H. (2006). Molecular phylogeny of treecreepers (*Certhia*) detects hidden diversity. *Ibis* **148**: 477-488.

Tjallingii, R., Claussen, M., Stuut, J-B. W., Fohlmeister, J., Jahn, A., Bickert, T., Lamy, F. and Rohl, U. (2008). Coherent high- and low-latitude control of northwest African hydrological balance. *Nature Geoscience* **1**: 670-675.

Töpfer, T, Haring E., Birkhead, T. R., Lopes, R. J. ,Liu Severinghaus, L., Martens, J. and Päckert, M. (2011). A molecular phylogeny of bullfinches *Pyrrhula* Brisson, 1760 (Aves: Fringillidae). *Molecular Phylogenetics and Evolution*, **58**, 271–282.

Treplin, S., Siegert, R., Bleidorn, C., Shokellu Thompson, H., Fotso, R. and Tiedemann, R. (2008). Molecular phylogeny of songbirds (Aves: Passeriformes) and the relative utility of common nuclear marker loci. *Cladistics* **24**: 328-349.

Turner, C. (2002). Formal Status and Vegetational Development of the Eemian Interglacial in Northwestern and Southern Europe. *Quaternary Research*, **58**, 41-44.

Tyrberg, T. (1998). *Pleistocene birds of the Palearctic: A Catalogue.* Publications Nuttall Ornithol. Club, **27**, Cambridge, Mass. 720 pp.

Tyrberg, T. (2008). http://web.telia.com/~u11502098/pleistocene.html

Tzedakis, P. C. (1994). Vegetation Change through Glacial-Interglacial Cycles: A Long Pollen Sequence Perspective. *Philosophical Transactions of the Royal Society, London, Series B.*, **345**, 403-432.

Tzedakis, P. C. (2005). Towards an understanding of the response of southern European vegetation to orbital and suborbital climate variability. *Quat. Sci. Rev.* **24**: 1585-1599.

Tzedakis, P. C., Frogley, M. R. & Heaton, T. H. E. (2002). Duration of Last Interglacial Conditions in Northwestern Greece. *Quaternary Research*, **58**, 53-55.

Väli, U. (2006). Mitochondrial DNA sequences support species status for the Indian Spotted Eagle *Aquila hastata*. *Bull. B. O. C.* **126**: 238-242.

van Andel, T. H. and Tzedakis, P. C. (1996). Palaeolithic landscapes of Europe and environs. *Quat. Sci. Rev.* **15**: 481-500.

van Beusekom, C. F. (1972). Ecological isolation with respect to food between sparrowhawk and goshawk. *Ardea* **60**: 72-96.

van Tuinen, M., Butvill, D. B., Kirsch, J. A. W. And Hedges, S. B. (2001). Convergence and divergence in the evolution of aquatic birds. *Proc. Roy. Soc. Lond. B* **268**: 1345-1350.

van Tuinen, M., Stidham, T. A. and Hadly, E. A. (2006). Tempo and mode of modern bird evolution observed with large-scale taxonomic sampling. *Hist. Biol.* **18**: 205-221.

Vandenberghe, J., Coope, R. & Kasse, K. (1998). Quantitative reconstructions of palaeoclimates during the last interglacial-glacial in western and central Europe: an introduction. *Journal of Quaternary Science*, **13**, 361-366.

Vignaud, P., Duringer, P., Mackaye, H. T. *et al.* (2002). Geology and Palaeontology of the Upper Miocene Toros-Menalla hominid locality, Chad. *Nature* **418**: 152-155.

Voelker, A. H. L. (2002). Global distribution of centennial-scale records for Marine Isotope Stage (MIS) 3: a database. *Quat. Sci. Rev.* **21**: 1185-1212.

Voelker, G. (1999a). Dispersal, Vicariance, and Clocks: Historical Biogeography and Speciation in a Cosmopolitan Passerine Genus (*Anthus*: Motacillidae). *Evolution* **53**: 1536-1552.

Voelker, G. (1999b). Molecular Evolutionary Relationships in the Avian Genus *Anthus* (Pipits: Motacillidae). *Mol. Phylog. Evol.* **11**: 84-94.

Voelker, G. (2002a). Molecular phylogenetics and the historical biogeography of dippers (*Cinclus*). *Ibis* **144**: 577-584.

Voelker, G. (2002b). Systematics and historical biogeography of wagtails: dispersal versus vicariance revisited. *Condor* **104**: 725-739.

Voelker, G. (2010). Repeated vicariance of Eurasian songbird lineages since the Late Miocene. *J. Biogeog.* **37**: 1251-1261.

Voelker, G. and Spellman, G. M. (2004). Nuclear and mitochondrial DNA evidence for polyphyly in the avian superfamily Muscicapoidea. *Mol. Phylog. Evol.* **30**: 386-394.

Voelker, G., Rohwer, S., Bowie, R. C. K. and Outlaw, D. C. (2007). Molecular systematics of a speciose, cosmopolitan songbird genus: Defining the limits of, and relationships among, the *Turdus* thrushes. *Mol. Phylog. Evol.* **42**: 422-434.

Waite, T. A. and Reeve, J. D. (1992). Caching behaviour in the Gray Jay and the source-departure decision for rate-maximizing scatterhoarders. *Behaviour* **120**: 51-68.

Walker, C. A. and Dyke, G. J. (2006). New records of fossil birds of prey from the Moicene of Kenya. *Hist. Biol.* **18**: 91-94.

Wang, Y., Deng, T. and Biasatti, D. (2006). Ancient diets indicate significant uplift of southern Tibet after ca. 7 Ma. *Geology* **34**: 309-312.

Weil, A. (2005). Living Large in the Cretaceous. *Nature* **433**: 116-117.

Weir, J. T. and Schluter, D. (2004). Ice sheets promote speciation in boreal birds. *Proc. Roy. Soc. Lond. B.* **271**: 1881-1887.

Willis, K. (1996). Where did all the flowers go? The fate of temperate European flora during glacial periods. *Endeavour* **20**: 110-114.

Willis, K., Ruder, E. and Sumegi, P. (2000). The full-glacial forests of central and southeastern Europe. *Quat. Res.* **55**: 203-213.

Wink, M. and Heidrich, P. (1999). Molecular evolution and systematics of owls (Strigiformes). In König, C., Weik, F. and Becking, J. H. (eds.). *Owls of the World.* Pica Press, Robertsbridge. pp. 39-57.

Wink, M., Heidrich, P. and Fentzloff, C. (1996). A mtDNA Phylogeny of Sea Eagles (genus *Haliaeetus*) Based on Nucleotide Sequences of the Cytochrome *b*-gene. *Biochem. Syst. Ecol.* **24**: 783-791.

Wink, M., Heidrich, P., Sauer-Gürth, H., Elsayed, A-A. and Gonzalez, J. (2009). Molecular Phylogeny and Systematics of Owls (Strigiformes). In König, C., Friedhelm, W. and Wink, M. (eds.). *Owls of the World.* Yale University Press, Yale. pp.42-63.

Wink, M. and Sauer-Gürth, H. (2004). Phylogenetic Relationships in Diurnal Raptors based on nucleotide sequences of mitochondrial and nuclear marker genes. In Chancellor, R. D. and Meyburg, B-U. (eds.). *Raptors Worldwide.* WWGBP/MME. pp. 483-498.

Wink, M., Sauer-Girth, H., Ellis, D. and Kenward, R. (2004a). Phylogenetic Relationships in the Hierofalco Complex (Saker-, Gyr-, Lanner-, Laggar Falcon). In Chancellor, R. D. and Meyburg, B.-U. (eds). *Raptors Worldwide.* WWGBP/MME. pp499-504.

Wink, M., Sauer-Gürth, H. and Pepler, D. (2004b). Phylogeographic Relationships of the Lesser Kestrel *Falco naumanni* in Breeding and Wintering Quarters, inferred from nucleotide sequences of the mitochondrial cytochrome b gene. In Chancellor, R. D. & Meyburg, B-U. (eds). *Raptors Worldwide,* WWGBP/MME. pp 505-510.

White, G. (1789). *The Natural History and Antiquities of Selborne.* Cassell and Co., London.

Wink, M., Sauer-Gürth, H. and Witt, H-H. (2004c). Phylogenetic Differentiation of the Osprey *Pandion haliaetus* inferred from nucleotide sequences of the mitochondrial cytochrome b gene. In Chancellor, R. D. & Meyburg, B-U. (eds). *Raptors Worldwide,* WWGBP/MME. pp 511-516.

Wink, M., Seibold, I., Lotfikhah, F. and Bednarek, W. (1998). Molecular Systematics of Holarctic Raptors (Order Falconiformes). In Chancellor, R. D., Meyburg, B.-U. and Ferrero, J. J. (eds.). *Holarctic Birds of Prey.* ADENEX-WWGBP. pp 29-48.

Woillard, G. (1979). Abrupt end of the last interglacial in north-east France. *Nature* **281**: 558-562.

Zagwijn, W. H. (1992). Migration of Vegetation during the Quaternary in Europe. *Courier Forschungs-Institut Senckenberg,* **153**, 9-20.

Zamora, J., Lowy, E., Ruíz-del-Valle, V., Moscoso, J., Serrano-Vela, J. I., Rivero-de-Aguilar, J. and Arnaiz-Villena, A. (2006). *Rhodopechys obsoleta* (desert finch): a pale ancestor of greenfinches (*Carduelis* spp.) according to molecular phylogeny. *J. Ornithol.* **147**: 448-456.

Zanazzi, A., Kohn, M. J., MacFadden, B. J. and Terry Jr., D. O. (2007). Large Temperature drop across the Eocene-Oligocene transition in Central North America. *Nature* **445**: 639-642.

Zink, R. M. and Klicka, J. (2006). The tempo of avian diversification: a comment on Johnson and Cicero. *Evolution* **60**: 411-412.

Zink, R. M., Drovetski, S. V. and Rohwer, S. (2006). Selective neutrality of mitochondrial ND2 sequences, phylogeography and species limits in *Sitta europaea. Mol. Phylog. Evol.* **40**: 679-686.

Zink, R. M., Drovetski, S. V., Questiau, S., Fadeev, I. V., Nesterov, E. V., Westberg, M. C. and Rohwer, S. (2003). Recent evolutionary history of the bluethroat (*Luscinia svecica*) across Eurasia. *Mol. Ecol.* **12**: 3069-3075.

Zink, R. M., Klicka, J. and Barber, B. R. (2004). The tempo of avian diversification during the Quaternary. *Phil. Trans Roy. Soc. Lond B* **359**: 215-220.

Zink, R. M., Pavlova, A., Rowher, S. and Drovetski, S. V. (2006). Barn swallows before barns: population histories and intercontinental colonization. *Proc. Roy. Soc. Lond. B* **273**: 1245-1251.

Zuccon, D. and Ericson, G. P. (2010). The *Monticola* rock-thrushes: Phylogeny and biogeography revisited. *J. Biogeog.* **55**: 901-910.

Zuccon, D., Cibois, A., Pasquet, E. and Ericson, P.G.P. (2006). Nuclear and mitochondrial sequence data reveal the major lineages of starlings, mynas and related taxa. *Mol. Phylog. Evol.* **41**: 333-344.

Index

D

N

O

P

U

V

W